AQA Biology

Exclusively endorsed by AQA

Glenn Toole
Susan Toole

Nelson Thornes

Published in 2008 by:
Nelson Thornes Ltd
Delta Place
27 Bath Road
CHELTENHAM
GL53 7TH
United Kingdom

10 11 12 / 10 9 8 7 6 5 4 3

A catalogue record for this book is available from the British Library

ISBN 978 0 7487 9813 1

Cover photograph by Getty Images/Blend Images

Illustrations include artwork drawn by Barking Dog Art

Page make-up by GreenGate Publishing, Kent

Printed and bound in China by 1010 Printing International Ltd

Contents

Unit 5
Control in cells and in organisms 140

Introduction

Nelson Thornes has worked in partnership with AQA to ensure this book and the accompanying online resources offer you the best support for your A level course.

All resources have been approved by senior AQA examiners so you can feel assured that they closely match the specification for this subject and provide you with everything you need to prepare successfully for your exams.

These print and online resources together **unlock blended learning**; this means that the links between the activities in the book and the activities online blend together to maximise your understanding of a topic and help you achieve your potential.

These online resources are available on kerboodle! which can be accessed via the internet at **http://www.kerboodle.com/live**, anytime, anywhere. If your school or college subscribes to this service you will be provided with your own personal login details. Once logged in, access your course and locate the required activity.

For more information and help visit **http://www.kerboodle.com**

Icons in this book indicate where there is material online related to that chapter. The following icons are used:

Learning activity

These resources include a variety of interactive and non-interactive activities to support your learning:

- Animations
- Simulations
- Maths skills
- Key diagrams
- Glossary

Progress tracking

These resources include a variety of tests that you can use to check your knowledge on particular topics (Test yourself) and a range of resources that enable you to analyse and understand examination questions (On your marks...). You will also find the answers to the examination-style questions online.

Research support

These resources include WebQuests, in which you are assigned a task and provided with a range of weblinks to use as source material for research.

These are designed as stretch and challenge resources to stretch you and broaden your learning, in order for you to attain the highest possible marks in your exams.

Weblinks

Our online resources feature a list of recommended weblinks, split by chapter. This will give you a head start, helping you to navigate to the best websites that will aid your learning and understanding of the topics in your course.

How science works

These resources are a mixture of interactive and non-interactive activities to help you learn the skills required for success in this new area of the specification.

Practical

This icon signals where there is a relevant practical activity to be undertaken, and support is provided online.

How to use this book

This book covers the specification for your course and is arranged in a sequence approved by AQA. It will cover all three of the Assessment Objectives required in your AQA A Level Biology course.

The main text of the book will cover AO1 – Knowledge and understanding. This consists of the main factual content of the specification. The other Assessment Objectives (AO2 – Application of knowledge and understanding and AO3 – How science works) make up around 50% of the assessment weighting of the specification, and as such will be covered in the textbook in the form of the feature 'Applications and How science works' (see below). You will **not** be asked to recall the information given under these headings for the purpose of examinations.

The book content is divided into the two theory units of the AQA Biology A2 specification; Unit 4 – Populations and environment and Unit 5 – Control in cells and in organisms. Units are then further divided into chapters, and then topics, making the content clear and easy to use.

Unit openers give you a summary of the content you will be covering, and a recap of ideas from GCSE and AS Level that you will need.

The features in this book include:

Learning objectives

At the beginning of each section you will find a list of learning objectives that contain targets linked to the requirements of the specification. The relevant specification reference is also provided.

Key terms

Terms that you will need to be able to define and understand are highlighted in bold type within the text, e.g. **niche**. Where terms are not explained within the same topic, they are highlighted in bold blue type, e.g. **hydrolysis**. You can look up these terms in the glossary. Sometimes a term appears in brackets. These are words that are useful to know but are not used on the specification. They therefore do not have to be learned for examination purposes.

Hint

Hints to aid your understanding of the content.

Link

Synoptic links are highlighted in the margin near the relevant text using the link icon accompanied by brief notes which include references to where the linked topics are to be found in this book or the AS book.

Applications and How science works

This feature may cover either or both of the assessment objectives AO2 – Application of knowledge and understanding and AO3 – How science works, both key parts of the new specification.

As with the specification, these objectives are integrated throughout the content of the book. This feature highlights opportunities to apply your knowledge and understanding and draws out aspects of 'How science works' as they occur within topics, so that it is always relevant to what you are studying. The ideas provided in these features intend to teach you the skills you will need to tackle this part of the course, and give you experience that you can draw upon in the examination. You will not be examined on the exact information provided in 'Applications and How science works'. For more information, see 'How science works' on page 1.

Summary questions

Short questions that test your understanding of the subject and allow you to apply the knowledge and skills you have acquired to different scenarios. These questions mostly cover material in the same topic but may sometimes include information from previous topics. Answers are supplied at the back of the book. These answers are more than just a mark scheme. They often include explanations of the answers to aid learning and understanding. The answers are not exhaustive and there may be acceptable alternatives.

AQA Examiner's tip

Hints from AQA examiners to help you with your studies and to prepare you for your exam.

AQA Examination-style questions

Questions from past AQA papers that are in the general style that you can expect in your exam. These occur at the end of each chapter to give practice in examination-style questions for a particular topic. They also occur at the end of each unit; the questions here may cover any of the content of the unit. These questions relate to earlier specifications but have been chosen because they are relevant to the new specification. Despite careful selection there may be certain terms that do not exactly match the new requirements. They should therefore be treated in the same way as Applications and used for examination practice and application of knowledge, rather than learning their content. When you answer the examination-style questions in this book, remember that quality of written communication (QWC) will be assessed in any question or part-question in the Unit 4 and 5 papers where extended descriptive answers are required.

Synopticity

Synoptic questions or part-questions are a key feature of your A2 examination papers. Such a question may require you to draw on knowledge, understanding and skills from AS Level that underpin the A2 topic which the question is about. They link knowledge, understanding and skills from topics in the A2 theory unit on which the question is set, perhaps in a new context, which is described in the question.

Stretch and challenge

Some of the questions in the papers for Units 4 and 5 are designed to test the depth of your knowledge and understanding of the subject. Such questions may require you to solve a problem where you have to decide on a suitable strategy and appropriate methods, possibly linking different ideas from within the unit, discuss in an extended written answer a controversial issue involving biology, perhaps in terms of advantages and disadvantages, that affects people or society at large.

The questions test your ability to think deeply and clearly about biology and to provide solutions and answers that are coherent and clear.

Answers to these questions are supplied online.

Web links in the book

As Nelson Thornes is not responsible for third party content online, there may be some changes to this material that are beyond our control. In order for us to ensure that the links referred to in the book are as up-to-date and stable as possible, the websites are usually homepages with supporting instructions on how to reach the relevant pages if necessary.

Please let us know at **kerboodle@nelsonthornes.com** if you find a link that doesn't work and we will do our best to redirect the link, or to find an alternative site.

Skills for studying A2 Biology

Welcome to biology at A Level.

This book aims to make your study of biology successful and interesting.

The book is written to cover the content of the A2 course for the AQA specification. Each chapter in the book corresponds exactly to the subdivisions of each unit of the specification.

A2 course structure (% of total A2 marks is shown in brackets)

Unit 4 Populations and environment
(33.33%) Chapters 1–8

Unit 5 Control in cells and in organisms
(46.67%) Chapters 9–16

Unit 6 Investigative and practical skills in A2 Biology
(20%) Online resources

Using the book

You will find that the A2 course builds on the skills and understanding you developed in your AS course. The topics in the A2 course will deepen your knowledge and understanding of biology. The course provides a solid biology foundation for those who intend to proceed to university to study biology or a related biological or medical science such as medicine, dentistry or veterinary science, where a good understanding of biology is necessary to fully appreciate key ideas, concepts and techniques.

New ideas are presented in the book in a careful step-by-step manner to enable you to develop a firm understanding of concepts and ideas. Biology at A2 Level will require you to describe and explain facts and processes in detail and with accuracy. However, the course is also about developing skills so that you can apply what you have learned. Examination papers will test skills such as interpreting new information, analysing experimental data, and evaluating information. In the AQA specification you will see sections which begin 'candidates should be able to ....' These are the sections which set out the skills you will need to develop to achieve success. You will find 'Application and How science works' features in the book. These features present relevant and challenging information which will enable you to develop these skills. The factual content of these sections is **not** required for examination purposes.

The AQA specification also emphasises how scientists work and how their work affects people in their everyday lives. For example, information is often presented in newspapers and on TV on science issues such as the possible side-effects of vaccines or drugs. Such reports may even contain conflicting evidence. The validity of evidence and the accuracy of conclusions is constantly questioned by scientists. Information in the text and in the accompanying resources will enable you to analyse evidence and data and to evaluate the way scientists obtain new evidence.

Checking your progress

You will find questions at the end of each chapter so that you can check your progress as you complete each section. Each chapter represents a manageable amount of learning so that you do not try to achieve too much too quickly. At the end of each unit there are questions written by AQA examiners in the same style that you will meet in examinations.

Investigative and practical skills

There are two routes for the assessment of Investigative and Practical Skills:

Either, **Route T:** Practical Skills Assessment (PSA) + Investigative Skills Assessment (ISA), which will be marked by your teacher.

Or, **Route X:** Practical Skills Verification (PSV) (assessed by your teacher) + Externally Marked Practical Assessment (EMPA), which is set and marked by an external AQA appointed examiner.

Both routes form 20% of the total A2 assessment and will involve carrying out practical work, collecting and processing data, and then using the data to answer questions in a written test. The resources which accompany the book provide examples of investigations so that you can develop your practical and investigative skills as you progress through the topics in Units 4 and 5.

The book and accompanying resources provide a wealth of material specifically written for your A2 Biology course. As well as helping you to achieve success, you should find the resources interesting and challenging.

How science works

You have already gained some skills through the 'How science works' component of your AS course. These skills will be developed further in your A2 course and are a key part of how every scientist works. Scientists use them to probe and test new theories and applications in whatever area they are working in. Now you will develop your scientific skills further and gain new ones as you progress through the course.

'How science works' is developed in this book through relevant features that are highlighted. The 'How science works' features will help you to develop the relevant 'How science works' skills necessary for examination purposes, but more importantly these features should give you a thorough grasp of how scientists work, as well as a deeper awareness of how science is used to improve the quality of life for everyone.

When carrying out their work scientists:

A Use their knowledge and understanding when observing objects or events, in defining a problem and when questioning the explanations of themselves or of other scientists.

B Make observations that lead to explanations in the form of hypotheses. In turn hypotheses lead to predictions that can be tested experimentally.

C Carry out experimental and investigative activities that involve making accurate measurements, and recording measurements methodically.

D Analyse and interpret data to look for patterns and trends, to provide evidence and identify relationships.

E Evaluate methodology, evidence and data, and resolve conflicting evidence.

F Appreciate that if evidence is reliable and reproducible and does not support a theory, the theory must be modified or replaced with a different theory.

G Communicate the findings of their research to provide opportunity for other scientists to replicate and further test their work.

H Evaluate, and report on the risks associated with new technology and developments.

I Consider ethical issues in the treatment of humans, other organisms and effects on the environment.

J Appreciate the role of the scientific community in validating findings and developments.

K Appreciate how society in general uses science to inform decision making.

UNIT

4

Populations and the environment

Chapters in this unit

1 Populations

2 ATP

3 Photosynthesis

4 Respiration

5 Energy and ecosystems

6 Nutrient cycles

7 Ecological succession

8 Inheritance and selection

Introduction

Living organisms live isolated lives but form part of interdependent communities. Each community interacts with other communities and with its non-living environment within ecosystems. While ecosystems as a whole remain relatively stable, their biotic and abiotic components are constantly changing. Ecosystems are maintained by light energy from the Sun that photosynthesising organisms convert to carbohydrates. These in turn are respired by all organisms to release the energy needed for survival. The Sun constantly provides energy, which flows through ecosystems. Other nutrients, by contrast, are finite and so have to be recycled.

This unit looks at how energy flows through ecosystems and how nutrients are recycled. It also covers the conflict between human needs and the conservation of natural resources. It explores how science works, using both information embedded in the text and specific applications designed to cover particular elements of this aspect of the specification. Each chapter in turn explores the following:

Populations looks at what populations are and how we can study them. In looking at the factors that affect population size, consideration is given to abiotic factors, competition and predation.

ATP covers energy and the synthesis, storage and role of ATP.

Photosynthesis describes the process by which plants convert the light energy of the Sun into ATP and how this is then used in the reduction of carbon dioxide to form carbohydrates. The factors affecting photosynthesis are also considered.

Respiration is an account of how carbohydrate is used by all organisms to release the energy required to carry out their activities and the different ways in which this is achieved, depending on whether free oxygen is available.

Energy and ecosystems explains how energy is transferred through ecosystems and the efficiency with which this is achieved. It goes on to compare natural ecosystems with those based on modern intensive farming.

Nutrient cycles explains how carbon and nitrogen are recycled in ecosystems and the role of microorganisms in these processes. It further considers the effect of human activities on these cycles, including the effects of using nitrogen fertilisers as well as the greenhouse effect and global warming.

Ecological succession discusses how bare land is colonised and its subsequent development into a climax community. How the management of succession is used to conserve habitats is also covered.

Inheritance and selection explains how characteristics are passed from generation to generation and how the alleles for these characters are distributed within populations. It concludes by looking at how these alleles are selected for and how geographical isolation can lead to the formation of new species.

What you already know

While the material in this unit is intended to be self-explanatory, there is certain information from GCSE that will prove very helpful to the understanding of the content of this unit. In addition to all AS material, a knowledge of the following elements of GCSE will be of assistance:

- Competition in plants and animals for the things they need.
- Photosynthesis and the factors that affect its rate.
- Aerobic respiration.
- Energy flow in ecosystems.
- The carbon cycle.
- The influence of humans on the environment.
- Principles of inheritance and how to construct genetic diagrams.
- Natural selection.

1 Populations

1.1 Populations and ecosystems

Learning objectives:

- What is meant by the terms 'environment', 'biotic', 'abiotic' and 'biosphere'?
- What is an ecosystem?
- What is meant by the terms 'population', 'community' and 'habitat'?
- What is a niche?

Specification reference: 3.4.1

Hint

Many students confuse the terms 'biotic' and 'abiotic'. 'Biotic' refers to living things, for example predation or competition between different species or individuals of one species. 'Abiotic' refers to non-living factors, such as temperature, rainfall or light intensity.

Figure 1 *Woodland ecosystem*

AQA Examiner's tip

'Population' is a very important concept that needs to be understood. It comes up in many areas of biology, including genetics and evolution.

In this chapter we shall look at how living organisms form communities within ecosystems through which energy is transferred and elements are recycled.

Ecology is the study of the inter-relationships between organisms and their environment. The environment includes both non-living (**abiotic**) components, such as temperature and rainfall, and living (**biotic**) components, such as competition and predation. Ecology is a complex area of study which includes most aspects of biology. It is, in effect, the study of the life-supporting layer of land, air and water that surrounds the Earth. This layer is called the **biosphere**.

Ecosystems

An ecosystem is made up of all the interacting biotic (living) and abiotic (non-living) features in a specific area. Ecosystems are more or less self-contained functional units. Within an ecosystem there are two major processes to consider:

- the flow of energy through the system
- the cycling of elements within the system.

An example of an ecosystem is a freshwater pond or lake. It has its own community of plants to collect the necessary sunlight energy to supply the organisms within it. Nutrients such as nitrates and phosphates are recycled within the pond or lake. There is little or no loss or gain between it and other ecosystems. Another example of an ecosystem is an oak woodland (Figure 1). Within each ecosystem, there are a number of species. Each species is made up of many groups of individuals that together make up a **population**.

Populations

A **population** is a group of interbreeding organisms of one **species** in a habitat. In the different habitats of an oak woodland there are populations of nettles, worms, woodpeckers, beetles, etc. The boundaries of a population are often difficult to define. In our oak woodland, for example, all the mature woodpeckers can breed with one another and so form a single population. However, the woodlice on a decaying log at one side of the wood can, in theory, breed with those on a log a kilometre or more away at the other side of the wood. In practice, the sheer distance makes interbreeding unlikely and therefore they can be considered as separate populations. Where exactly the boundary lies between these two populations is, however, unclear. Populations of different species form a **community**.

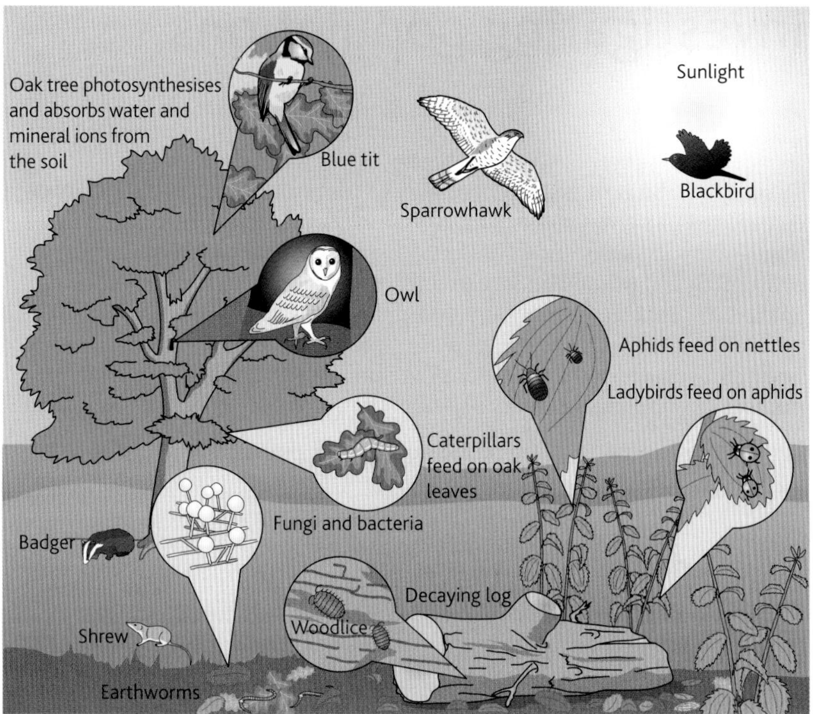

Figure 2 *Part of an oak woodland ecosystem*

Community

A **community** is defined as all the populations of different organisms living and interacting in a particular place at the same time. Within an oak woodland, a community may include a large range of organisms, such as oak trees, hazel shrubs, bluebells, nettles, sparrowhawks, blue tits, ladybirds, aphids, woodlice, earthworms, fungi and bacteria (see Figure 2).

Habitat

A **habitat** is the place where a community of organisms lives. Within an ecosystem there are many habitats. For example, in an oak woodland, the leaf canopy of the trees may be a habitat for blue tits while a decaying log is the habitat for woodlice. A stream flowing through the woodland provides a very different habitat, within which aquatic plants and water beetles live. For a water vole, the stream and its banks are its habitat. Within each habitat there are smaller units, each with their own microclimate. These are called **microhabitats**. The mud at the bottom of the stream may be the microhabitat for a bloodworm while a crevice on the bark of an oak tree may be the microhabitat for a lichen.

Ecological niche

A **niche** describes how an organism fits into the environment. A niche refers to where an organism lives and what it does there. It includes all the biotic and abiotic conditions required for an organism to survive, reproduce and maintain a viable population. Some species may appear very similar, but their nesting habits or other aspects of their behaviour will be different, or they may show different levels of tolerance to environmental factors, such as a pollutant or a shortage of oxygen or nitrates. No two species occupy exactly the same niche.

Hint

Organisms are found in places where the local environmental conditions fall within the range that their adaptations enable them to cope with.

AQA Examiner's tip

Make sure that you can accurately define the basic ecological terms described in this topic. They turn up frequently in examinations.

Figure 3 *This lake is an example of a habitat*

Summary questions

In the following passage, state the word that best replaces each of the numbers in brackets.

The study of the inter-relationships between organisms and their environment is called (1). The layer of land, air and water that surrounds the Earth is called the (2). An ecosystem is a more or less self-contained functional unit made up of all the living or (3) features and non-living or (4) features in a specific area. Within each ecosystem are groups of different organisms, called a (5), which live and interact in a particular place at the same time. A group of interbreeding organisms occupying the same place at the same time is called a (6), and the place where they live is known as a (7).

1.2 Investigating populations

Learning objectives:

- What factors should be considered when using a quadrat?
- How is a transect used to obtain quantitative data about changes in communities along a line?
- How is the abundance of different species measured?
- How can the mark-release-recapture method be used to measure the abundance of mobile species?

Specification reference: 3.4.1

To study a **habitat**, it is often necessary to count the number of individuals of a species in a given space. This is known as **abundance**. It is virtually impossible to identify and count every organism. To do so would be time-consuming and would almost certainly cause damage to the habitat being studied. For this reason only small samples of the habitat are usually studied in detail. As long as these samples are representative of the habitat as a whole, any conclusion drawn from the findings will be valid. There are a number of sampling techniques used in the study of habitats. These include:

- random sampling using frame quadrats or point quadrats
- systematic sampling along transects.

Quadrats

There are three factors to consider when using quadrats:

- **the size of quadrat to use**. This will depend upon the size of the plants or animals being counted and how they are distributed within the area. Larger species require larger quadrats. Where a species occurs in a series of groups rather than being evenly distributed throughout the area, a large number of small quadrats will give more representative results than a small number of large ones.
- **the number of sample quadrats to record within the study area**. The larger the number of sample quadrats the more reliable the results will be. As the recording of species within a quadrat is a time-consuming task a balance needs to be struck between the validity of the results and the time available. The greater the number of different species present in the area being studied, the greater the number of quadrats required to produce valid results.
- **the position of each quadrat within the study area**. To produce statistically significant results a technique known as random sampling must be used.

Random sampling

It is important that sampling is random to avoid any bias in collecting data. Avoiding bias ensures that the data obtained are valid. How do we make sampling random?

Suppose we wish to investigate the effects of grazing animals on the species of plants growing in a field. We begin by choosing two fields as close together as possible in order to minimise soil, climatic and other abiotic differences. One field is regularly grazed by animals such as sheep, whereas the other has not been grazed for many years. We then take random samples at many sites in each field by placing the quadrat on the ground and recording the names and numbers of every species found within the area of the quadrat.

But how do we get a truly random sample? We could simply stand in one of the fields and throw the quadrat over our shoulder. However, even with the best of intentions, it is difficult not to introduce an element of personal bias using this method. For example, are we more likely to stand in a dry area than a wet muddy one? Will we deliberately try to avoid an area covered in sheep droppings or full of nettles?

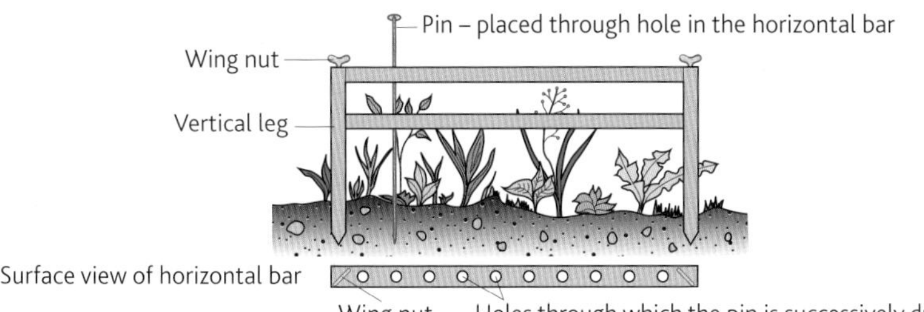

Figure 1 *A point quadrat*

A better method of random sampling is to:

1 Lay out two long tape measures at right angles, along two sides of the study area.
2 Obtain a series of coordinates by using random numbers taken from a table or generated by a computer.
3 Place a quadrat at the intersection of each pair of coordinates and record the species within it.

Systematic sampling along transects

It is sometimes more informative to measure the abundance and distribution of a species in a systematic rather than a random manner. This is particularly important where some form of transition in the communities of plants and animals takes place. For example, the distribution of organisms on a tidal seashore is determined by the relative periods of time that they spend under water and exposed to the air, that is, by their vertical height up the shore. The stages of zonation are especially well shown using transects. A line transect comprises a string or tape stretched across the ground in a straight line. Any organism over which the line passes is recorded. A belt transect is a strip, usually a metre wide, marked by putting a second line parallel to the first. The species occurring within the belt between the lines are recorded.

Measuring abundance

Random sampling with quadrats and counting along transects are used to obtain measures of **abundance**. Abundance is the number of individuals of a species within a given space. It can be measured in several ways, depending upon the size of the species being counted and the habitat. Examples include:

- **frequency**, which is the likelihood of a particular species occurring in a quadrat. If, for example, a species occurs in 15 out of 30 quadrats, the frequency of its occurrence is 50 per cent. This method is useful where a species, such as grass, is hard to count. It gives a quick idea of the species present and their general distribution within an area. However, it does not provide information on the density and detailed distribution of a species.
- **percentage cover**, which is an estimate of the area within a quadrat that a particular plant species covers. It is useful where a species is particularly abundant or is difficult to count. The advantages in these situations are that data can be collected rapidly and individual plants do not need to be counted. It is less useful where organisms occur in several overlapping layers (more probably plants).

To obtain reliable results, it is necessary to ensure that the sample size is large, that is, many quadrats are used and the mean of all the samples is obtained. The larger the number of samples, the more representative of the community as a whole will be the results.

How science works

Types of quadrat

A point quadrat consists of a horizontal bar supported by two legs. At set intervals along the horizontal bar are ten holes, through each of which a long pin may be dropped (Figure 1). Each species that the pin touches is then recorded.

A frame quadrat is a square frame divided by string or wire into equally sized subdivisions (Figure 2). It is often designed so that it can be folded to make it more compact for storage and transport. The quadrat is placed in different locations within the area being studied. The abundance of each species within the quadrat is then recorded.

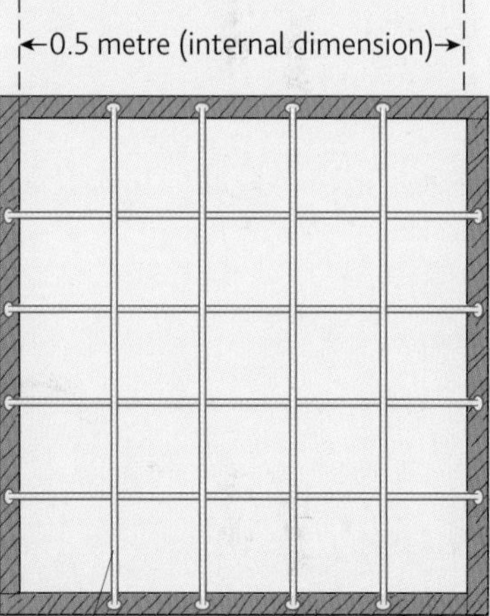

Figure 2 *A frame quadrat*

Figure 3 *Students carrying out fieldwork*

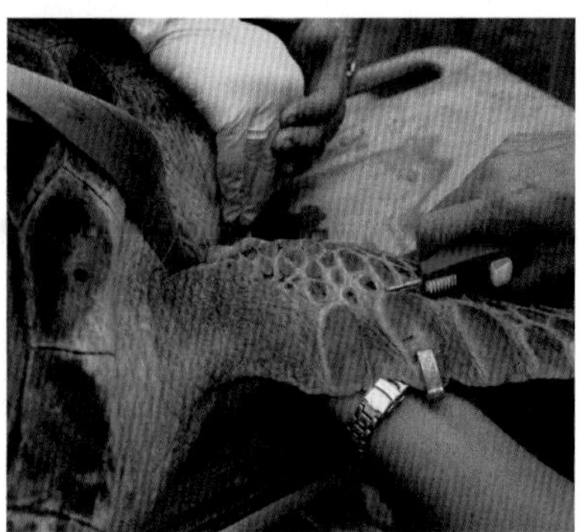

Figure 4 *A green sea turtle with a flipper tag used in estimating population size by the mark-release-recapture technique*

Mark-release-recapture techniques

The methods of measuring abundance described above work well with plant communities but not with most animals. Most animals are mobile and move away when approached. Others are often hidden and are therefore difficult to find and identify. To estimate the abundance of most animals requires an altogether different technique.

A known number of animals are caught, marked in some way, and then released back into the community. Some time later, a given number of individuals is collected randomly and the number of marked individuals is recorded. The size of the population is then calculated as follows:

$$\text{estimated population size} = \frac{\text{total number of individuals in the first sample} \times \text{total number of individuals in the second sample}}{\text{number of marked individuals recaptured}}$$

This technique relies on a number of assumptions:

- The proportion of marked to unmarked individuals in the second sample is the same as the proportion of marked to unmarked individuals in the population as a whole.
- The marked individuals released from the first sample distribute themselves evenly amongst the remainder of the population and have sufficient time to do so.
- The population has a definite boundary so that there is no immigration into or emigration out of the population.
- There are few, if any, deaths and 'births' within the population.
- The method of marking is not toxic to the individual nor does it make the individual more conspicuous and therefore more liable to predation.
- The mark or label is not lost or rubbed off during the investigation.

Link

Standard deviations form part of 3.2.1 of the AS specification and are covered in Topic 7.2 of the *AS Biology* book.

Analysing data

The quantitative data collected from an ecological survey then have to be analysed and interpreted. The first stage is usually to present the data in the form of a table or graph. This makes it easier to compare data, for example, from two different locations. Such comparisons can be made more precisely using statistical analysis of the data. Comparing the mean of two sets of data is helpful but, as we saw during the AS Biology course, this tells us nothing of the spread of the data about the mean. It is therefore useful to calculate the standard deviation as well.

Only tentative conclusions may be drawn from comparing two sets of data as there are many factors that can contribute to these differences. One such factor, that chance alone is the reason for the difference, can be checked statistically. This can be done using one of the various methods for testing the significance of differences between two sets of data (e.g. the *t*-test). The results show whether these differences are due to some particular factor or are just a matter of chance.

Data can also be analysed for possible correlations and causes. You should remember from the AS course that two factors are said to be correlated when they vary in relation to each other. A positive correlation is where an increase in the value of one variable is accompanied by an increase in the value of the other variable, for example, the rate of photosynthesis increases with an increase in light intensity. A negative correlation is where an increase in the value of one variable is accompanied by a decrease in the value of the other variable.

Statistical tests can be used to calculate the strength and direction of any correlation between two variables. One example is the Spearman rank correlation. You may be required to use this in your practical work. Showing that two variables correlate statistically does not prove that one causes the other. The numbers of two species in a population may correlate very well but it is possible that both of them are affected by the same environmental factor, for example, a rise in temperature.

Application and How science works

Ethics and fieldwork

Understanding the complex inter-relationships between organisms in a community helps us to minimise the effect of human activities on the environment and ensure these communities are conserved. For us to understand communities we need to collect data about them. However, the very collection of such data may be harmful to the communities that we are trying to conserve. To minimise the impact of an ecological investigation on the environment a number of basic procedures should be observed:

- Where possible, the organisms should be studied *in situ*. If it is necessary to remove them, the numbers taken should be kept to the absolute minimum.
- Any organisms removed from a site should be returned to their original habitat as soon as possible. This applies even if they are dead.
- A sufficient period of time should elapse before a site is used for future studies.
- Disturbance and damage to the habitat should be avoided. Trampling, overturning stones, permanently removing organisms, etc. can all adversely affect a habitat.

Field studies often lead to varying degrees of damage to habitats and some of the organisms they contain. The important thing is that there is an appropriate balance between the damage done and the value of the information gained. This an example of How science works (HSW: J).

1 Suggest a reason why even dead organisms should be returned to the habitat from which they came.

2 Suggest why it is beneficial to a habitat that further investigations are not carried out too soon after an initial study.

3 In the study of a seashore, students turn over large stones to record the numbers of different organisms on their underside. Suggest reasons why it is important that these stones are replaced the same way up as they were originally.

4 In the case of experienced ecologists obtaining data that enables habitats to be conserved, the benefits usually outweigh any damage that they cause to the habitats. This makes their work ethically justifiable. It might be said that the same is not true of school or college students performing field studies. Give reasons for and against A-level students carrying out ecological investigations in the field.

AQA Examiner's tip

You will not be required to know tests of significance, such as the Spearman rank correlation or the chi-squared (χ^2) test, for theory examination purposes. However, you **will** be required to select and apply such tests when analysing the results of experiments and investigations carried out as part of your ecological fieldwork.

Summary questions

1 An ecologist was estimating the population of sandhoppers on a beach. One hundred sandhoppers were collected, marked and released again. A week later 80 sandhoppers were collected, of which five were marked. Calculate the estimated size of the sandhopper population on the beach. Show your working.

2 When using the mark-release-recapture technique, explain how each of the following might affect the final estimate of a population.

a The marks put on the individuals captured in the first sample make them more easily seen by predators and so proportionately more are eaten than unmarked individuals.

b Between the release of marked individuals and the collection of a second sample an increased 'birth' rate leads to a very large increase in the population.

c Between the release of marked individuals and the collection of a second sample, disease kills large numbers of all types of individual.

1.3 Variation in population size

Learning objectives:

- What factors determine the size of a population?
- Which abiotic factors affect the size of a population?
- How do each of these factors influence population size?

Specification reference: 3.4.1

Figure 1 *A population of lesser flamingos*

Hint

Humans exist in populations just like other species and therefore the rules also apply to us.

Hint

Algae are a group of mostly aquatic photosynthetic organisms belonging to the kingdom Protoctista. Although many are unicellular, some seaweeds can be up to 45 m long.

A population is a group of interbreeding individuals of the same species in a habitat. The number of individuals in a population is the **population size**. We saw in Topic 1.1 that all the populations of the different organisms that live and interact together are known as a community.

Population growth curves

The usual pattern of growth for a natural population has three phases (see Figure 2):

1. a period of slow growth as the initially small number of individuals reproduce to slowly build up their numbers
2. a period of rapid growth where the ever-increasing number of individuals continue to reproduce. The population size doubles during each interval of time, as seen by the gradient of the curve in Figure 2, which becomes increasingly steep
3. a period when the population growth declines until its size remains more or less stable. The decline may be due to the food supply limiting numbers or to increased predation. The graph therefore levels out with only cyclic fluctuations due to variations in factors such as food supply or the population size of predators.

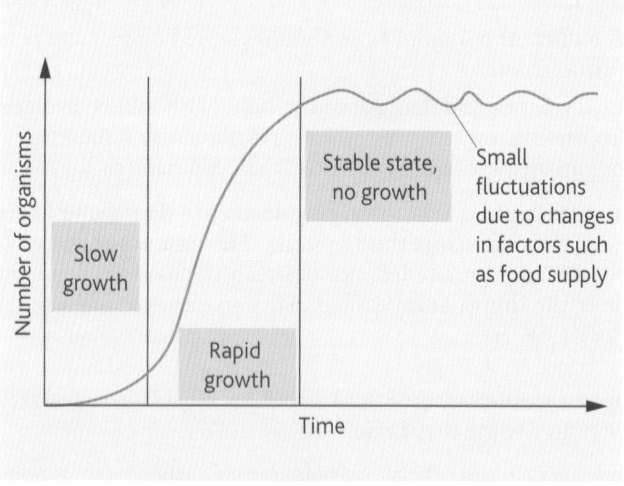

Figure 2 *Growth curve of most populations*

Population size

Imagine a situation in which a single algal cell, capable of asexual reproduction, is placed in a newly created pond. It is summer and so there is plenty of light and the temperature of the water is around 12 °C; mineral nutrients have been added to the water. In these circumstances the algal cell divides rapidly because all the factors needed for the growth of the population are present. There are no **limiting factors**. In time, however, things change. For example:

- Mineral ions are used up as the population becomes larger.

- The population becomes so large that the algae at the surface prevent light reaching those at deeper levels.
- Other species are introduced into the pond, carried by animals or the wind, and some of these species may use the algae as food or compete for light or minerals.
- Winter brings much lower temperatures and lower light intensity of shorter duration.

In short, the good life ends and the going gets tough. As a result the growth of the population slows, and possibly ceases altogether, and the population size may even diminish. Ultimately the population is likely to reach a relatively constant size. There are many factors, living (biotic) and non-living (abiotic), which affect this ultimate size. Changes in these factors will influence the rate of growth and the final size of the population.

In summary, no population continues to grow indefinitely because certain factors limit growth, for example, the availability of food, light, water, oxygen and shelter, and the accumulation of toxic waste, disease and predators. Each population has a maximum size that can be sustained over a relatively long period and this is determined by these limiting factors.

The various limiting factors that affect the size of a population are of two basic types:

- **Abiotic factors** are concerned with the non-living part of the environment.
- **Biotic factors** are concerned with the activities of living organisms and include, for example, competition and predation. We shall look at the effects of biotic factors on population size in Topics 1.4. and 1.5.

Abiotic factors

The abiotic conditions that influence the size of a population include:

- **temperature**. Each species has a different optimum temperature at which it is best able to survive. The further away from this optimum, the smaller the population that can be supported. In plants and cold-blooded animals, as temperatures fall below the optimum, the enzymes work more slowly and so their metabolic rate is reduced. Populations therefore grow more slowly. At temperatures above the optimum, the enzymes work less efficiently because they gradually undergo **denaturation**. Again the population grows more slowly.

  The warm-blooded animals, that is, birds and mammals, can maintain a relatively constant body temperature regardless of the external temperature. Therefore you might think that their population growth and size would be unaffected by temperature. However, the further the temperature of the external environment gets from their optimum temperature, the more energy these organisms expend in trying to maintain their normal body temperature. This leaves less energy for individual growth and so they mature more slowly and their reproductive rate slows. The population size therefore gets smaller.
- **light**. As the ultimate source of energy for **ecosystems**, light is a basic necessity of life. The rate of photosynthesis increases as light intensity increases. The greater the rate of photosynthesis, the faster plants grow and the more spores or seeds they produce. Their population

Figure 3 *The collared dove only arrived in Britain in the 1950s but its population has increased rapidly since then*

Figure 4 *A population of migrating birds, like these terns, fluctuates seasonally*

Hint

Remember that the growth of any population is eventually slowed by a limiting factor.

Hint

A species can only live within a certain range of abiotic factors and this range differs from species to species.

Link

To remind yourself of the effects of temperature and pH on enzyme action revisit Topic 2.7 in the *AS Biology* book.

Figure 5 *This cactus is adapted to survive in conditions where water is scarce. Its population in dry regions is therefore relatively large as there is little competition from other species, most of which are not adapted to survive in such conditions*

growth and size is therefore potentially greater. In turn, the population of animals that feed on plants is potentially larger.

- **pH**. This affects the action of enzymes. Each enzyme has an optimum pH at which it operates most effectively. A population of organisms is larger where the appropriate pH exists and smaller, or non-existent, where the pH is very different from the optimum.
- **water and humidity**. Where water is scarce, populations are small and consist only of species that are well adapted to living in dry conditions. Humidity affects the **transpiration** rates in plants and the evaporation of water from the bodies of animals. Again, in dry air conditions, the populations of species adapted to tolerate low humidity will be larger than those with no such adaptations.

Application

The influence of abiotic factors on plant populations

Species X and species Y are two species of flowering plants. Each is able to tolerate different temperatures and different pHs. The chart (Figure 6) below illustrates the way each species is able to tolerate each of these two abiotic factors.

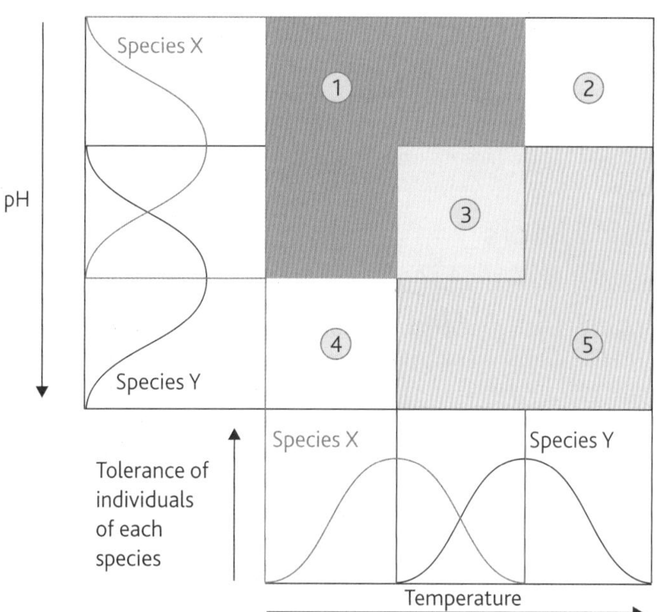

Figure 6

1. State the numbered box that best fits each of the descriptions below.
   a. Only a population of species X is found.
   b. Both temperature and pH allow a population of both species to exist.
   c. The temperature is too high for a population of species X and the pH is too low for a population of species Y.
   d. There is competition between species X and species Y.
2. Explain why there is no population of either species in box 4.

Summary questions

1. Why do populations never grow indefinitely?
2. Distinguish between biotic and abiotic factors.
3. Suggest the level and type of abiotic factor that is most likely to limit the population size of the organisms and their habitats given below.
   a. Ground plants on a forest floor
   b. Hares in a sandy desert
   c. Bacteria on the summit of a high mountain.

1.4 Competition

Learning objectives:

- What is intraspecific competition?
- What factors do different species compete for?
- What is interspecific competition?
- How does interspecific competition influence population size?

Specification reference: 3.4.1

Where two or more individuals share any resource (e.g. light, food, space, oxygen) that is insufficient to satisfy all their requirements fully, then competition results. Where such competition arises between members of the same species it is called **intraspecific competition**. Where it arises between members of different species it is termed **interspecific competition**.

Intraspecific competition

Intraspecific competition occurs when individuals of the *same* species compete with one another for resources such as food, water, breeding sites, etc. It is the availability of such resources that determines the size of a **population**. The greater the availability, the larger the population. The lower the availability, the smaller the population. Examples of intraspecific competition include:

- limpets competing for algae, which is their main food. The more algae available, the larger the limpet population becomes.
- oak trees competing for resources. In a large population of small oak trees some will grow larger and restrict the availability of light, water and minerals to the rest, which then die. In time the population will be reduced to relatively few large dominant oaks.
- robins competing for breeding territory. Female birds are normally only attracted to males who have established territories. Each territory provides adequate food for one family of birds. When food is scarce, territories have to become larger to provide enough food. There are therefore fewer territories in a given area and fewer breeding pairs, leading to a smaller population size.

Interspecific competition

Interspecific competition occurs when individuals of *different* species compete for resources such as food, light, water, etc. Where populations of two species initially occupy the same **niche**, one will normally have a competitive advantage over the other. The population of this species will gradually increase in size while the population of the other will diminish. If conditions remain the same, this will lead to the complete removal of one species. This is known as the competitive exclusion principle.

This principle states that where two species are competing for limited resources, the one that uses these resources most effectively will ultimately eliminate the other. In other words, no two species can occupy the same niche indefinitely when resources are limiting. Two species of sea birds, shags and cormorants, appear to occupy the same niche, living and nesting on the same type of cliff face and eating fish from the sea. Analysis of their food, however, shows that shags feed largely on sand eels and herring, whereas cormorants eat mostly flat fish, gobies and shrimps. They therefore occupy different niches.

To show how a factor influences the size of a population it is necessary to link it to the birth rate and death rate of individuals in a population. For example, an increase in food supply does not necessarily mean there will be more individuals; it could just result in bigger individuals. It is therefore important to show how a factor, such as a change in food

Hint

Which of two species in a niche has the competitive advantage depends upon the conditions at any point in time. If one species can tolerate a higher temperature than another, a rise in environmental temperature will favour it. If however there is a fall in environmental temperature, the other species is more likely to become dominant.

Summary questions

1 Distinguish between intraspecific competition and interspecific competition.

2 Name any two resources that species compete for.

supply, affects the number of individuals in a population. For example, a decrease in food supply could lead to individuals dying of starvation and directly reduce the size of a population. An increase in food supply means that more individuals are likely to survive and so there is an increased probability that they will produce offspring and the population will increase. This effect therefore takes longer to influence population size.

Application

The effects of interspecific competition on population size

The red squirrel is native to the British Isles and exclusively occupied a particular niche until around 130 years ago, when the grey squirrel was introduced from North America. Since then the two species have been competing for food and territory. There are now an estimated 2.5 million grey squirrels and just 160 000 red squirrels in the British Isles. The red squirrel population occurs mostly in Wales and Scotland, with smaller groups in northeastern England and on islands such as Anglesey and the Isle of Wight. Figure 1 illustrates the changes in red and grey squirrel populations in Wales and Scotland between 1970 and 1990.

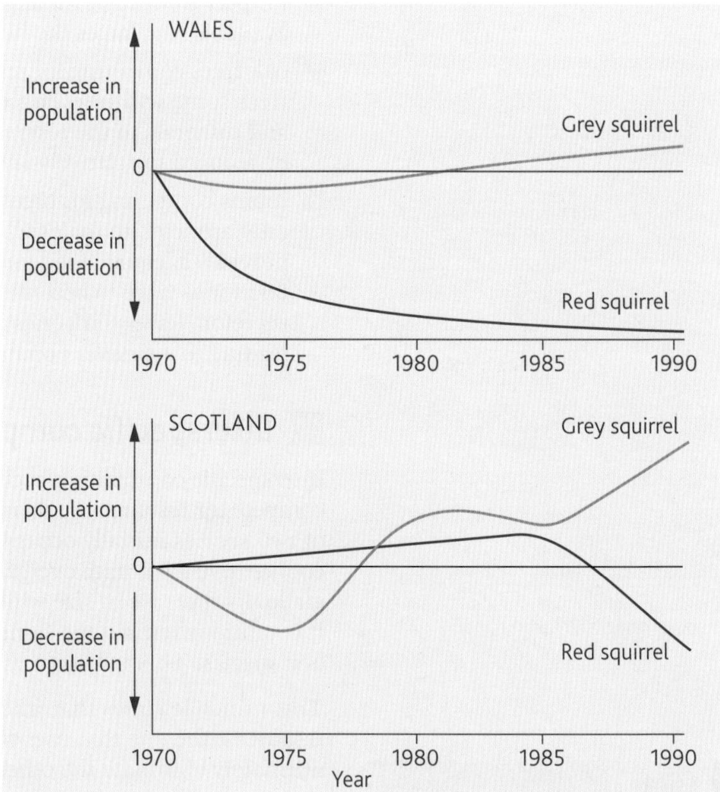

Figure 1 *Changes in red and grey squirrel populations in Wales and Scotland between 1970 and 1990. The lines show changes in comparison with the 1970 population*

In many cases we suspect that competition is the reason for variations in population. In practice it is difficult to prove for a number of reasons:

- There are many other factors that influence population size, such as abiotic factors.
- A causal link has to be established to show that competition is the cause of an observed correlation.

- There is a time lag in many cases of competition and so a population change may be due to competition that took place many years earlier.
- Data on natural population sizes are hard to obtain and not always reliable.

Study Figure 1 and answer the following questions.

1 State one piece of evidence from the graph for Scotland which shows that changes in the red squirrel population are due to competition from the grey squirrel.

2 In Wales the populations of both grey and red squirrels declined between 1970 and 1975. Suggest a possible reason for this.

3 Both types of squirrels eat nuts, seeds and fruit as part of their diet. Grey squirrels spend more time foraging on the forest floor than red squirrels. Suggest how this behaviour might give the grey squirrel a competitive advantage over the red squirrel.

4 Suggest an explanation why islands such as Anglesey and the Isle of Wight still have significant red squirrel populations while they have disappeared from much of the rest of England and Wales.

Figure 2 *Red squirrel*

Figure 3 *Grey squirrel*

Application and How science works

Competing to the death

In an experiment, two species of a genus of unicellular organism called *Paramecium* were grown separately in different test tubes that contained yeast as a source of food. The two species were then grown together in the same test tube – again with yeast as a food source. In each case the populations of both species were measured over a period of 20 days. The results are shown in the graph in Figure 4. This is an example of How science works (HSW: E).

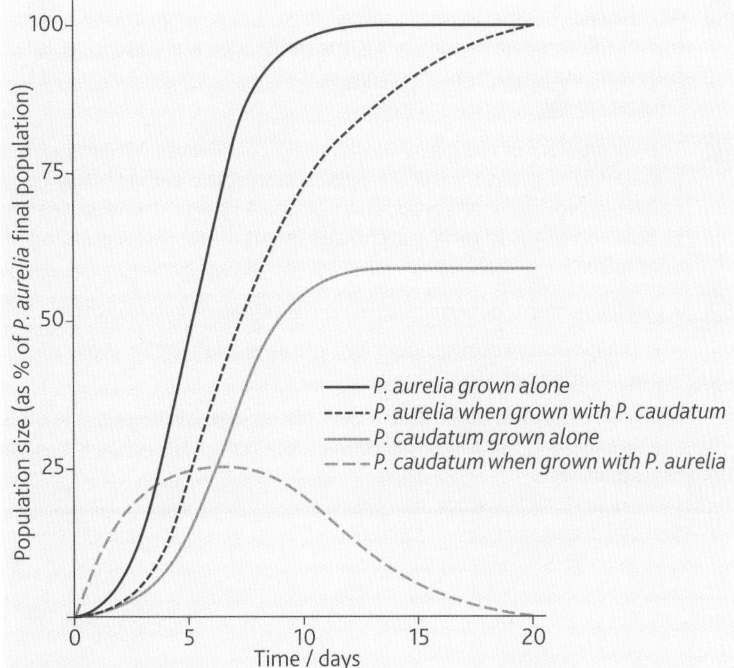

Figure 4 *Population growth of* Paramecium aurelia *and* P. caudatum *grown separately and together*

Hint

Although the population of one species may increase as another decreases, this does not prove that this is due to direct competition between them. To be certain, it is necessary to establish a causal link for the observed correlation.

1 Describe the population growth curve of *P. caudatum* when grown alone over the 20-day period.

2 Compare the population growth curve of *P. caudatum* when grown with *P. aurelia* to the curve when *P. caudatum* is grown alone.

3 Suggest an explanation for the difference in the final population size of *P. caudatum* when grown with *P. aurelia* compared to when it is grown alone.

4 Suggest why the growth rate of *P. aurelia* is slower in the presence of *P. caudatum* than when grown alone.

5 Suggest why, after 20 days, the population size of *P. aurelia* grown with *P. caudatum* is the same as that when *P. aurelia* is grown alone.

Application

Effects of abiotic and biotic factors on population size

Oak trees produce acorns in the autumn. Deer mice feed on acorns. Table 1 shows the dry mass of acorns produced per hectare (ha) from 1992 to 1997 in an area of woodland. It also shows the estimated population size of deer mice per hectare of the same area of woodland in the spring of each year from 1993 to 1998.

Table 1

Year	Dry mass of acorns/kg ha^{-1} produced in autumn	Estimated deer mice population/ number ha^{-1} in spring
1992	28	–
1993	131	260
1994	318	550
1995	211	1320
1996	726	990
1997	39	3440
1998	–	340

1 Suggest a method by which the population of deer mice might be estimated.

2 Calculate the mean annual growth rate in deer mice population over the period 1993 to 1995. Show your working.

3 With reference to the data in the table, describe the relationship between acorn production in autumn and the deer mice population the following spring.

4 Acorn seeds begin to form in spring. It has been suggested that the higher the temperature in spring, the more acorns are produced the following autumn. From the table, state which year probably had the coldest spring.

5 The caterpillars of the gypsy moth feed on oak leaves. When the population of gypsy moth caterpillars is large, the damage they cause to oak trees reduces acorn production. Suggest how and why a rise in the population of gypsy moth caterpillars might affect the population of deer mice.

6 As well as acorns, deer mice also eat the pupae of gypsy moths.

a Explain how a warm spring might result in a fall in the gypsy moth population the following year.

b Owls are natural predators of deer mice. Suggest the possible effect of an increase in the owl population on the production of acorns. Explain your answer.

1.5 Predation

Learning objectives:

- What is predation?
- How does the predator–prey relationship affect the population size of the predator and prey?

Specification reference: 3.4.1

In Topic 1.4 we looked at interspecific competition. We shall now turn our attention to another type of interspecific relationship: the predator–prey relationship. A **predator** is an organism that feeds on another organism, known as their **prey**.

As predators have evolved they have become better adapted for capturing their prey: faster movement, more effective camouflage, better means of detecting prey. Prey have equally become more adept at avoiding predators: better camouflage, more protective features such as spines, concealment behaviour. In other words the predator and the prey have evolved together. If either of them had not matched the improvements of the other, it would most probably have become extinct.

Predation

Predation occurs when one organism is consumed by another. When a population of a predator and a population of its prey are brought together in a laboratory, the prey is usually exterminated by the predator. This is largely because the range and variety of the habitat provided is normally limited to the confines of the laboratory. In nature the situation is different. The area over which the population can travel is far greater and the variety of the environment is much more diverse. In particular, there are many more potential refuges. In these circumstances some of the prey can escape predation, so although the prey population falls to a low level, it rarely becomes extinct.

Evidence collected on predator and prey populations in a laboratory does not necessarily reflect what happens in the wild. At the same time, it is difficult to obtain reliable data on natural populations because it is not possible to count all the individuals in a natural population. Its size can only be estimated from sampling and surveys. These are only as good as the techniques used, none of which guarantee complete accuracy. We must therefore treat all data produced in this way with caution.

Effect of predator–prey relationship on population size

The relationship between predators and their prey and its effect on population size can be summarised as follows:

- Predators eat their prey, thereby reducing the population of prey.
- With fewer prey available the predators are in greater competition with each other for the prey that are left.
- The predator population is reduced as some individuals are unable to obtain enough prey for their survival.
- With fewer predators left, fewer prey are eaten.
- The prey population therefore increases.
- With more prey now available as food, the predator population in turn increases.

This general predator–prey relationship is illustrated in Figure 1. In natural **ecosystems**, however, organisms eat a range of foods and therefore the fluctuations in population size shown in the graph are often less severe.

AQA Examiner's tip

When asked to describe predator–prey relationships from a graph in examinations, candidates often just write 'one goes up and then the other one does'. To gain credit you should use names to describe precisely the changes taking place.

Summary questions

1 Explain why a predator population often exterminates its prey population in a laboratory but rarely does so in natural habitats.

2 Explain how a fall in the population of a predator can lead to a rise in its prey population.

3 A species of mite (A) is fed on oranges in a laboratory tank until its population is stable. A second mite species (B), that preys on species A, is introduced into the tank. Sketch a graph of the likely cycle of population change that the two species will undergo. Explain the changes that the graph illustrates.

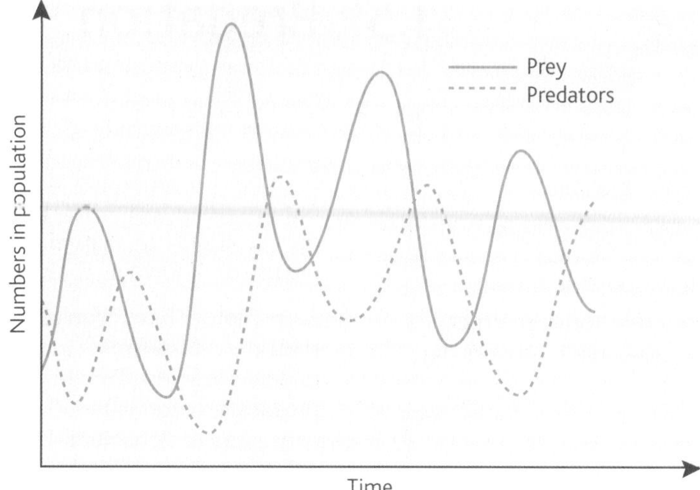

Figure 1 *Relationships between prey and predator populations*

Although predator–prey relationships are significant reasons for cyclic fluctuations in populations, they are not the only reasons; disease and climatic factors also play a part. These periodic population crashes are important in evolution as they create a **selection pressure** whereby only those individuals who are able to escape predators, or withstand disease or an adverse climate, will survive to reproduce. The population therefore evolves to be better adapted to the prevailing conditions.

Application and How science works

The Canadian lynx and the snowshoe hare

The long-term study of the predator–prey relationship of the Canadian lynx and the snowshoe hare was made possible because records exist of the number of furs traded by companies such as the Hudson Bay Company in Canada over 200 years. By analysing these records the relative population size of the Canadian lynx and the snowshoe hare can be determined. The data collected are shown as a graph in Figure 3.

1 What assumption is being made if we use the number of each type of fur traded as a measure of the population size of each species?

2 Describe the changes that occur in the populations of Canadian lynx and snowshoe hare.

3 Explain the changes that you have described.

It has long been observed that the population of snowshoe hares fluctuates in cycles. The question is whether these fluctuations are due mostly to predation by the lynx, mostly to changes in the food supply or mostly to a combination of both. To find out, ecologists fenced off 1 km^2 areas of coniferous forest in Canada where the hares lived. Separate areas were treated in four different ways:

- In the first set of areas, the hares were given extra food.
- In the second set of areas, lynx were excluded.
- In the third set of areas, the hares were given extra food and lynx were excluded.
- In the fourth set of areas, conditions were left unaltered as a control.

The results of the experiment are shown in Figure 4. This is an example of How science works (HSW: B).

Figure 2 *Canadian lynx catching a snowshoe hare*

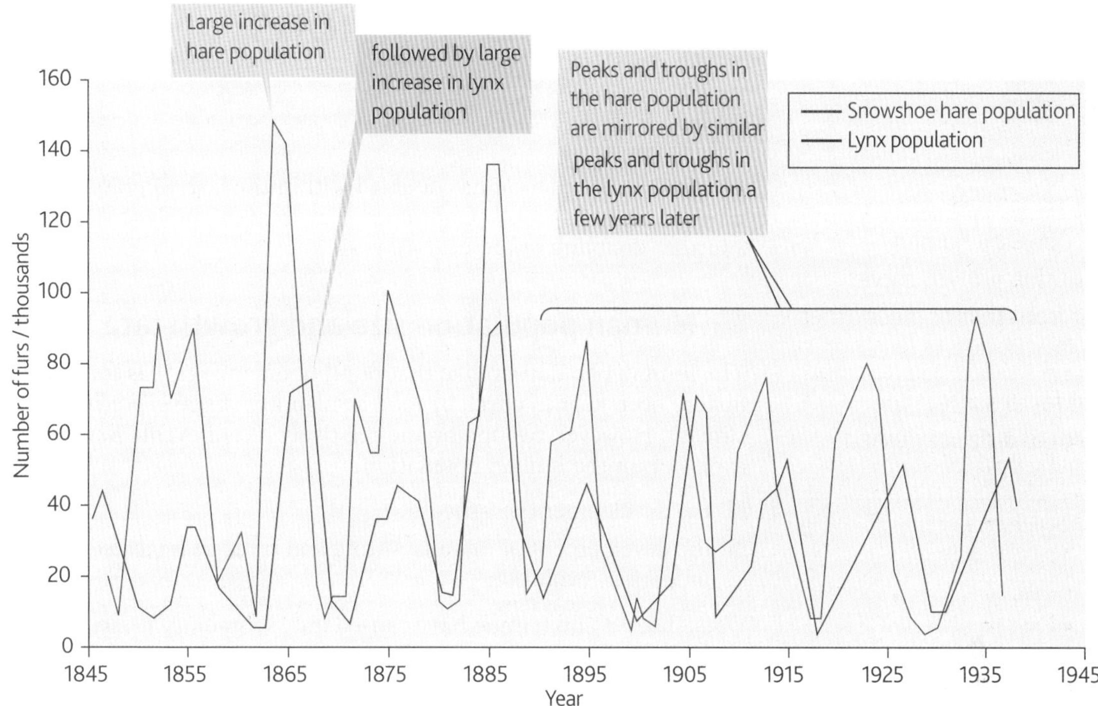

Figure 3 *The predator–prey relationship illustrated by the number of snowshoe hare and lynx trapped for the Hudson Bay Company between 1845 and 1940*

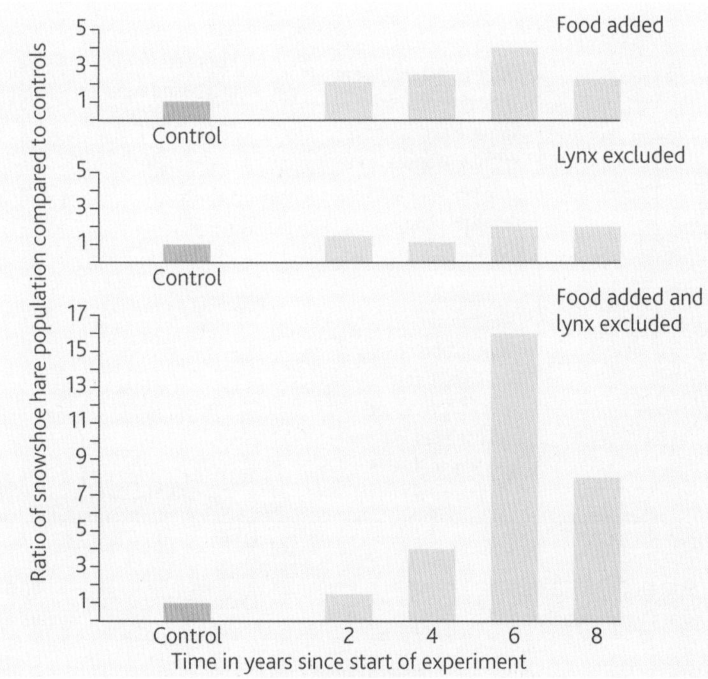

Figure 4 *Snowshoe hare population experiment*

4 By how many times had the addition of food increased the population after 6 years compared to the control?

5 Which had the greater influence on the population of hares: the addition of food or the exclusion of the lynx? Give a reason for your answer.

6 What conclusions can be drawn from this experiment?

1.6 Human populations

Learning objectives:

- How does the human population growth curve differ from that of most other organisms?
- What factors affect the growth and size of human populations?

Specification reference: 3.4.1

The human population has doubled within the last 50 years and now exceeds 6 billion people. Continued growth at the present rate will lead to a further doubling in the next 40 years.

Human population size and growth rate

The human population, like that of other organisms, has for most of our history been kept in check by food availability, disease, predators and climate. However, two major and relatively recent events have led to an explosion in the human population.

- the development of agriculture
- the development of manufacturing and trade that created the industrial revolution.

Wars, disease and famine have caused only temporary reversals in this upward trend. Consequently, the usual sigmoid population growth curve (see Topic 1.3, Figure 1), which is typical of other populations, is not followed by the human population. The exponential phase, in which the population grows rapidly, continues rather than gives way to the stationary phase in which the population stabilises (Figure 1).

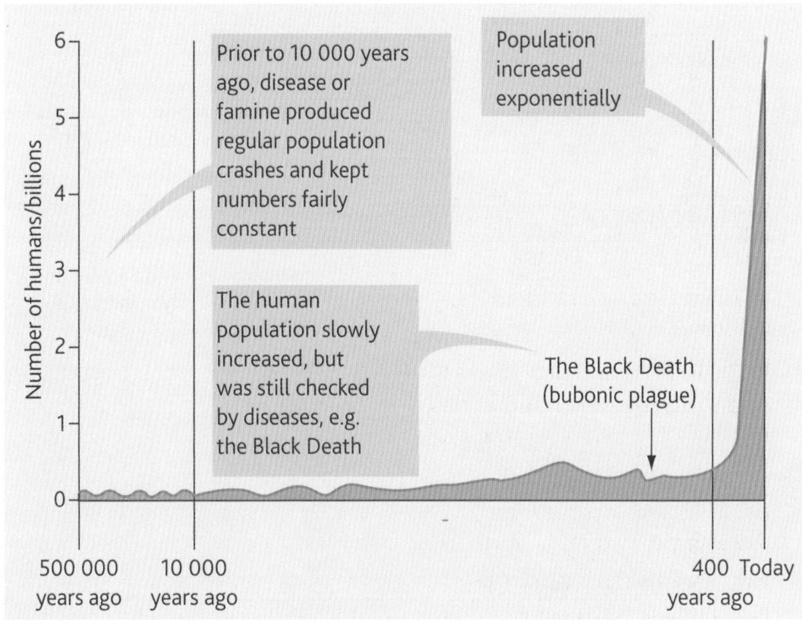

Figure 1 *The human population growth curve*

Factors affecting the growth and size of human populations

The basic factors that affect the growth and size of human populations are the **birth rate** and the **death rate**. It is the balance between these two factors that determines whether a human population increases, decreases or remains the same.

Individual populations are further affected by **migration**, which occurs when individuals move from one population to another. There are two types of migration:

- **immigration**, where individuals join a population from outside
- **emigration**, where individuals leave a population.

Again it is a balance between these two components that affects population size.

$$\text{population growth} = (\text{births} + \text{immigration}) - (\text{deaths} + \text{emigration})$$

$$\text{percentage population growth rate (in a given period)} = \frac{\text{population change during the period}}{\text{population at the start of the period}} \times 100$$

Figure 2 *The human population*

Factors affecting birth rates

Birth rates are affected by:

- **economic conditions**. Countries with a low per capita income tend to have higher birth rates.
- **cultural and religious backgrounds**. Some countries encourage larger families and some religions are opposed to birth control.
- **social pressures and conditions**. In some countries a large family improves social standing.
- **birth control**. The extent to which contraception and abortion are used markedly influences the birth rate.
- **political factors**. Governments influence birth rates through education and taxation policies.

$$\text{birth rate} = \frac{\text{number of births per year}}{\text{total population in the same year}} \times 1000$$

Factors affecting death rates

Death rates are affected by:

- **age profile**. The greater the proportion of elderly people in a population, the higher the death rate is likely to be.
- **life expectancy at birth**. The residents of economically developed countries live longer than those of economically less developed countries.
- **food supply**. An adequate and balanced diet reduces death rate.
- **safe drinking water and effective sanitation**. Reduce death rate by reducing the risk of contracting water-borne diseases such as cholera.
- **medical care**. Access to healthcare and education reduces the death rate.
- **natural disasters**. The more prone a region is to drought, famine or disease, the higher its death rate.
- **war**. Deaths during wars produce an immediate drop in population and a longer term fall as a result of fewer fertile adults.

$$\text{death rate} = \frac{\text{number of deaths per year}}{\text{total population in the same year}} \times 1000$$

Population structure

In most economically well-developed nations, such as those of North America and Europe, there has been a large increase in life expectancy. This has led to a change in societies, from those where life expectancy is short and birth rates are high, to those where life expectancy is long and birth rates are low. This change is called **demographic transition**.

Hint

It is not always possible to make predictions based on what went before. In the case of human populations, previously unknown factors can arise that affect populations in unknown ways, for example, HIV and AIDS, or obesity.

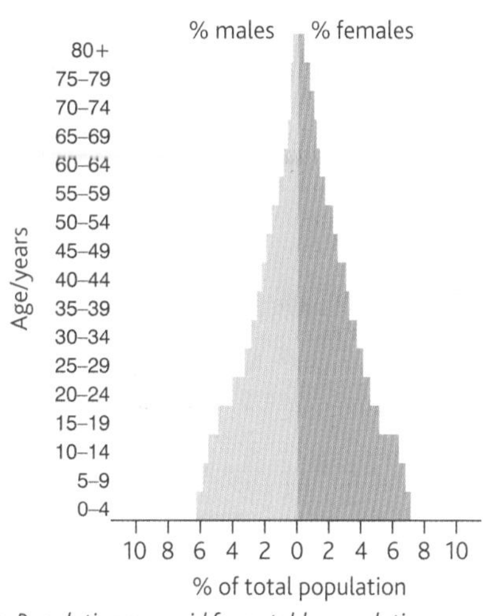

a *Population pyramid for a stable population*

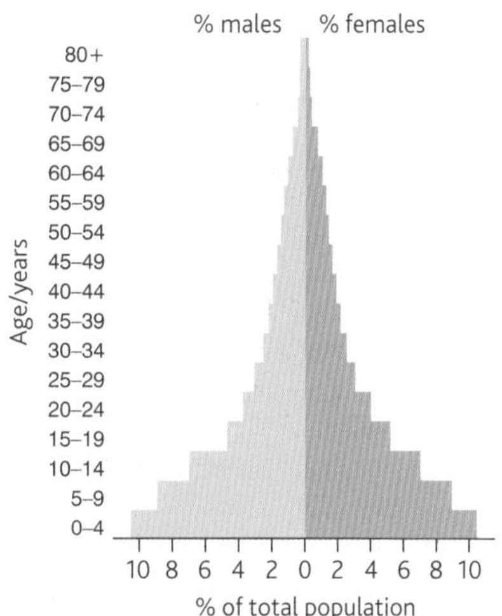

b *Population pyramid for an increasing population*

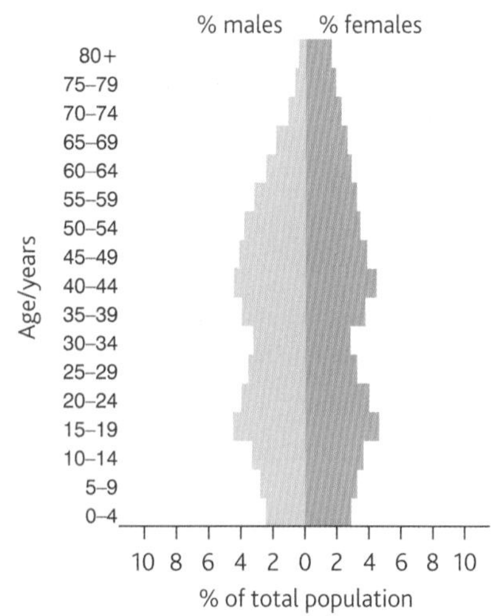

c *Population pyramid for a decreasing population*

Figure 3 *Age population pyramids*

It leads to a levelling-off of the population and the re-establishment of the typical sigmoid population growth curve. It is uncertain, however, whether demographic transition will affect all countries. It does not automatically follow that what happens in one population will happen in another as they develop economically. There are many different factors that influence human population size.

As the future size of a human population depends upon the number of females of child-bearing age, it is useful to know the age and gender profile of a population. This is displayed graphically by a series of stacked bars representing the percentages of males and females in each age group. These graphs, called **age population pyramids**, give useful information on the future trends of different populations. Three typical types of population are represented in the age population pyramids in Figure 3. These are:

- **stable population** (Figure 3a), where the birth rate and death rate are in balance and so there is no increase or decrease in the population size.
- **increasing population** (Figure 3b), where there is a high birth rate, giving a wider base to the population pyramid (compared to a stable population) and fewer older people, giving a narrower apex to the pyramid. This type of population is typical of economically less developed countries.
- **decreasing population** (Figure 3c), where there is a lower birth rate (narrower base of the population pyramid) and a lower mortality rate leading to more elderly people (wider apex to pyramid). This type of population occurs in certain economically more developed countries, such as Japan.

Survival rates and life expectancy

A survival curve plots the number of people alive as a function of time. Typically it plots the percentage of a population still alive at different ages but it can also be used to plot the percentage of a population still alive following a particular event, such as a medical operation or the onset of a disease. Figure 6 opposite shows the survival curves for two different populations.

The **average life expectancy** is the age at which 50 per cent of the individuals in a particular population are still alive. It follows that life expectancy can be calculated from a survival curve.

Application

The demographic transition of the human population

As countries have developed economically their human populations have, so far, displayed a pattern of growth known as demographic transition. This pattern can be divided into four stages depending on the birth rate, death rate and total population size. The relationship between these four stages and the birth rates, death rates and total population are illustrated in Figure 4.

1 Using Figure 4, suggest which of the four stages (1, 2, 3 or 4) best applies to each of the descriptions below.

a A country that has a rapidly falling birth rate and a relatively low death rate.

b A country in which there is a high birth rate but much starvation and periodic epidemic disease.

c A country where there have been many years of improved nutrition, far less infectious disease and a large number of children.

d Britain 20 000 years ago when famine and disease led to regular population crashes.

e Britain today.

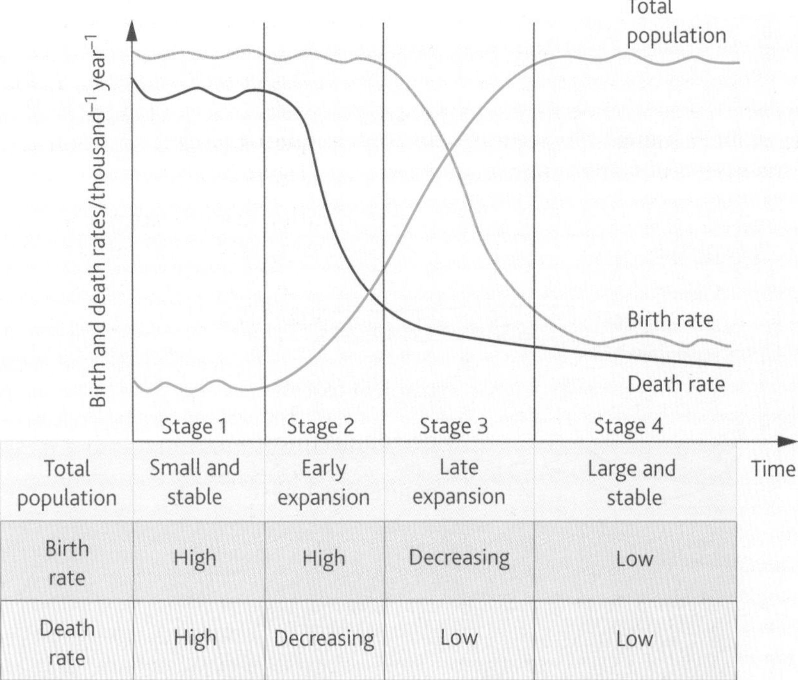

Total population	Small and stable	Early expansion	Late expansion	Large and stable
Birth rate	High	High	Decreasing	Low
Death rate	High	Decreasing	Low	Low

Figure 4 *Demographic transition*

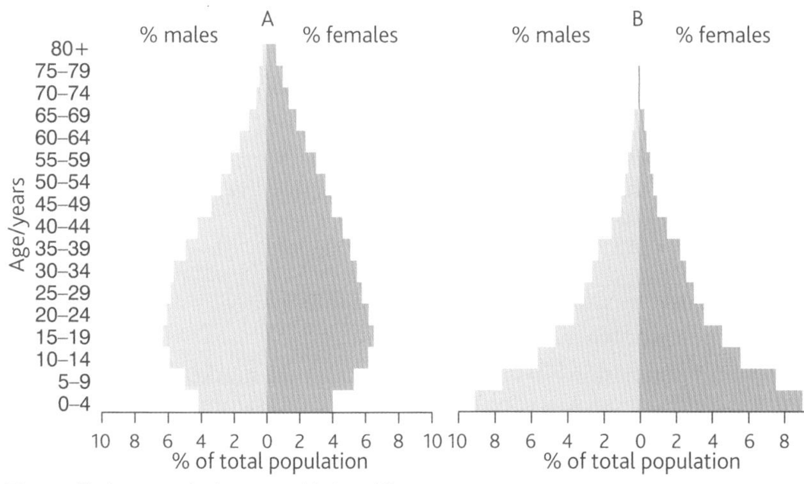

Figure 5 *Age population pyramids A and B*

2 Figure 5 shows two age population pyramids: A and B. Which stage of the demographic transition model shown in Figure 4 does each pyramid represent? Give reasons for your answer in each case.

AQA Examiner's tip

Make certain that you practise simple calculations – it is amazing how many A-level candidates get them wrong in examinations.

Summary questions

1 Which **two** events have contributed most to the human population explosion?

2 The figures below show some population statistics for a country.

Total population at the start of 2007 = 1 000 000

Birth rate in 2007 = 25 per 1000 of population

Death rate in 2007 = 20 per 1000 of population

a How many births were there during the year?

b How many deaths were there during the year?

c Calculate the percentage population growth for this country in 2007. Show your working.

3 Using the survival curves shown in Figure 6, calculate the life expectancy for each of the countries shown.

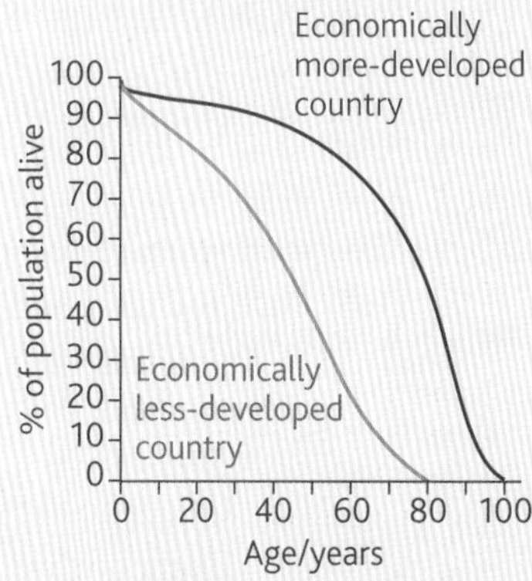

Figure 6 *Survival curves for an economically more-developed country and an economically less-developed country*

AQA Examination-style questions

1 An ecosystem can be described as a dynamic system involving interaction of biotic and abiotic components. Within an ecosystem populations of organisms occupy ecological niches.

(a) Explain what is meant by the following terms.

(i) population

(ii) ecological niche *(3 marks)*

(b) **Figure 1** shows the ranges of mean annual temperatures and precipitation (water falling as rain or snow) for six types of ecosystem.

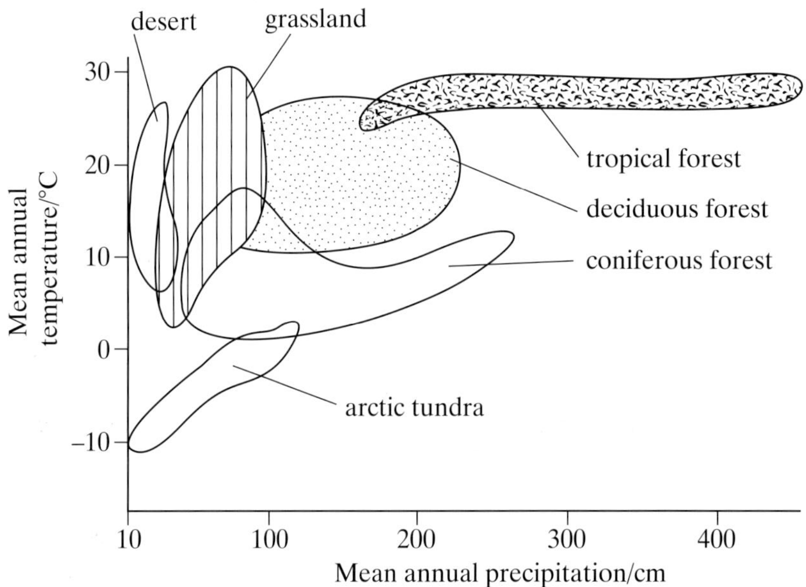

Figure 1

Arctic tundra is considered to be an extreme environment whereas tropical forest is physically less hostile to living organisms.

(i) Explain how the information in the diagram supports this view.

(ii) Describe and explain the difference in the effect of abiotic factors on the diversity of organisms in the tundra and in tropical forest. *(3 marks)*

AQA, 1998

2 The mark-release-recapture technique may be used to estimate population size.

(a) Give **two** assumptions that must be made when using this technique. *(2 marks)*

(b) In estimating the size of a ladybird population, 70 ladybirds were trapped, marked and released. A week later, a second sample was captured. In this second sample, 27 were marked and 13 were not marked.

(i) Calculate the estimated size of the ladybird population. Show your working.

(ii) Explain why it is important that the samples contain as many ladybirds as possible. *(3 marks)*

AQA, 2003

3 (a) In a demographic transition, give one factor that might cause:

(i) an increase in the birth rate;

(ii) a decrease in the death rate. *(2 marks)*

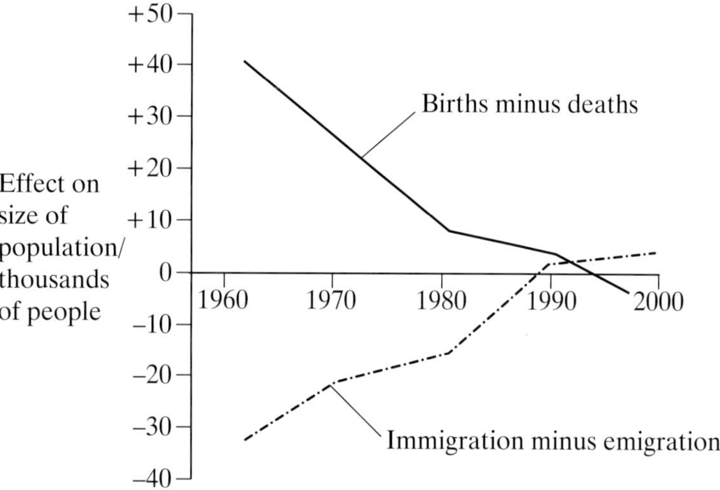

Figure 2

(b) Births, deaths and migration affect population growth. **Figure 2** shows the effects of these factors on a human population between 1960 and 2000. During this period the death rate was almost constant.

(i) From the information given, what does the graph show about changes in birth rate between 1960 and 1980? Explain your answer.

(ii) Describe the effect of immigration and emigration on the growth of this population between 1960 and 2000. *(4 marks)*

AQA, 2006

4 (a) The population pyramids in **Figure 3** show the age distribution in two countries in 2000.

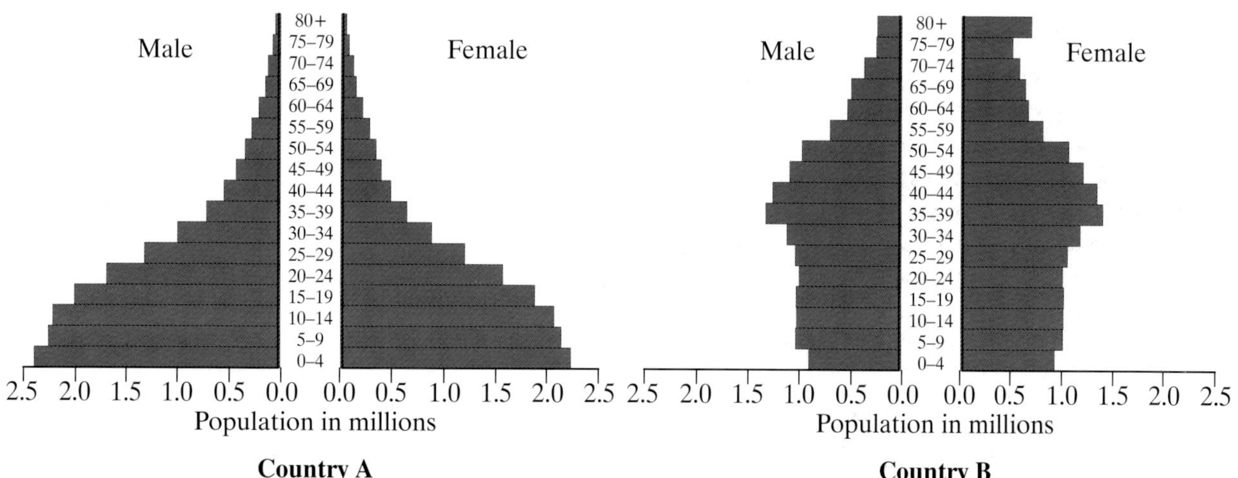

Figure 3

(i) Describe the pattern of age distribution in each country.

(ii) The population size of the two countries is about the same. In which country is the population growing more rapidly? Explain your answer. *(3 marks)*

(b) What information is required in order to calculate the growth rate of a population? *(2 marks)*

AQA, 2004

5 Tree and canyon lizards are found in desert areas in Texas. Both species eat insects. In an investigation the population of both species of lizard were measured over a four-year period in:

- control areas, where both species lived;
- experimental areas, from which one of the species had been removed.

The results for tree lizards are shown in **Figure 4** and the results for canyon lizards are shown in **Figure 5**.

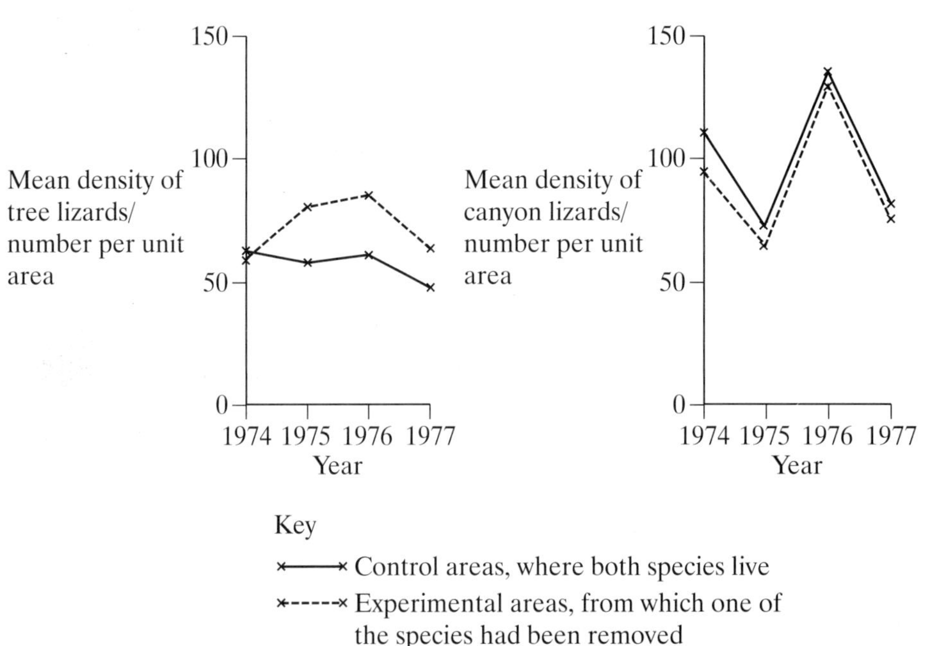

(a) Name the type of interaction being studied in the control areas. *(1 mark)*

(b) Describe how the population density of each lizard species is affected by the presence of the other. Give evidence from **Figure 4** and **Figure 5** to support your answer.

(i) Tree lizard

(ii) Canyon lizard *(2 marks)*

(c) The investigators concluded that, during the four-year period, an abiotic factor had a greater effect on the canyon lizards than on the tree lizards. What evidence from **Figure 4** and **Figure 5** supports this conclusion? *(1 mark)*

(d) Adult tree lizards were found to have shorter lives in experimental areas than in control areas. Using **Figure 4**, suggest an explanation for this. *(1 mark)*

AQA, 2007

6 Scientists used a line transect to find the distribution of three species of *Ranunculus* (buttercup) in a field. The field consists of a series of ridges and furrows. **Figure 6** shows the distribution of the species of *Ranunculus* along the line transect.

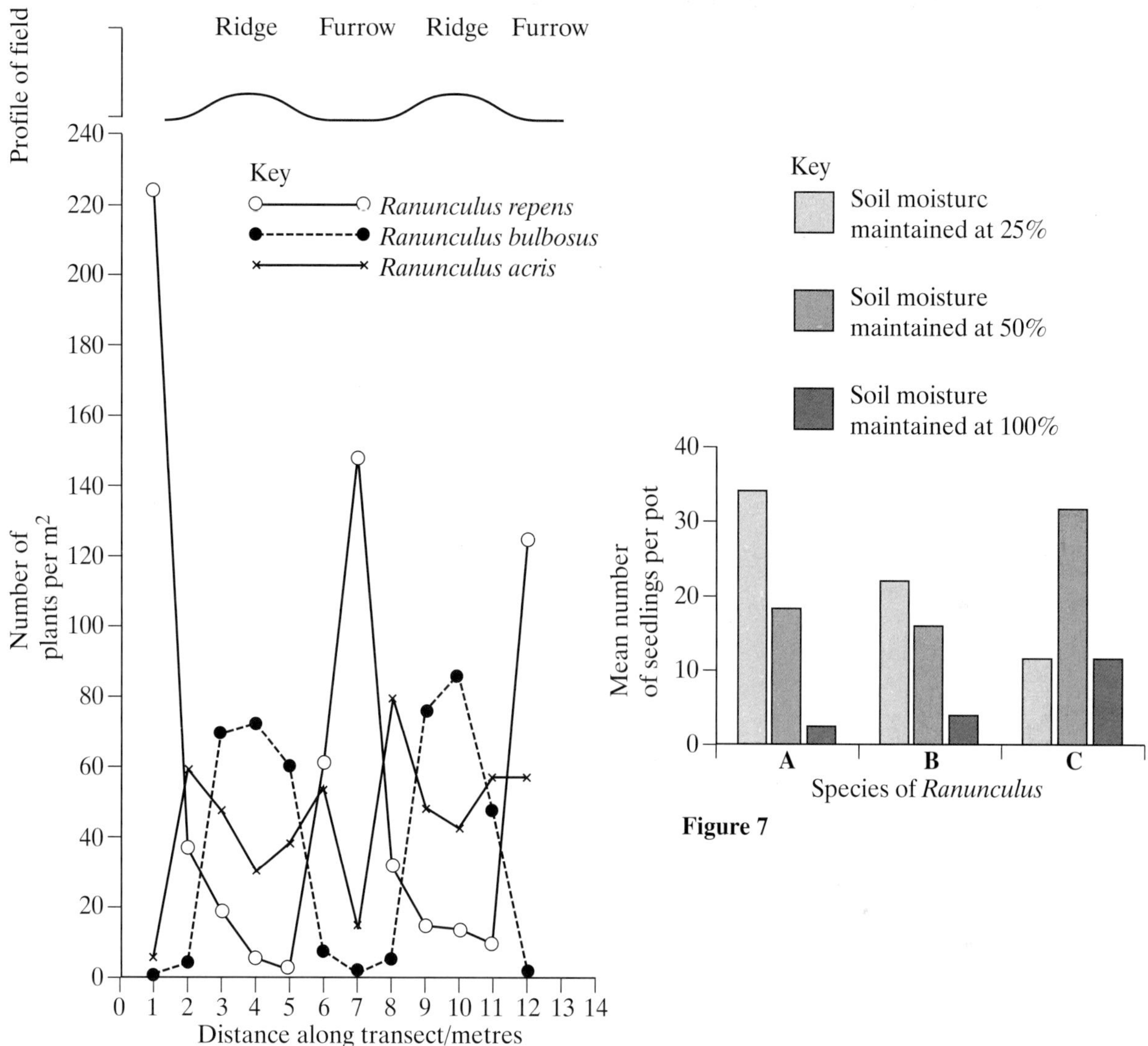

Figure 7

Figure 6

(a) Describe how you would use a transect to obtain data on the distribution of the *Ranunculus* species as shown in **Figure 6**. *(2 marks)*

(b) Other than soil moisture, give **one** abiotic factor and explain how it could lead to the abundance of *Ranunculus repens* in the furrows of this field. *(2 marks)*

(c) The scientists then investigated the effect of soil moisture on seed germination of the three species of *Ranunculus*. They planted seeds of species **A** in three sets of pots. The soil in one set of pots was maintained at 25% water content, the soil in the second set was maintained at 50% water content and the soil in the third set was maintained at 100% water content. They repeated this with seeds of species **B** and species **C**. Four weeks later, the scientists recorded the mean number of seedlings in the pots in each set. Their results are shown in **Figure 7**.

(i) Suggest **two** factors which should be controlled during this investigation.

(ii) Use information from **Figure 6** and **Figure 7** to identify each species of *Ranunculus* **A–C**.

(iii) Describe how you could determine whether two *Ranunculus* plants belong to the same species. *(5 marks)*

AQA, 2007

2 ATP

2.1 Energy and ATP

Learning objectives:

- What is energy and why do organisms need it?
- How does ATP store energy?
- How is ATP synthesised?
- What is the role of ATP in biological processes?

Specification reference: 3.4.2

All living organisms require energy in order to remain alive. This energy comes initially from the Sun. Plants use solar energy to combine water and carbon dioxide into complex organic molecules by the process of photosynthesis. Both plants and animals then break down these organic molecules to make adenosine triphosphate (ATP), which is used as the energy source to carry out processes that are essential to life.

What is energy?

Energy is defined as 'the ability to do work'. Some facts about energy are listed below:

- It takes a variety of different forms, for example, light, heat, sound, electrical, magnetic, mechanical, chemical and atomic energy.
- It can be changed from one form into another.
- It cannot be created or destroyed.
- It is measured in joules (J).

Why do organisms need energy?

Without some input of energy, natural processes tend to break down in disorder. Living organisms are highly ordered systems that require a constant input of energy to prevent them becoming disordered – a condition that would lead to their death. More particularly, energy is needed for:

- **metabolism** – all the reactions that take place in living organisms involve energy
- **movement** both within an organism (e.g. the circulation of blood) and of the organism itself (e.g. locomotion)
- **active transport** of ions and molecules against a concentration gradient across plasma membranes
- **maintenance, repair and division** of cells and of organelles within the cells
- **production of substances** used within organisms, for example, enzymes and hormones
- **maintenance of body temperature** in birds and mammals. These organisms are **endothermic** and need energy to replace that lost as heat to the environment.

Energy and metabolism

The flow of energy through living systems occurs in three stages:

1. Light energy from the Sun is converted by plants into chemical energy during photosynthesis.
2. The chemical energy from photosynthesis, in the form of organic molecules, is converted into ATP during respiration in all cells.
3. ATP is used by cells to perform useful work.

AQA Examiner's tip

You cannot 'make' energy or 'produce' energy. Energy can only be transformed from one type to another and it can be transferred – you need to think and write in these terms. It is a common error in examinations for candidates to state that energy is produced during respiration.

Hint

When it comes to energy requirements, think of a cell as a house. If energy is put into the house, in the form of maintenance, then it remains in an orderly condition and continues to stand, that is, it survives. Without this input of energy, tiles fall off the roof, water penetrates and eventually the house falls down.

How ATP stores energy

Adenosine triphosphate (ATP), as the name suggests, has three phosphate groups. These are the key to how ATP stores energy. The bonds between these phosphate groups are unstable and so have a low **activation energy**, which means they are easily broken. When they do break they release a considerable amount of energy. Usually in living cells it is only the terminal phosphate that is removed, according to the equation:

ATP	+	(H_2O)	$\longrightarrow$	ADP	+	P_i	+	E
adenosine triphosphate		water		adenosine diphosphate		inorganic phosphate		energy

As water is used to convert ATP to ADP, this is known as a **hydrolysis** reaction.

Figure 1 *Birds and mammals need energy to maintain a constant body temperature*

Synthesis of ATP

The conversion of ATP to ADP is a reversible reaction and therefore energy can be used to add an inorganic phosphate to ADP to re-form ATP according to the reverse of the equation above. As water is removed in this process, the reaction is known as a **condensation reaction**. Figure 2 summarises the interconversion of ATP and ADP.

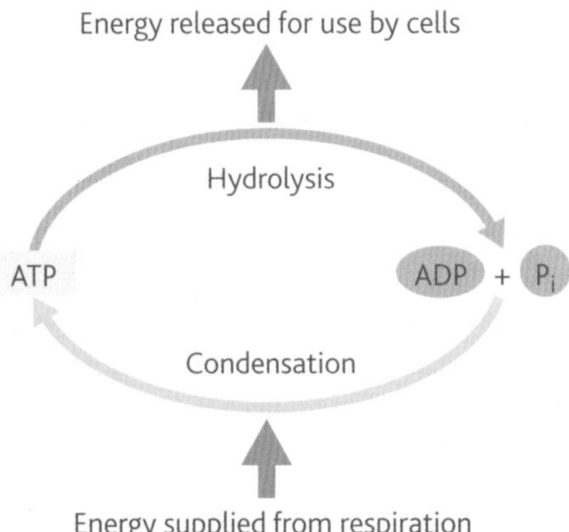

Figure 2 *Interconversion of ATP and ADP*

The synthesis of ATP from ADP involves the addition of a phosphate molecule to ADP. It occurs in three ways:

- **photophosphorylation**, which takes place in chlorophyll-containing plant cells during photosynthesis
- **oxidative phosphorylation**, which occurs in the mitochondria of plant and animal cells during the process of electron transport (see Topic 4.3)
- **substrate-level phosphorylation**, which occurs in plant and animal cells when phosphate groups are transferred from donor molecules to ADP to make ATP. For example, in the formation of pyruvate at the end of glycolysis (see Topic 4.1).

In the first two cases, ATP is synthesised using energy released during the transfer of electrons along a chain of **electron carrier molecules** in either the chloroplasts or the mitochondria.

Hint

Think of the unstable bonds that link the phosphates in ATP as coiled springs. Due to these spring-like bonds the end phosphate is straining to break away from its nearest partner. Any small addition of energy and the end phosphate springs away releasing all the energy that is stored in the 'spring', that is, stored in the bond.

Roles of ATP

The same feature that makes ATP a good energy donor, namely the instability of its phosphate bonds, is also the reason it is not a good long-term energy store. Fats, and carbohydrates such as glycogen, serve this purpose far better. ATP is therefore the **immediate energy source** of a cell. As a result, cells do not store large quantities of ATP, but rather just maintain a few seconds' supply. This is not a problem, as ATP is rapidly re-formed from ADP and inorganic phosphate (P_i) and so a little goes a long way. ATP is a better immediate energy source than glucose for the following reasons:

- Each ATP molecule releases less energy than each glucose molecule. The energy for reactions is therefore released in smaller, more manageable quantities rather than the much greater, and therefore less manageable, release of energy from a glucose molecule.

Hint

ATP is synthesised during reactions that **release** energy and it is hydrolysed to provide energy for reactions that **require** it.

- The **hydrolysis** of ATP to ADP is a single reaction that releases immediate energy. The breakdown of glucose is a long series of reactions and therefore the energy release takes longer.

ATP cannot be stored and so has to be continuously made within the mitochondria of cells that need it. Cells, such as muscle fibres and the epithelium of the small intestine, which require energy for movement and active transport respectively, possess many large mitochondria.

ATP is the source of energy for:

- **metabolic processes**. ATP provides the energy needed to build up macromolecules from their basic units, for example:
  - polysaccharide synthesis from monosaccharides
  - polypeptide synthesis from amino acids
  - DNA/RNA synthesis from nucleotides.
- **movement**. ATP provides the energy for muscle contraction. In muscle contraction, ATP provides the energy for the filaments of muscle to slide past one another and therefore shorten the overall length of a muscle fibre (see Topic 11.2).
- **active transport**. ATP provides the energy to change the shape of carrier proteins in plasma membranes. This allows molecules or ions to be moved against a concentration gradient.
- **secretion.** ATP is needed to form the lysosomes necessary for the secretion of cell products.
- **activation of molecules**. When a phosphate molecule is transferred from ATP to another molecule it makes it more reactive and so lowers the activation energy of that molecule. ATP therefore allows enzyme-catalysed reactions to occur more readily, e.g. the addition of phosphate to glucose molecules at the start of **glycolysis** (see Topic 4.1).

Summary questions

1 ATP is sometimes referred to as 'an immediate energy source'. Explain why.

2 Explain how ATP can make an enzyme-catalysed reaction take place more readily.

3 State **three** roles of ATP in plant cells.

AQA Examiner's tip

It is a common mistake of candidates to write about ATP as a 'high-energy' substance. ATP is an 'intermediate energy' substance that is used to transfer energy.

Link

The roles of ATP link to a number of topics in the *AS Biology* book:

- the synthesis of polysaccharides (Topic 2.3), proteins (Topic 2.5) and DNA (Topic 11.1)
- active transport (Topic 3.8)
- lysosomes (Topic 3.3)
- activation energy and enzymes (Topic 2.7).

It would help to remind yourself of how exactly ATP functions in each of these processes.

3 Photosynthesis

3.1 Overview of photosynthesis

Learning objectives:

- How is the plant leaf adapted to carry out photosynthesis?
- What are the main stages of photosynthesis?

Specification reference: 3.4.3

Link

The structure of a leaf and of the chloroplast were considered in Topics 13.4 and 10.4 of the *AS Biology* book respectively. A review of these topics will help you to follow how both are linked in the process of photosynthesis as described here.

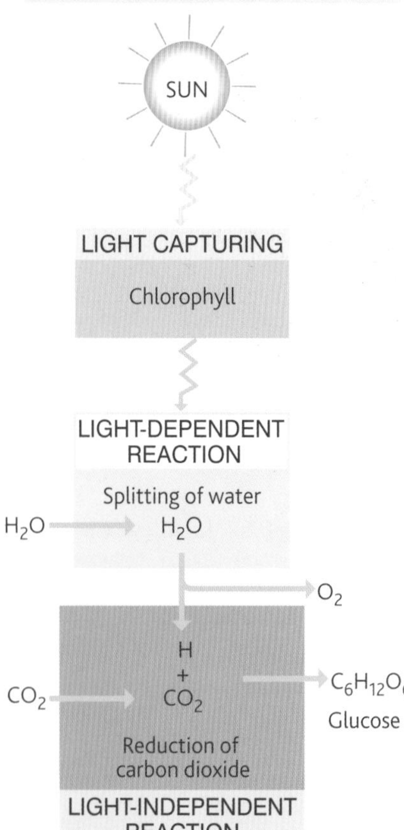

Figure 1 *Overview of photosynthesis*

Humans, along with almost every other living organism, owe their continued existence to photosynthesis. The energy we use, whether it comes from food when we respire or from the wood, coal, oil or gas that we burn in our homes, has been captured by photosynthesis from sunlight. Photosynthesis likewise produces the oxygen we breathe by releasing it from water molecules.

Energy flows though all organisms. How this energy enters an organism depends on its type of nutrition. In plants, light energy is transformed into the chemical energy of the molecules formed during photosynthesis. These molecules are used by the plant to produce **ATP** during respiration. Non-photosynthetic organisms feed on the molecules produced by plants and then also use them to make ATP during respiration.

Site of photosynthesis

The leaf is the main photosynthetic structure. The chloroplasts are the cellular organelles within the leaf where photosynthesis takes place.

Structure of the leaf

Photosynthesis takes place largely in the leaf, the structure of which is shown in Figure 2. Leaves are adapted to bring together the three raw materials of photosynthesis (water, carbon dioxide and light) and remove its products (oxygen and glucose). These adaptations include:

- a large surface area that collects as much sunlight as possible
- an arrangement of leaves on the plant that minimises overlapping and so avoids the shadowing of one leaf by another
- thin, as most light is absorbed in the first few millimetres of the leaf and the diffusion distance is thus kept short
- a transparent **cuticle** and epidermis that let light through to the photosynthetic mesophyll cells beneath
- long, narrow upper mesophyll cells packed with chloroplasts that collect sunlight
- numerous stomata for gaseous exchange
- stomata that open and close in response to changes in light intensity
- many air spaces in the lower mesophyll layer to allow diffusion of carbon dioxide and oxygen
- a network of xylem that brings water to the leaf cells, and phloem that carries away the sugars produced in photosynthesis.

An outline of photosynthesis

The overall equation for photosynthesis is:

$$\underset{\text{carbon dioxide}}{6CO_2} + \underset{\text{water}}{6H_2O} \xrightarrow{\text{light}} \underset{\text{glucose}}{C_6H_{12}O_6} + \underset{\text{oxygen}}{6O_2}$$

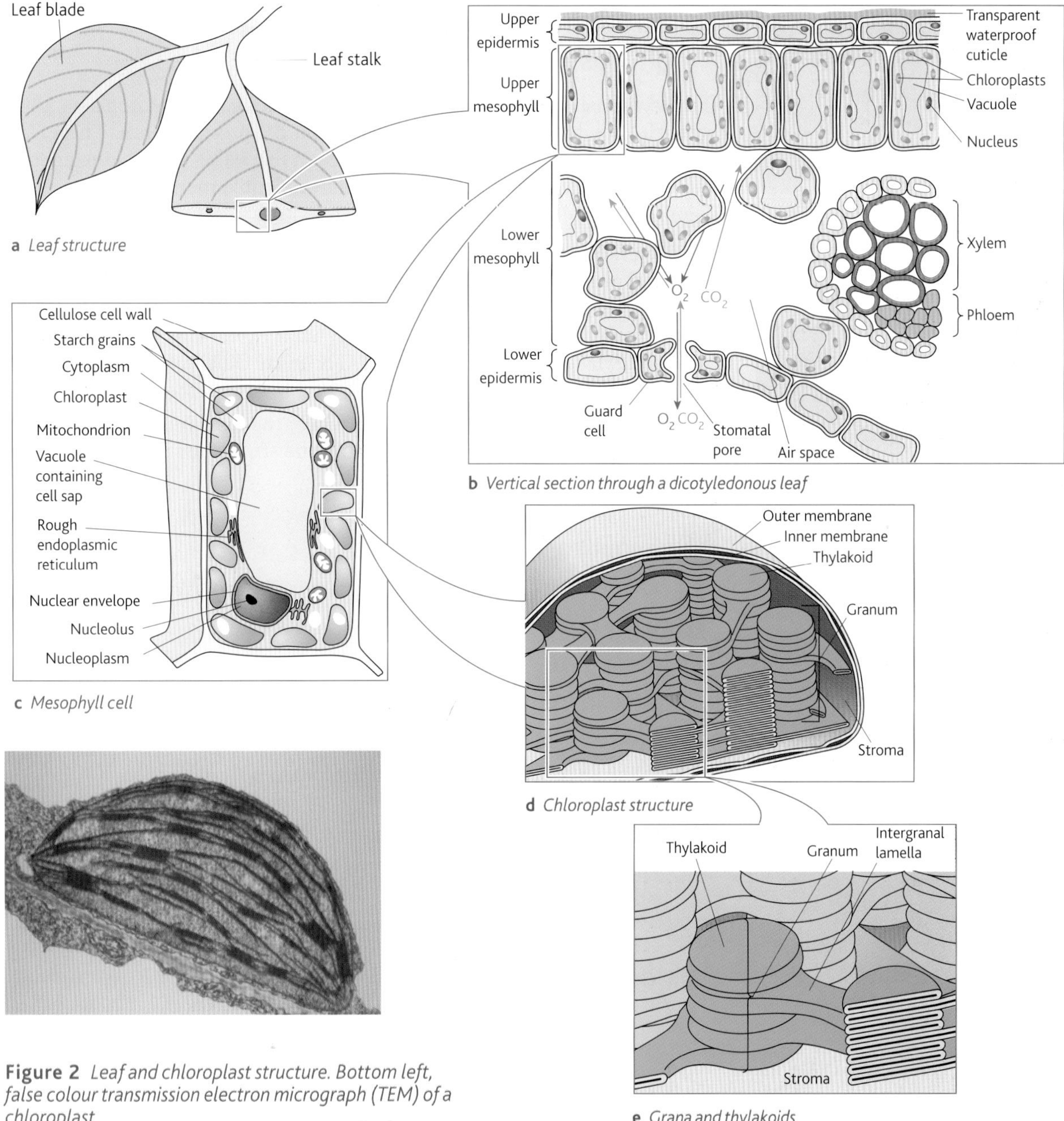

Figure 2 *Leaf and chloroplast structure. Bottom left, false colour transmission electron micrograph (TEM) of a chloroplast*

The equation shown is highly simplified. Photosynthesis is a complex metabolic pathway involving many intermediate reactions. It is a process of energy transformation in which light energy is firstly changed into electrical energy and then into chemical energy. There are three main stages to photosynthesis, see Figure 1 on the previous page:

1 **capturing of light energy** by chloroplast pigments such as chlorophyll
2 **the light-dependent reaction**, in which light energy is converted into chemical energy. During the process an electron flow is created by the effect of light on chlorophyll and this causes water to split (**photolysis**) into **protons**, **electrons** and oxygen. The products are reduced NADP, ATP and oxygen.
3 **the light-independent reaction**, in which these protons (hydrogen ions) are used to reduce carbon dioxide to produce sugars and other organic molecules.

AQA Examiner's tip

Candidates frequently forget that all plant cells respire all the time, while only those plant cells with chloroplasts carry out photosynthesis – and then only in the light.

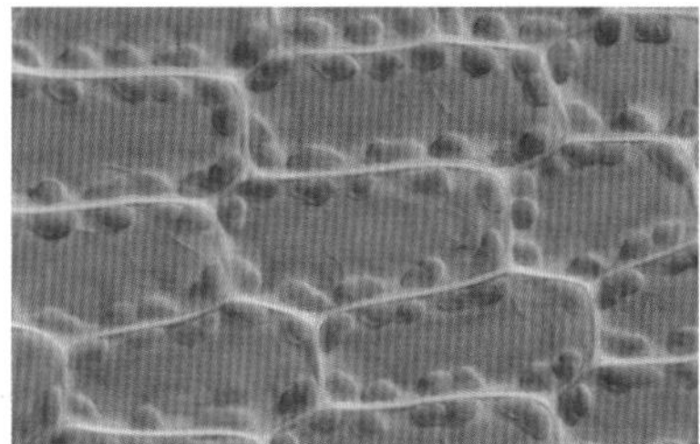

Figure 3 *Photomicrograph of a moss leaf showing cells that contain chloroplasts (green) around their margins*

Structure and role of chloroplasts in photosynthesis

Photosynthesis takes place within cell organelles called chloroplasts, the structure of which is shown in Figure 2d on the previous page. These vary in shape and size but are typically disc-shaped, 2–10 µm long and 1 µm in diameter. They are surrounded by a double membrane. Inside the chloroplast membranes are two distinct regions:

- **The grana** are stacks of up to 100 disc-like structures called **thylakoids** where the light-dependent stage of photosynthesis takes place. Within the thylakoids is the photosynthetic pigment called chlorophyll. Some thylakoids have tubular extensions that join up with thylakoids in adjacent grana. These are called inter-granal lamellae.
- **The stroma** is a fluid-filled matrix where the light-independent stage of photosynthesis takes place. Within the stroma are a number of other structures such as starch grains.

Summary questions

1 Which **two** molecules are the raw materials of photosynthesis?

2 Which **two** molecules are the products of photosynthesis?

3 In which parts of the chloroplast does each of the following occur?

a the light-dependent reaction

b the light-independent reaction

4 What are the products of each of the following?

a the light-dependent reaction

b the light-independent reaction

3.2 The light-dependent reaction

Learning objectives:

- What are oxidation and reduction?
- How is ATP made during the light-dependent reaction?
- What is the role of photolysis in the light-dependent reaction?
- How are chloroplasts adapted to carry out the light-dependent reaction?

Specification reference: 3.4.3

Hint

Oxidation and reduction can each be described in three ways:

Oxidation – loss of electrons or loss of hydrogen or gain of oxygen.

Reduction – gain of electrons or gain of hydrogen or loss of oxygen.

AQA Examiner's tip

Make sure you know the three ways in which something can be oxidised or reduced.

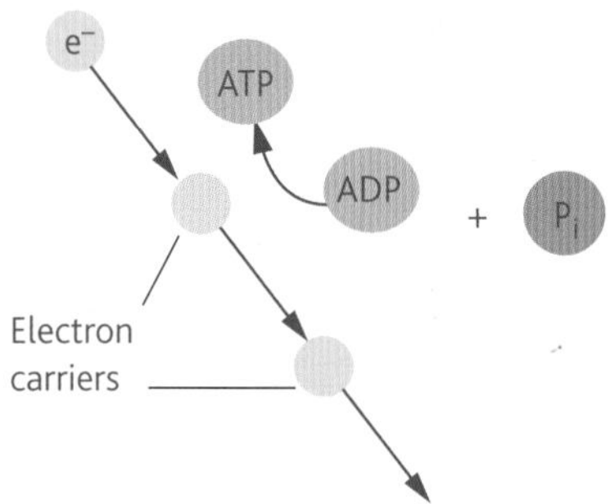

Figure 1 *Electron transfer chain and the making of ATP*

The light-dependent reaction of photosynthesis involves the capture of light whose energy is used for two purposes:

- to add an inorganic phosphate (P_i) molecule to ADP, thereby making **ATP**
- to split water into H^+ ions (protons) and OH^- ions. As the splitting is caused by light, it is known as **photolysis**.

Oxidation and reduction

Before we look at what happens in the light-dependent reaction, it is necessary to understand what oxidation and reduction are.

When a substance combines with oxygen the process is called **oxidation**. The substance to which oxygen has been added is said to be oxidised. When one substance gains oxygen from another, the one losing the oxygen is said to be reduced, the process being known as **reduction**. In practice, when a substance is oxidised it loses **electrons** and when it is reduced it gains electrons. This is the more usual way to define oxidation and reduction. Oxidation results in energy being given out, whereas reduction results in it being taken in. Oxidation and reduction always take place together.

The making of ATP

When a chlorophyll molecule absorbs light energy, it boosts the energy of a pair of electrons within this chlorophyll molecule, raising them to a higher energy level. These electrons are said to be in an excited state. In fact the electrons become so energetic that they leave the chlorophyll molecule altogether. The electrons that leave the chlorophyll are taken up by a molecule called an **electron carrier**. Having lost a pair of electrons, the chlorophyll molecule has been oxidised. The electron carrier, which has gained electrons, has been reduced.

The electrons are now passed along a number of electron carriers in a series of oxidation-reduction reactions. These electron carriers form a transfer chain that is located in the membranes of the **thylakoids**. Each new carrier is at a slightly lower energy level than the previous one in the chain, and so the electrons lose energy at each stage. This energy is used to combine an inorganic phosphate molecule with an ADP molecule in order to make ATP. This process is summarised in Figure 1.

Photolysis of water

The loss of electrons when light strikes a chlorophyll molecule leaves it short of electrons. If the chlorophyll molecule is to continue absorbing light energy, these electrons must be replaced. The replacement electrons are provided from water molecules that are split using light energy. This photolysis of water also yields hydrogen ions (protons). The equation for this process is:

$2H_2O$	$\longrightarrow$	$4H^+$	+	$4e^-$	+	O_2
water		protons		electrons		oxygen

Hint

Reduced NADP is the most important product of the light-dependent reaction.

Hint

To picture how thylakoids are arranged in the grana, think of a thylakoid as a coin and the grana as a stack of many such coins, one on top of the other.

These hydrogen ions (protons) are taken up by an electron carrier called NADP. On taking up the hydrogen ions (protons) the NADP becomes reduced. The reduced NADP then enters the light-independent reaction (see Topic 3.3) along with the electrons from the chlorophyll molecules. The reduced NADP is important because it is a further potential source of chemical energy to the plant. The oxygen by-product from the photolysis of water is either used in respiration or diffuses out of the leaf as a waste product of photosynthesis.

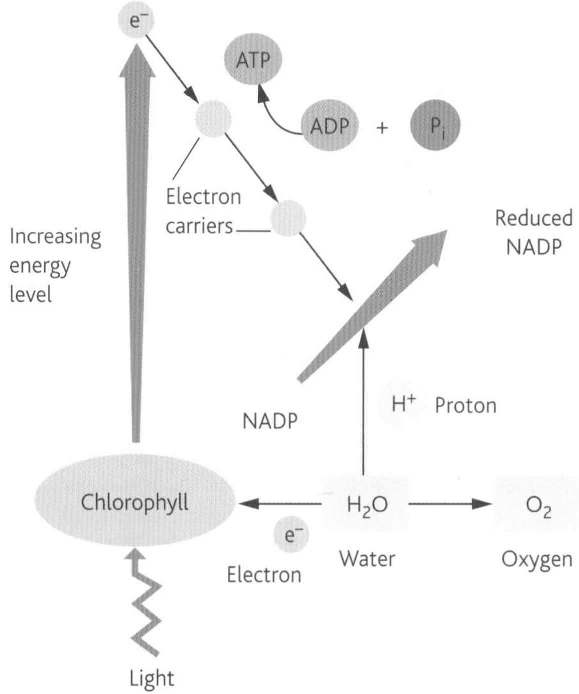

Figure 2 *Summary of the light-dependent reaction of photosynthesis*

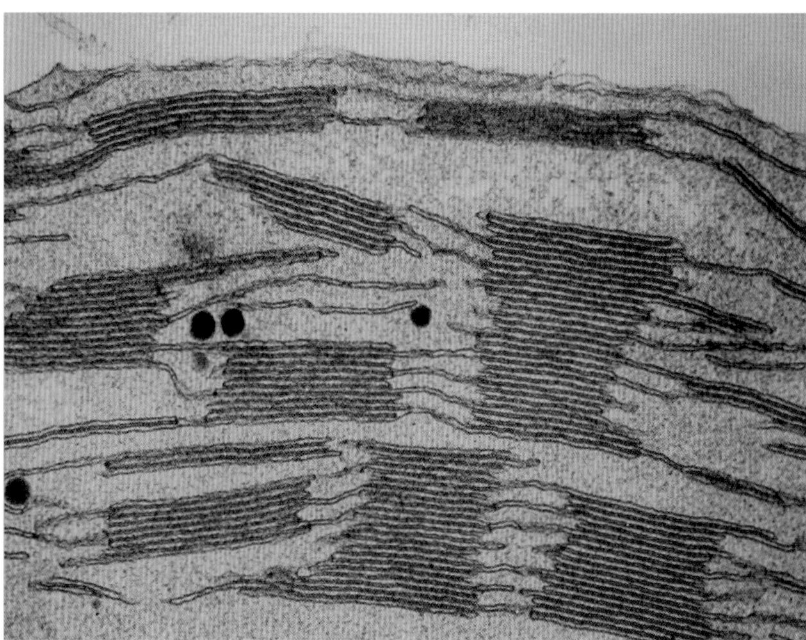

Figure 3 *False-colour TEM of grana in a chloroplast from a leaf of maize. The grana are made up of disc-like thylakoids where the light-dependent reaction of photosynthesis takes place*

Site of the light-dependent reaction

The light-dependent reaction of photosynthesis (see Figure 2) takes place in the thylakoids of chloroplasts. The thylakoids are disc-like structures that are stacked together in groups called grana.

Chloroplasts are structurally adapted to their function of capturing sunlight and carrying out the light-dependent reaction of photosynthesis in the following ways:

- The thylakoid membranes provide a large surface area for the attachment of chlorophyll, electron carriers and enzymes that carry out the light-dependent reaction.
- A network of proteins in the grana hold the chlorophyll in a very precise manner that allows maximum absorption of light.
- The granal membranes have enzymes attached to them, which help manufacture ATP.
- Chloroplasts contain both DNA and ribosomes so they can quickly and easily manufacture some of the proteins needed for the light dependent-reaction.

Application

Chloroplasts and the light-dependent reaction

Figure 4 shows the structure of a chloroplast.

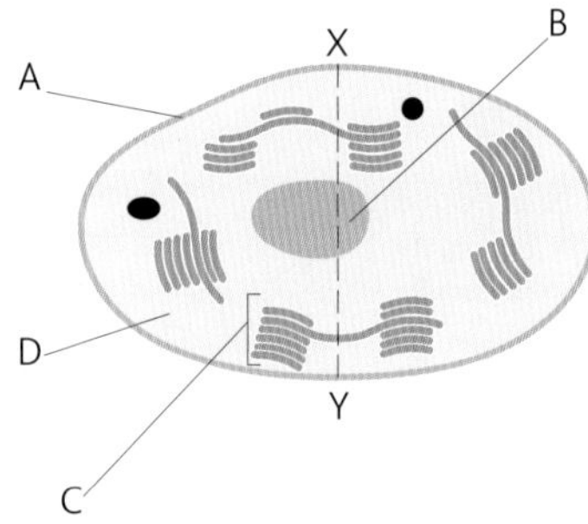

Figure 4

1. Name the parts labelled A, C and D.
2. In which of these labelled parts does the light-dependent reaction take place?
3. Structure B is used for storage. Suggest the name of the substance likely to be stored in B.
4. ATP is produced in the light-dependent reaction of photosynthesis. Suggest **two** reasons why plants cannot use this as their only source of ATP.
5. The actual length of X–Y in this chloroplast is 2 µm. What is the magnification used in Figure 4? Show your working.

Summary questions

1. Where precisely within a plant cell are the electron carriers involved in the light-dependent reaction found?
2. Describe what happens in the photolysis of water.
3. In each of the following, state whether the process involves oxidation or reduction of the molecule named.
   - a An unsaturated fat molecule gains a hydrogen atom.
   - b Oxygen is lost from a carbon dioxide molecule.
   - c Light causes an electron to leave a chlorophyll molecule.

3.3 The light-independent reaction

Learning objectives:

- How is the carbon dioxide absorbed by plants incorporated into organic molecules?
- What are the roles of ATP and reduced NADP in the light-independent reaction?
- What is the Calvin cycle?

Specification reference: 3.4.3

The products of the light-dependent reaction of photosynthesis, namely **ATP** and reduced **NADP**, are used to reduce carbon dioxide in the second stage of photosynthesis. Unlike the first stage, this stage does not require light directly and, in theory, occurs whether or not light is available. It is therefore called the light-independent reaction. In practice, it requires the products of the light-dependent stage and so rapidly ceases when light is absent. The light-independent reaction takes place in the stroma of the chloroplasts. The details of this stage were worked out by Melvin Calvin and his co-workers and so it is often referred to as the Calvin cycle.

The Calvin cycle

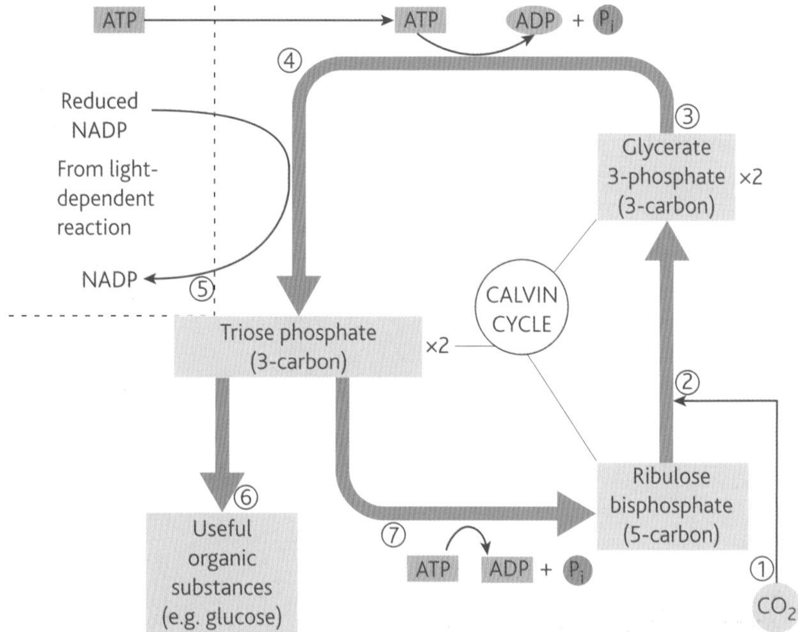

Figure 1 *Summary of the light-independent reaction of photosynthesis (or Calvin cycle)*

In the following account of the Calvin cycle, the numbered stages are illustrated in Figure 1.

1. Carbon dioxide from the atmosphere diffuses into the leaf through **stomata** and dissolves in water around the walls of the mesophyll cells. It then diffuses through the plasma membrane, cytoplasm and chloroplast membranes into the **stroma** of the chloroplast.
2. In the stroma, the carbon dioxide combines with the 5-carbon compound **ribulose bisphosphate (RuBP)** using an enzyme.
3. The combination of carbon dioxide and RuBP produces two molecules of the 3-carbon **glycerate 3-phosphate (GP)**.
4. ATP and reduced NADP from the light-dependent reaction are used to reduce the activated glycerate 3-phosphate to **triose phosphate (TP)**.
5. The NADP is re-formed and goes back to the light-dependent reaction to be reduced again by accepting more hydrogen.

6 Some triose phosphate molecules are converted to useful organic substances, such as glucose.

7 Most triose phosphate molecules are used to regenerate ribulose bisphosphate using ATP from the light-dependent reaction.

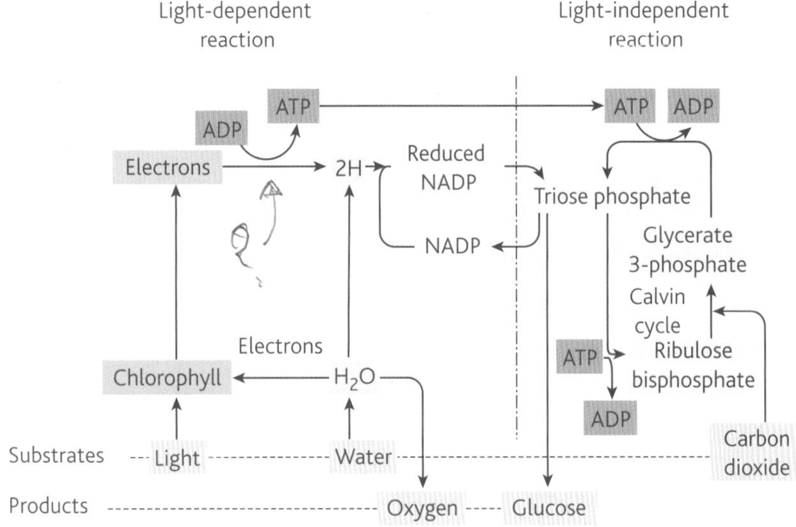

Figure 2 *Summary of photosynthesis*

Site of the light-independent reaction

The light-independent reaction of photosynthesis takes place in the stroma of the chloroplasts.

The chloroplast is adapted to carrying out the light-independent reaction of photosynthesis in the following ways:

- The fluid of the stroma contains all the enzymes needed to carry out the light-independent reaction (reduction of carbon dioxide).
- The stroma fluid surrounds the grana and so the products of the light-dependent reaction in the grana can readily diffuse into the stroma.
- It contains both DNA and ribosomes so it can quickly and easily manufacture some of the proteins needed for the light-independent reaction.

Application and How science works

Using a lollipop to work out the light-independent reaction

The details of the light-independent reaction were worked out by Melvin Calvin and his co-workers using his 'lollipop' experiment. It was so called because the apparatus, shown in Figure 3, resembled a lollipop.

In the experiment, single-celled algae are grown under light in a thin transparent 'lollipop'. Radioactive hydrogencarbonate is injected into the 'lollipop'. This supplies radioactive carbon dioxide to the algae. At 5-second intervals, samples of the photosynthesising algae are dropped into hot methanol to stop chemical reactions instantly. The compounds in the algae are then separated out and those that are radioactive are identified. The results are given in Table 1. This is an example of How science works (HSW: D).

Hint

Any substance whose name ends in 'ose' is a sugar. The ending 'ate' usually means that the substance is an acid (in solution).

Summary questions

1 What is the role of ribulose bisphosphate (RuBP) in the Calvin cycle?

2 How is the reduced NADP from the light-dependent reaction used in the light-independent reaction?

3 Apart from reduced NADP, which other product of the light-dependent reaction is used in the light-independent reaction?

4 Where precisely in a plant cell are the enzymes involved in the Calvin cycle found?

5 Light is not required for the Calvin cycle to take place. Explain therefore why the Calvin cycle cannot take place for long in the absence of light.

Table 1

Time / s	Substances found to be radioactive
0	carbon dioxide
5	glycerate 3-phosphate
10	glycerate 3-phosphate + triose phosphate
15	glycerate 3-phosphate + triose phosphate + glucose
20	glycerate 3-phosphate + triose phosphate + glucose + ribulose bisphosphate

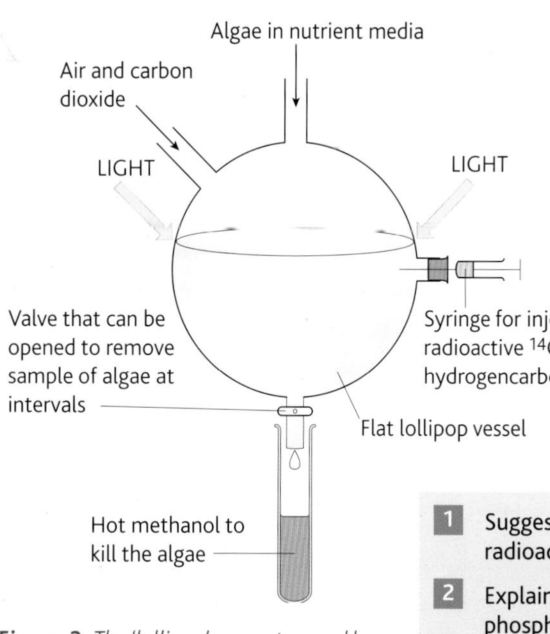

Figure 3 *The 'lollipop' apparatus used by Melvin Calvin*

- Algae are grown under light in the thin transparent lollipop.
- Radioactive ^{14}C in the form of hydrogencarbonate is injected.
- At intervals (seconds to minutes) samples of the photosynthesising algae are dropped into the hot methanol to stop chemical reactions instantly.
- The compounds in the algae are separated by two-way chromatography.
- The radioactive compounds are identified and the pathway determined by the time at which each first appeared. The first to appear is the first in the pathway, etc.

1 **Suggest a reason why the carbon dioxide supplied to the algae was radioactively labelled.**

2 **Explain how information in Table 1 provides evidence that glycerate 3-phosphate is converted into triose phosphate.**

3 **Suggest an explanation of how the hot methanol might stop further chemical reactions taking place.**

In a further experiment, samples of algae were collected at 1-minute intervals over a period of 5 minutes. The quantities of glycerate 3-phosphate (GP) and ribulose bisphosphate (RuBP) were measured. At the beginning of the experiment, the concentration of carbon dioxide supplied was high. After 2 minutes the concentration of carbon dioxide was reduced. The graph in Figure 4 shows the results of this experiment.

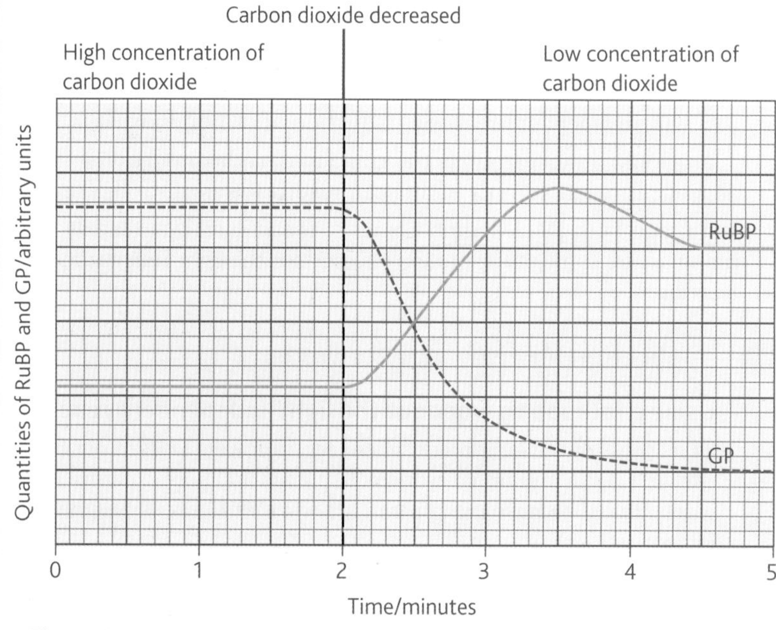

Figure 4

4 **Describe the effects on the quantities of GP and RuBP of the decrease in carbon dioxide after 2 minutes.**

5 **Suggest explanations for these changes to the levels of GP and RuBP.**

3.4 Factors affecting photosynthesis

Learning objectives:

- What is meant by the concept of limiting factors?
- How can photosynthesis be measured?
- How do temperature, carbon dioxide concentration and light intensity affect the rate of photosynthesis?

Specification reference: 3.4.3

We have seen that photosynthesis is the process that captures light energy and thereby makes life on Earth possible. It follows that an understanding of those factors which influence the rate of photosynthesis can help ensure that an adequate supply of energy (food) is available, not only to ourselves, but also to those other organisms with which we share this planet. Before we consider some of these factors, it is necessary to understand the concept of limiting factors.

Limiting factors

In any complex process such as photosynthesis, the factors that affect its rate all operate together. However, the rate of the process at any given moment is not affected by all the factors, but rather by the one whose level is at the least favourable value. This factor is called a **limiting factor** because it limits the rate at which the process can take place. Changing the levels of the other factors will not alter the rate of the process.

To take the example of light intensity limiting the rate of photosynthesis:

- In complete darkness, it is the absence of light alone that prevents photosynthesis occurring. No matter how much we raise or lower the temperature or change the concentration of carbon dioxide, there will be no photosynthesis. Light, or rather the absence of it, is the factor determining the rate of photosynthesis at that moment.
- If we provide light, however, the rate of photosynthesis will increase.
- As we add more light, the rate increases further. This does not continue indefinitely, however, because there comes a point at which further increases in light intensity have no effect on the rate of photosynthesis.
- At this point some other factor, such as the concentration of carbon dioxide, is in short supply and so limits the process. Carbon dioxide is now the limiting factor and only an increase in its level will increase the rate of photosynthesis.
- As with light, providing more carbon dioxide will lead to more photosynthesis.
- Further increases in carbon dioxide level will have no effect on the rate of photosynthesis.
- At this point a different factor, such as temperature, is the limiting factor and only an alteration in its level will affect the rate of photosynthesis.

These events are illustrated in Figure 1.

Processes such as photosynthesis are made up of a series of small reactions. It is the slowest of these reactions that determines the overall rate of photosynthesis. In turn, it is the level of factors such as temperature and the supply of raw materials that determines the speed of each step.

The law of limiting factors can therefore be expressed as:

At any given moment, the rate of a physiological process is limited by the factor that is at its least favourable value.

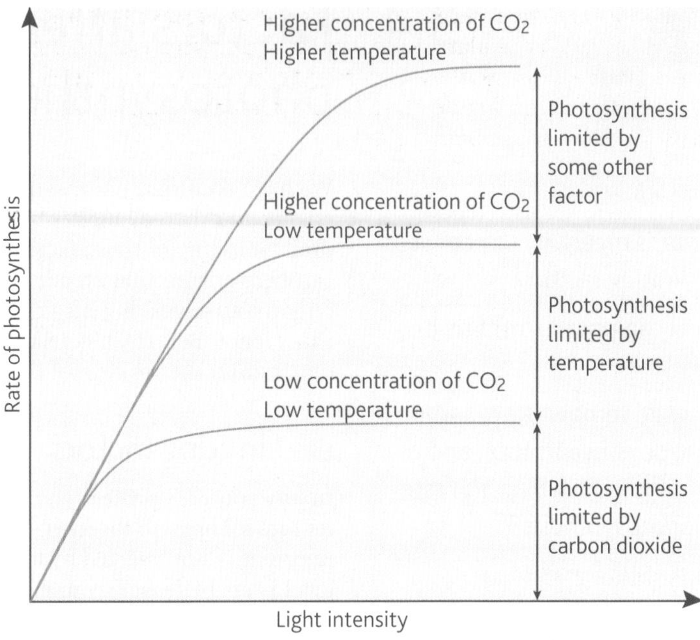

Figure 1 *Concept of limiting factors as illustrated by the effects of the levels of different conditions on the rate of photosynthesis*

Effect of light intensity on the rate of photosynthesis

The rate of photosynthesis is usually measured in one of two ways:

- the volume of oxygen released by a plant
- the volume of carbon dioxide taken up by a plant.

When light is the limiting factor, the rate of photosynthesis is directly proportional to light intensity. As light intensity is increased, the volume of oxygen produced and carbon dioxide absorbed due to photosynthesis will increase to a point at which it is exactly balanced by the oxygen absorbed and the carbon dioxide produced by cellular respiration. At this point there will be no net exchange of gases into or out of the plant. This is known as the light **compensation point**. Further increases in light intensity will cause a proportional increase in the rate of photosynthesis and increasing volumes of oxygen will be given off and carbon dioxide taken up. A point will be reached at which further increases in light intensity will have no effect on photosynthesis. At this point some other factor, such as carbon dioxide concentration or temperature, is limiting the reaction. These events are illustrated in Figure 2.

Effect of carbon dioxide concentration on the rate of photosynthesis

Carbon dioxide is present in the atmosphere at a concentration of around 0.04 per cent. This level continues to increase as the result of human activities such as burning fossil fuels and the clearing of rain forests. It is still one of the rarest gases present and is often the factor that limits the rate of photosynthesis under normal conditions. The optimum concentration of carbon dioxide for a consistently high rate of photosynthesis is 0.1 per cent and growers of some greenhouse crops, such as tomatoes, enrich the air in the greenhouses with more carbon dioxide to provide higher yields. Carbon dioxide concentration affects enzyme activity, in particular the enzyme that catalyses the combination of ribulose bisphosphate with carbon dioxide in the light-independent reaction (see Topic 3.3). Figures 1 and 2 illustrate the effect of different carbon dioxide levels on photosynthesis.

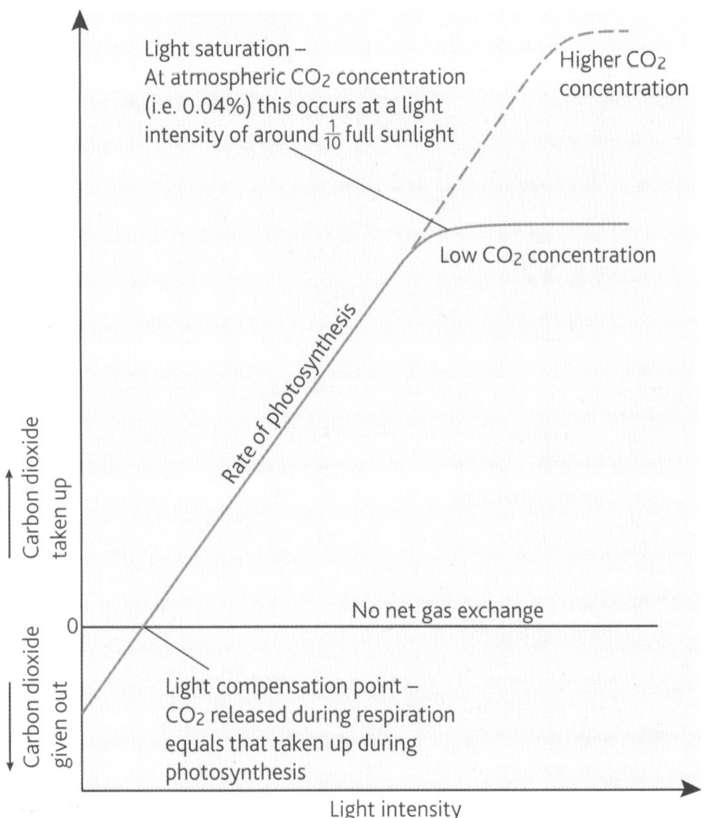

Figure 2 *Graph showing the effect of light intensity on the rate of photosynthesis as measured by the amount of CO_2 exchange*

Effect of temperature on the rate of photosynthesis

Provided that other factors are not limiting, the rate of photosynthesis increases in direct proportion to the temperature. Between the temperatures of 0 °C and 25 °C the rate of photosynthesis is approximately doubled for each 10 °C rise in temperature. In many plants, the optimum temperature is 25 °C above which the rate levels off and then declines – largely as a result of enzyme denaturation. Purely photochemical reactions are not usually affected by temperature, and so the fact that photosynthesis is temperature-sensitive suggested to early researchers that there was also a totally chemical process involved as well as a photochemical one. We now know that this chemical process is the light-independent reaction (see Topic 3.3).

Application

Measuring photosynthesis

The rate of photosynthesis in an aquatic plant such as Canadian pondweed (*Elodea*) can be found by measuring the volume of oxygen produced by using the apparatus (called a photosynthometer) illustrated in Figures 4 and 5 overleaf.

Summary questions

Figure 3 illustrates the influence of light intensity, carbon dioxide and temperature on the rate of photosynthesis.

1 0.1% carbon dioxide at 25 °C
2 0.04% carbon dioxide at 35 °C
3 0.04% carbon dioxide at 25 °C
4 0.04% carbon dioxide at 15 °C

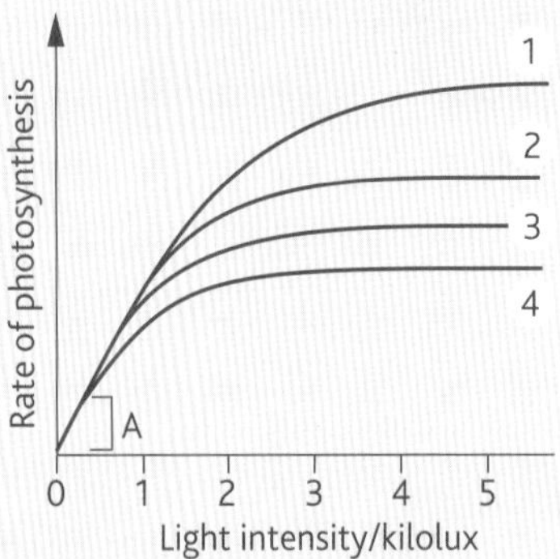

Figure 3

1 State **one** measurement that could be taken to determine the rate of photosynthesis in this experiment?

2 Name the factor that is limiting the rate of photosynthesis over the region marked A on the graph. Explain your answer.

3 In the spring a commercial grower of tomatoes keeps his greenhouses at 25 °C and at a carbon dioxide concentration of 0.04%. The light intensity is 4 kilolux at this time of year. Using the graph, predict whether the tomato plants would grow more if the carbon dioxide level was raised to 0.1% or if the temperature were increased to 35 °C. Explain your answer.

4 Why is there no point in the grower heating his greenhouses on a dull day?

5 Using your knowledge of the light-independent reaction, explain why, at 25 °C, raising the level of carbon dioxide from 0.04% to 0.1% increases the amount of glucose produced.

Figure 4 *Student using a photosynthometer to measure the rate of photosynthesis*

Hint

The light-independent reaction is enzyme controlled. Temperature affects enzyme activity and so must affect the light-independent reaction.

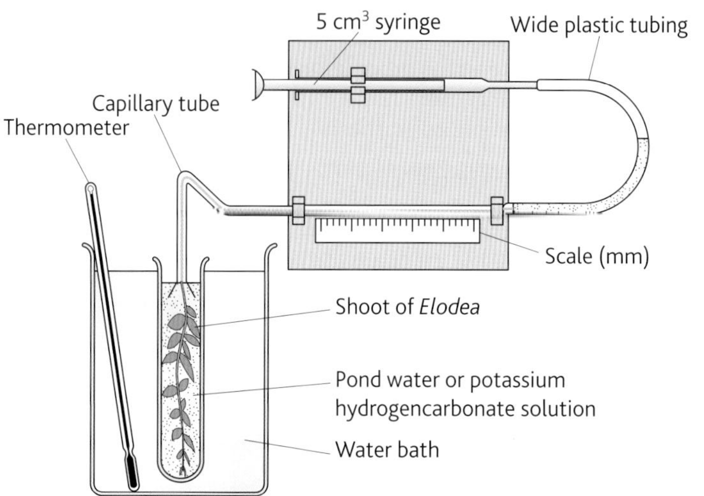

Figure 5 *Apparatus used to measure the rate of photosynthesis under various conditions*

- The apparatus is set up as in Figure 5, taking care not to introduce any air bubbles into it and that the apparatus is completely air-tight.
- The water bath is used to maintain a constant temperature throughout the experiment and can be adjusted as necessary.
- Potassium hydrogencarbonate solution is used around the plant to provide a source of carbon dioxide.
- A source of light, whose intensity can be adjusted, is arranged close to the apparatus, which is kept in an otherwise dark room.
- The apparatus is kept in the dark for 2 hours before the experiment begins.
- The light source is switched on and timing is begun.
- Oxygen produced by the plant during photosynthesis collects in the funnel end of the capillary tube above the plant.
- After 30 minutes this oxygen is drawn up the capillary tube by gently withdrawing the syringe until its volume can be measured on the scale, which is calibrated in mm^3.
- The gas is drawn up into the syringe, which is then depressed again before the process is repeated at the same light intensity four or five times and the average volume of oxygen produced per hour is calculated.
- The apparatus is left in the dark for 2 hours before the procedure is repeated with the light source set at a different light intensity.

1 Why does the apparatus need to be airtight?

2 Why does the temperature of the water bath need to be kept constant?

3 Suggest an advantage of providing an additional source of carbon dioxide.

4 Suggest a reason for carrying out the experiment in a room that is dark except for the light source.

5 Suggest why the plant is kept in the dark before the experiment begins.

6 Suggest a reason why measuring the volume of oxygen produced by the plant in this experiment may not be an accurate measure of photosynthesis.

AQA Examination-style questions

1 (a) **Figure 1** summarises some of the light-dependent reactions of photosynthesis.

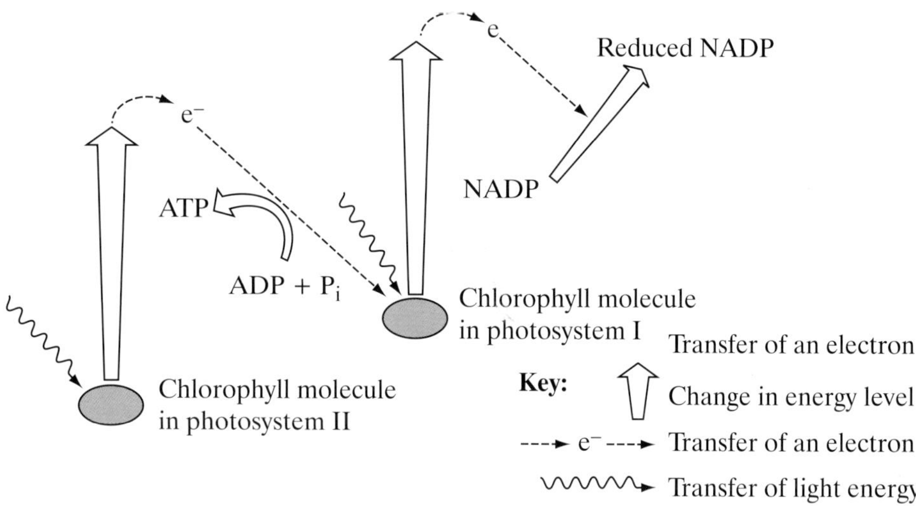

Figure 1

(i) Use **Figure 1** to describe what happens to a molecule of chlorophyll in photosystem II when it absorbs a photon of light.

(ii) Molecules of ATP are formed as electrons are transferred from photosystem II to photosystem I. Explain how this is possible. *(3 marks)*

(b) Reduced NADP produced during the light-dependent reactions of photosynthesis is used in the light-independent reactions. Explain how. *(2 marks)*

AQA, 2005

2 (a) Describe how NADP is reduced in the light-dependent reaction of photosynthesis. *(2 marks)*

(b) In an investigation of the light-independent reaction, the amounts of glycerate 3-phosphate (GP) and ribulose bisphosphate (RuBP) in photosynthesising cells were measured under different environmental conditions.

Figure 2 shows the effect of reducing the carbon dioxide concentration on the amounts of glycerate 3-phosphate and ribulose bisphosphate in photosynthesising cells.

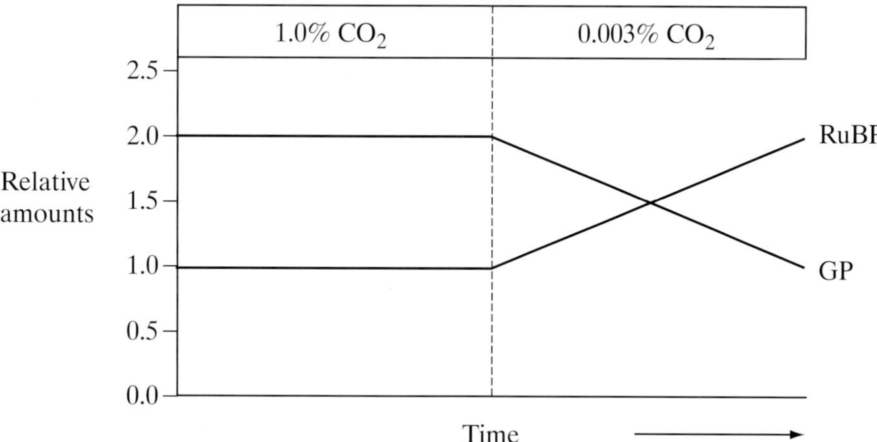

Figure 2

(i) Explain why there is twice the amount of glycerate 3-phosphate as ribulose bisphosphate when the carbon dioxide concentration is high.

(ii) Explain the rise in the amount of ribulose bisphosphate after the carbon dioxide concentration is reduced. *(2 marks)*

(c) **Figure 3** shows the results of an experiment in which photosynthesising cells were kept in the light and then in darkness.

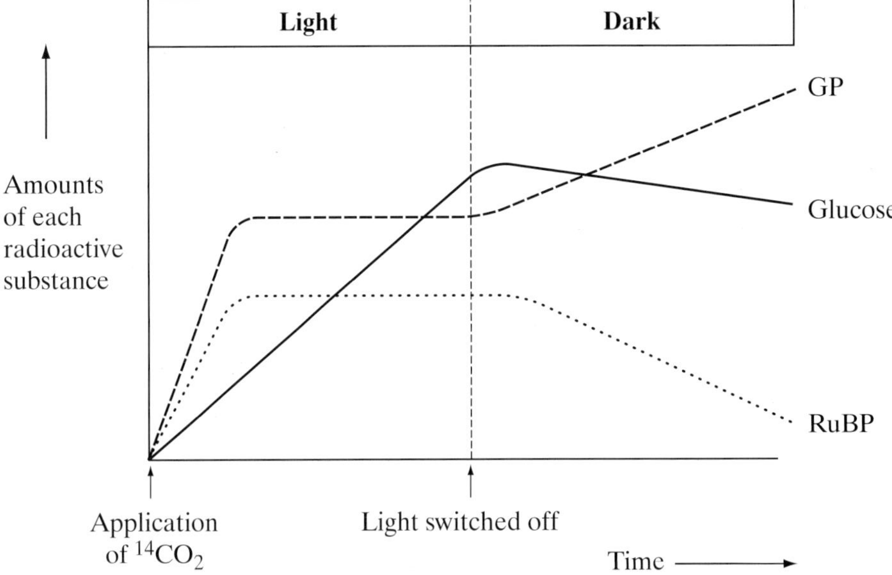

Figure 3

(i) In the experiment the cells were supplied with radioactively labelled $^{14}CO_2$. Explain why the carbon dioxide used was radioactively labelled.

(ii) Explain how lack of light caused the amount of radioactively labelled glycerate 3-phosphate to rise.

(iii) Explain what caused the amount of radioactively labelled glucose to decrease after the light was switched off. *(4 marks)*

AQA, 2006

3 **Figure 4** shows the absorption of different wavelengths of light by three photosynthetic pigments in a red seaweed.

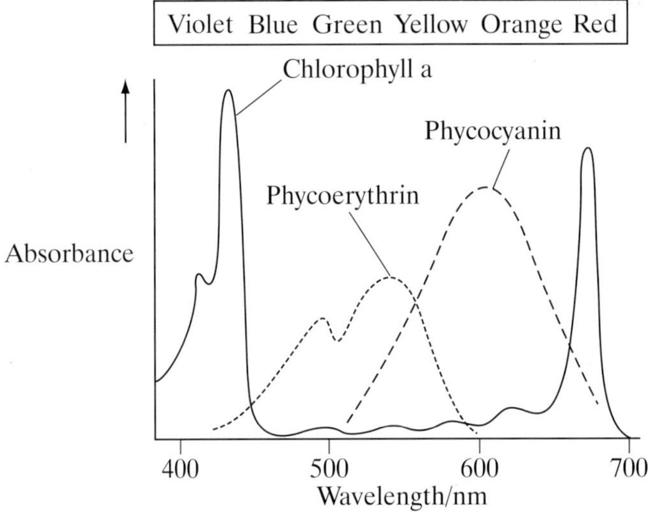

Figure 4

(a) (i) Describe what the graph shows about the properties of chlorophyll a.
(ii) Describe the part played by chlorophyll in photosynthesis. *(4 marks)*

(b) The red seaweed lives under water at a depth of 2 metres. Suggest an advantage to the red seaweed of having other pigments in addition to chlorophyll a. *(2 marks)*

AQA, 2005

4 Different wavelengths of light are used to illuminate a tube containing a suspension of photosynthetic algal cells. The percentage of light absorbed and the rate of photosynthesis are measured using the apparatus shown in **Figure 5.**

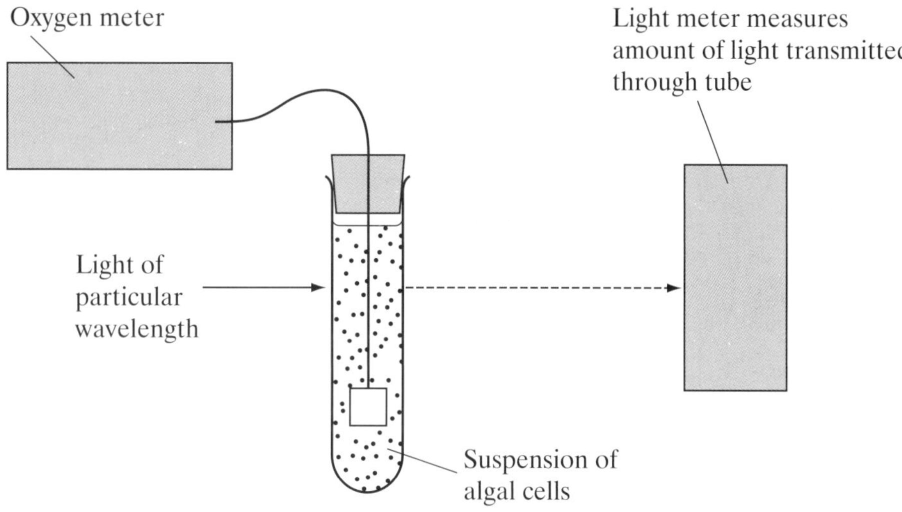

Figure 5

(a) (i) The light meter is calibrated to read 100% using a glass tube containing water but no algal cells. Explain how the percentage of light absorbed by the algal cells is calculated.
(ii) What measurements should be taken to determine the rate of photosynthesis? *(2 marks)*

(b) (i) Apart from temperature and pH, give **one** factor that should be kept constant during the investigation.
(ii) A buffer is used to maintain a constant pH. Explain why the pH of the suspension would increase during photosynthesis in the absence of a buffer. *(2 marks)*

(c) Explain why temperature has little influence on the absorption of light by photosynthetic organisms. *(2 marks)*

AQA, 2007

5 (a) **Figure 6** summarises the pathways involved in photosynthesis. Name molecule Z. *(1 mark)*

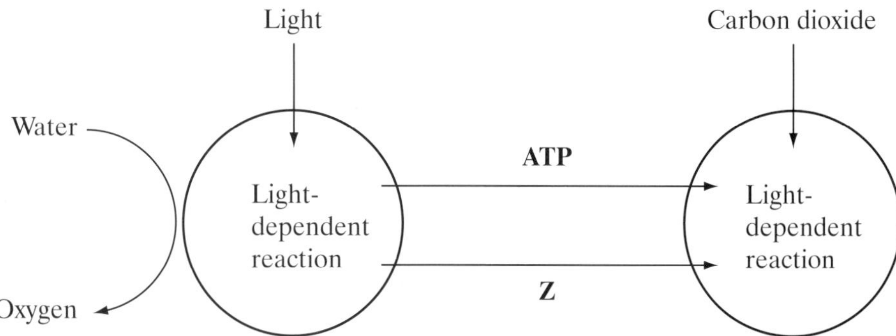

Figure 6

(b) Under some conditions oxygen reacts with ribulose bisphosphate to give glycerate 3-phosphate and phosphoglycolate. This reaction is summarised in the equation.

RuBP + oxygen ⟶ glycerate 3-phosphate + phosphoglycolate

Phosphoglycolate takes no part in the light-independent reaction.

(i) Give the number of carbon atoms in one molecule of phosphoglycolate.

(ii) The production of phosphoglycolate could lead to a reduction in the rate of photosynthesis. Explain how.

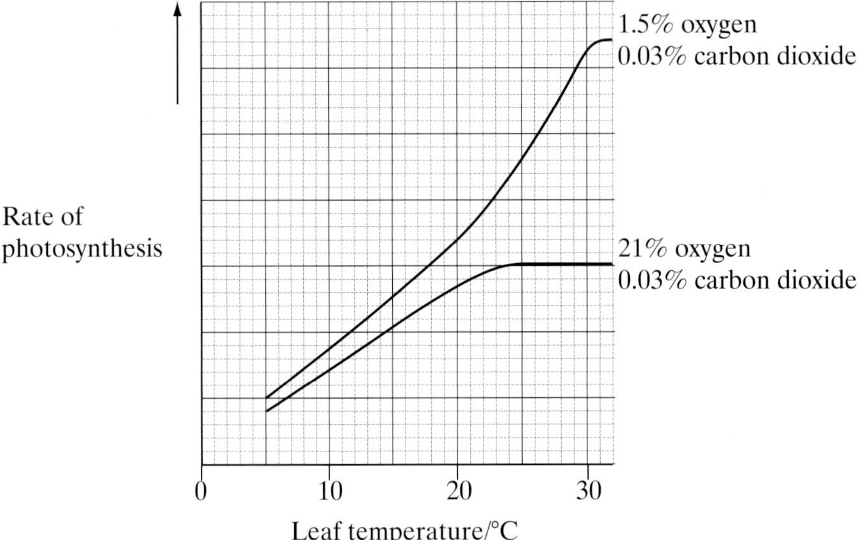

Figure 7

(iii) An investigation was carried out on the effect of temperature and oxygen concentration on the rate of photosynthesis in leaves. The results are shown in **Figure 7**. Describe and explain the effect of oxygen concentration on the rate of photosynthesis. *(6 marks)*

AQA, 2007

6 The percentage of light absorbed by an aquatic plant was measured when it was exposed to different wavelengths. The rate of photosynthesis was also measured at each wavelength of light. The results are shown in **Figure 8**.

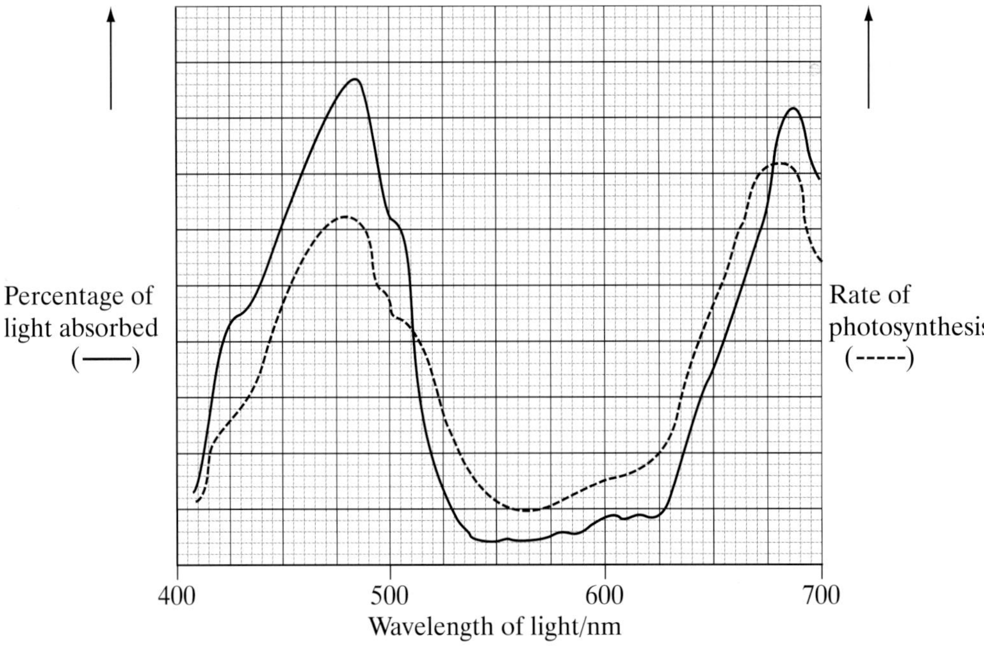

Figure 8

(a) Describe and explain the relationship between light absorption and the rate of photosynthesis for the wavelengths of light between 410 nm and 500 nm. *(2 marks)*

(b) Give **one** dependent variable you could measure in order to determine the rate of photosynthesis in an aquatic plant. *(1 mark)*

(c) Use **Figure 8** to identify the range of wavelengths of light that would be green in colour. Give a reason for your answer. *(2 marks)*

(d) A suspension of chloroplasts was isolated from an aquatic plant and a reagent was added. The reagent is blue when oxidised and is colourless when reduced.

(i) The suspension of chloroplasts in blue reagent was exposed to sunlight. The blue colour disappeared. Use your knowledge of the light-dependent reactions of photosynthesis to explain why.

(ii) Another suspension of chloroplasts was set up as before. Small quantities of ADP and phosphate ions were added and then the tube was exposed to light. The blue colour disappeared more quickly. Explain why. *(4 marks)*

AQA, 2004

7 (a) **Figure 9** shows the effect of different conditions on the rate of photosynthesis in wheat.

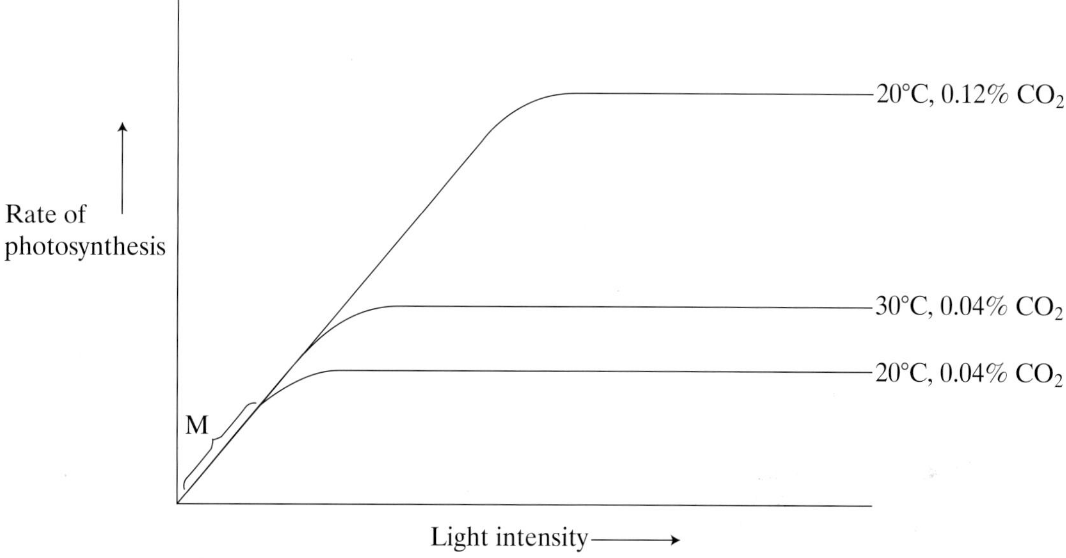

Figure 9

(i) Name the factor which is limiting photosynthesis at **M**.

(ii) Give a reason for your answer.

(iii) Copy the graph and mark with **X** a point where carbon dioxide concentration is limiting the rate of photosynthesis.

(iv) Explain why you have chosen this point. *(4 marks)*

(b) Suggest how carbon dioxide concentration could be increased in a glasshouse (greenhouse). *(1 mark)*

AQA, 2004

4 Respiration

4.1 Glycolysis

Learning objectives:

- Where does glycolysis fit into the overall process of respiration?
- What are the main stages of glycolysis?
- What are the products of glycolysis?

Specification reference: 3.4.4

We have seen in Chapter 3 that photosynthesis converts energy in the form of sunlight into the chemical energy of carbohydrates such as glucose. We also saw, in Chapter 2, that this glucose cannot be used directly by cells as a source of energy. Instead, cells use ATP as their immediate energy source. The conversion of glucose into ATP takes place during the process of cellular respiration. There are two different forms of cellular respiration depending upon whether oxygen is available or not:

- **Aerobic respiration** requires oxygen and produces carbon dioxide, water and much ATP.
- **Anaerobic respiration (fermentation)** takes place in the absence of oxygen and produces lactate (in animals) or ethanol and carbon dioxide (in plants) but only a little ATP in both cases.

Aerobic respiration can be divided into four stages:

1 **glycolysis** – the splitting of the 6-carbon glucose molecule into two 3-carbon pyruvate molecules

2 **link reaction** – the conversion of the 3-carbon pyruvate molecule into carbon dioxide and a 2-carbon molecule called acetylcoenzyme A

3 **Krebs cycle** – the introduction of acetylcoenzyme A into a cycle of oxidation-reduction reactions that yield some ATP and a large number of electrons

4 **electron transport chain** – the use of the electrons produced in the Krebs cycle to synthesise ATP with water produced as a by-product.

The main respiratory pathways are summarised in Figure 1.

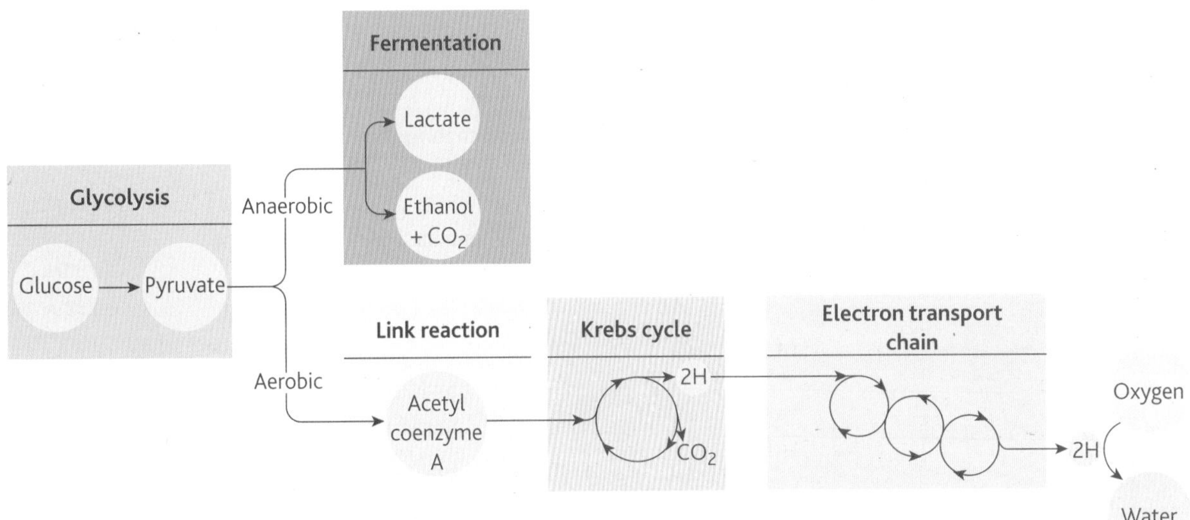

Figure 1 *Summary of respiratory pathways*

Glycolysis is the initial stage of both aerobic and anaerobic respiration. It occurs in the cytoplasm of all living cells and is the process by which a hexose (6-carbon) sugar, usually glucose, is split into two molecules of the 3-carbon molecule, pyruvate. Although there are a number of smaller enzyme-controlled reactions in glycolysis, these can be conveniently grouped into four stages:

1 **activation of glucose by phosphorylation**. Before it can be split into two, glucose must first be made more reactive by the addition of two phosphate molecules (phosphorylation). The phosphate molecules come from the **hydrolysis** of two ATP molecules to ADP. This provides the energy to activate glucose (lowers the **activation energy** for the enzyme-controlled reactions that follow).

2 **splitting of the phosphorylated glucose**. Each glucose molecule is split into two 3-carbon molecules known as triose phosphate.

3 **oxidation of triose phosphate**. Hydrogen is removed from each of the two triose phosphate molecules and transferred to a hydrogen-carrier molecule known as NAD to form reduced NAD.

4 **the production of ATP**. Enzyme-controlled reactions convert each triose phosphate into another 3-carbon molecule called pyruvate. In the process, two molecules of ATP are regenerated from ADP.

The events of glycolysis are summarised in Figure 2.

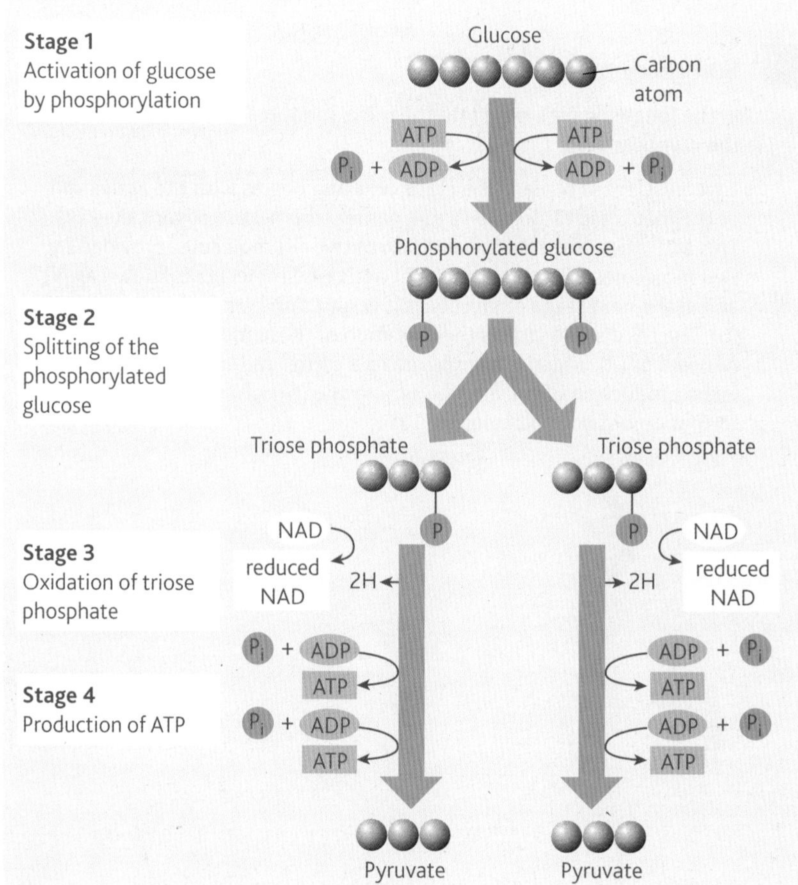

Figure 2 *Summary of glycolysis*

Hint

It must be remembered that for each molecule of glucose at the start of the process two molecules of triose phosphate are produced. Therefore the yields must be doubled, i.e. four molecules of ATP and two molecules of reduced NAD.

Hint

Gluc**ose**, a sugar, is oxidised to pyruv**ate**, an acid.

Energy yields from glycolysis

The overall yield from one glucose molecule undergoing glycolysis is therefore:

- two molecules of ATP (four molecules of ATP are produced, but two were used up in the initial phosphorylation of glucose and so the net increase is two molecules)
- two molecules of reduced NAD (these have the potential to produce more ATP as we shall see in Topic 4.3)
- two molecules of pyruvate.

Glycolysis is a universal feature of every living organism. The enzymes for the glycolytic pathway are found in the cytoplasm of cells and so glycolysis does not require any organelle or membrane for it to take place. It does not require oxygen and therefore it can take place whether or not it is present. In the absence of oxygen the pyruvate produced by glycolysis can be converted into either lactate or ethanol by a process called anaerobic respiration. This is necessary in order to re-oxidise NAD so that glycolysis can continue. This is explained, along with details of the reactions, in Topic 4.4. Anaerobic respiration, however, yields only a small fraction of the potential energy stored in the pyruvate molecule. In order to release the remainder of this energy, most organisms use oxygen to break down pyruvate further in a process called the Krebs cycle.

Summary questions

In the following passage, state the most suitable word to replace each of the numbers 1–10.

Glycolysis takes place in the (1) of cells and begins with the activation of the main respiratory substrate, namely the hexose sugar called (2). This activation involves the addition of two (3) molecules provided by two molecules of (4). The resultant activated molecule is known as (5) and in the next stage of glycolysis it is split into two molecules called (6). The third stage entails the oxidation of these molecules by the removal of (7), which is transferred to a carrier called (8). The final stage is the production of the 3-carbon molecule (9), which also results in the formation of two molecules of (10).

4.2 Link reaction and Krebs cycle

Learning objectives:

- What is the link reaction?
- What happens during the Krebs cycle?
- What are hydrogen carrier molecules and what is their role in the Krebs cycle?

Specification reference: 3.4.4

 Link

It will be helpful in this section to remind yourself of the structure of mitochondria. Details can be found in Topic 3.3 of the *AS Biology* book.

AQA Examiner's tip

Most candidates in examinations **wrongly** think that the carbon dioxide produced in respiration is formed by combination with molecular oxygen. It is, in fact, formed directly from molecules involved in the link reaction and the Krebs cycle.

The pyruvate molecules produced during glycolysis possess potential energy that can only be released using oxygen in a process called the Krebs cycle. Before they can enter the Krebs cycle, these pyruvate molecules must first be **oxidised** in a procedure known as the **link reaction**. In **eukaryotic cells** both the Krebs cycle and the link reaction take place exclusively inside mitochondria.

The link reaction

The pyruvate molecules produced in the cytoplasm during **glycolysis** are actively transported into the matrix of mitochondria. Here pyruvate undergoes a series of reactions during which the following changes take place:

- The pyruvate is oxidised by removing hydrogen. This hydrogen is accepted by NAD to form reduced NAD, which is later used to produce ATP (see Topic 4.4).
- The 2-carbon molecule, called an acetyl group, that is thereby formed combines with a molecule called coenzyme A (CoA) to produce a compound called **acetylcoenzyme A**.
- A carbon dioxide molecule is formed from each pyruvate.

The overall equation can be summarised as:

$$\text{pyruvate} + \text{NAD} + \text{CoA} \longrightarrow \text{acetyl CoA} + \text{reduced NAD} + CO_2$$

The Krebs cycle

The Krebs cycle was named after the British biochemist, Hans Krebs, who worked out its sequence. The Krebs cycle involves a series of oxidation-reduction reactions that take place in the matrix of mitochondria. Its events are illustrated in Figure 1 and can be summarised as follows:

- The 2-carbon acetylcoenzyme A from the link reaction combines with a 4-carbon molecule to produce a 6-carbon molecule.
- This 6-carbon molecule loses carbon dioxide and hydrogens to give a 4-carbon molecule and a single molecule of ATP produced as a result of **substrate-level phosphorylation** (see Topic 2.1).
- The 4-carbon molecule can now combine with a new molecule of acetylcoenzyme A to begin the cycle again.

For each molecule of pyruvate, the link reaction and the Krebs cycle therefore produce:

- **reduced coenzymes** such as NAD and FAD (see below). These have the potential to produce ATP molecules (see Topic 4.4) and are therefore the important products of Krebs cycle
- one molecule of ATP
- three molecules of carbon dioxide.

As two pyruvate molecules are produced for each original glucose molecule, the yield from a single glucose molecule is double the quantities above.

A summary of the link reaction and the Krebs cycle is shown in Figure 1.

Hint

Only a small amount of ATP is formed directly by the Krebs cycle. The vast majority of energy is carried away from the Krebs cycle by reduced NAD and reduced FAD and only later converted to ATP.

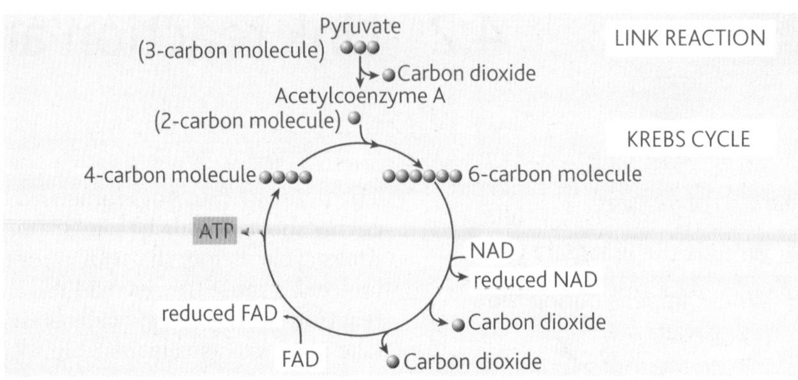

Figure 1 *Summary of the link reaction and the Krebs cycle*

Coenzymes

Coenzymes are molecules that some enzymes require in order to function. Coenzymes play a major role in photosynthesis and respiration where they carry hydrogen atoms from one molecule to another. Examples include:

- **NAD**, which is important throughout respiration
- **FAD**, which is important in the Krebs cycle
- **NADP**, which is important in photosynthesis (see Topic 3.2).

In respiration, NAD is the most important carrier. It works with dehydrogenase enzymes that catalyse the removal of hydrogen ions from substrates and transfer them to other molecules such as the hydrogen carriers involved in oxidative phosphorylation (see Topic 4.4).

The significance of the Krebs cycle

The Krebs cycle performs an important role in the cells of organisms for four reasons:

- It breaks down macromolecules into smaller ones; pyruvate is broken down into carbon dioxide.
- It produces hydrogen atoms that are carried by NAD to the electron transport chain for oxidative phosphorylation. This leads to the production of ATP that provides metabolic energy for the cell.
- It regenerates the 4-carbon molecule that combines with acetylcoenzyme A, which would otherwise accumulate.
- It is a source of intermediate compounds used by cells in the manufacture of other important substances such as fatty acids, amino acids and chlorophyll.

Summary questions

1. How many carbon molecules are there in a single molecule of pyruvate?
2. What 2-carbon molecule is pyruvate converted to during the link reaction?
3. State precisely in which part of the cell the Krebs cycle takes place.
4. Table 1 lists statements about some biochemical processes in a plant cell. State whether each of the numbers 1–18 represents 'true' or 'false'.

Table 1

Statement	Glycolysis	Krebs cycle	Light-dependent reaction of photosynthesis
ATP is produced	1	2	3
ATP is needed	4	5	6
NAD is reduced	7	8	9
NADP is reduced	10	11	12
CO_2 is produced	13	14	15
CO_2 is needed	16	17	18

Application and How science works

Coenzymes in respiration

Coenzymes such as NAD are important in respiration. They help enzymes to function by carrying hydrogen atoms from one molecule to another. Scientists can model the way coenzymes work in cells using a blue dye called methylene blue. It can accept hydrogen atoms and so become reduced. Reduced methylene blue is colourless. This is an example of How science works (HSW: A).

methylene blue + **hydrogen** ⟶ **reduced methylene blue**
(blue colour) **(colourless)**

In an investigation into respiration in yeast, three test tubes were set up as follows:

Tube A	Tube B	Tube C
2 cm^3 yeast suspension	2 cm^3 distilled water	2 cm^3 yeast suspension
2 cm^3 glucose solution	2 cm^3 glucose solution	2 cm^3 distilled water
1 cm^3 methylene blue	1 cm^3 methylene blue	1 cm^3 methylene blue

All three tubes were incubated at a temperature of 30 °C. The colour of each tube was recorded at the start of the experiment and after 5 and 15 minutes. The results are shown in the table below:

Time/ min	Colour of tube contents		
	Tube A	Tube B	Tube C
0	blue	blue	blue
5	colourless	blue	blue
15	colourless	blue	pale blue

1. Tube B acts as a control. Explain why this control was necessary in this investigation.
2. Using your knowledge of respiration, suggest an explanation for the colour change after 15 minutes in:
   a tube A
   b tube C.
3. How might the results in tube A after 15 minutes have been different if the experiment had been carried out at 70 °C? Explain your answer.
4. After 20 minutes the contents of tube A were mixed with air by shaking it vigorously, turning the methylene blue back to a blue colour. Suggest a reason for this colour change.
5. Suggest why conclusions made only on the basis of the results of this experiment may not be reliable.

4.3 Electron transport chain

Learning objectives:

- Where does the electron transport chain take place?
- How is ATP synthesised in the electron transport chain?
- What is the role of oxygen in aerobic respiration?

Specification reference: 3.4.4

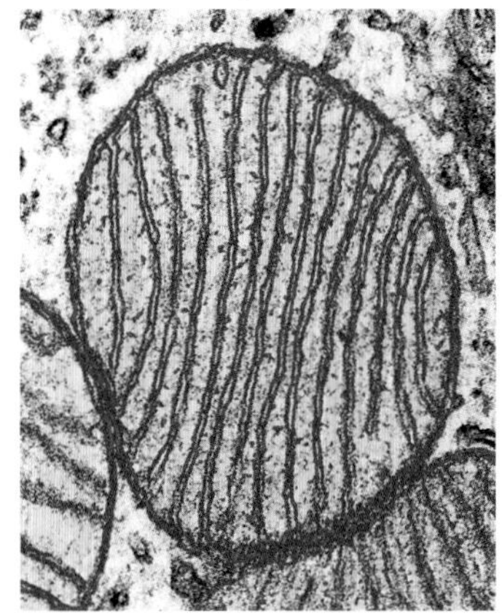

Figure 1 *Coloured TEM of a sectioned mitochondrion (red and yellow). It has two membranes: an outer surrounding membrane and an inner membrane that forms folds called* cristae*, seen here as red lines. The cristae are the sites of the electron transport chain*

Link

Re-reading about the structure of mitochondria in Topic 3.3 of the *AS Biology* book will help you follow the processes of the electron transport chain.

Hint

Remember that a single hydrogen atom is made up of one proton (H^+) and one electron (e^-).

So far in the process of **aerobic** respiration, we have seen how hexose sugars such as glucose are split (glycolysis) and how the 3-carbon pyruvate that results is fed into the Krebs cycle to yield carbon dioxide and hydrogen atoms. The carbon dioxide is a waste product and is removed during the process of gaseous exchange. The hydrogen atoms (or more particularly the electrons they possess) are valuable as a potential source of energy. These hydrogen atoms are carried by the coenzymes NAD and FAD into the next stage of the process: the **electron transport chain**. This is the mechanism by which the energy of the **electrons** within the hydrogen atoms is converted into a form that cells can use, namely **adenosine triphosphate** (**ATP**).

The electron transport chain and mitochondria

Mitochondria are rod-shaped organelles that are found in **eukaryotic cells**. We saw in Topic 3.3 of the *AS Biology* book that each mitochondrion is bounded by a smooth outer membrane and an inner one that is folded into extensions called cristae. The inner space, or matrix, of the mitochondrion is made up of a semi-rigid material of protein, lipids and traces of DNA.

Mitochondria are the sites of the electron transport chain. Attached to the inner folded membrane (cristae) are the enzymes and other proteins involved in the electron transport chain and hence ATP synthesis.

As mitochondria play such a vital role in respiration and the release of energy, it is hardly surprising that they occur in greater numbers in metabolically active cells, such as those of the muscles, liver and epithelial cells, which carry out active transport. The mitochondria in these cells also have more densely packed cristae which provide a greater surface area for the attachment of enzymes and other proteins involved in electron transport.

The electron transport chain and the synthesis of ATP

ATP is synthesised using the electron transport chain as follows:

- The hydrogen atoms produced during glycolysis and the Krebs cycle combine with the coenzymes NAD and FAD that are attached to the cristae of the mitochondria.
- The reduced NAD and FAD donate the electrons of the hydrogen atoms they are carrying to the first molecule in the electron transport chain.
- This releases the **protons** from the hydrogen atoms and these protons are actively transported across the inner mitochondrial membrane.
- The electrons meanwhile, pass along a chain of electron transport carrier molecules in a series of **oxidation-reduction** reactions. The electrons lose energy as they pass down the chain and some of this is used to combine ADP and inorganic phosphate to make ATP. The remaining energy is released in the form of heat.
- The protons accumulate in the space between the two mitochondrial membranes before they diffuse back into the mitochondrial matrix through special protein channels.
- At the end of the chain the electrons combine with these protons and oxygen to form water. Oxygen is therefore the final acceptor of electrons in the electron transport chain.

These events are summarised in Figure 2.

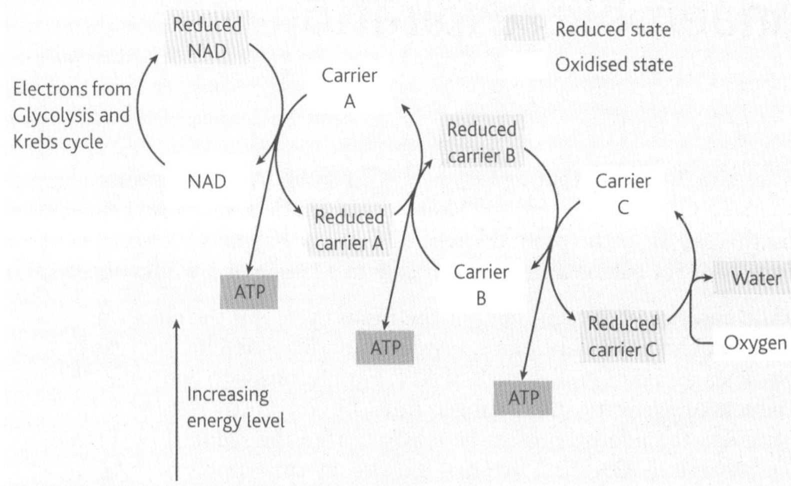

Figure 2 *Summary of electron transport chain*

The importance of oxygen in respiration is to act as the final acceptor of the hydrogen atoms produced in glycolysis and the Krebs cycle. Without its role in removing hydrogen atoms at the end of the chain, the hydrogen ions (protons) and electrons would 'back up' along the chain and the process of respiration would come to a halt. This point is illustrated by the effect of cyanide on respiration. Most people are aware that cyanide is a very potent poison that causes death rapidly. It is lethal because it is a non-competitive inhibitor of the final enzyme in the electron transport chain. This enzyme catalyses the addition of the hydrogen ions and electrons to oxygen to form water. Its inhibition causes hydrogen ions and electrons to accumulate on the carriers, bringing cellular respiration to a halt.

Application

Sequencing the chain

The order in which the carrier molecules of the electron transport chain are arranged can be determined experimentally. The experiments rely on the fact that each transfer of electrons between one molecule and the next is catalysed by a specific enzyme. In a series of experiments, three different inhibitors, 1, 2 and 3, are added to four electron transport molecules, A, B, C and D. Table 1 shows whether the molecules A–D are oxidised or reduced after the inhibitor is added.

Table 1

Inhibitor added	Electron transport molecules			
	A	B	C	D
1	reduced	oxidised	reduced	oxidised
2	oxidised	oxidised	reduced	oxidised
3	reduced	oxidised	reduced	reduced

1 Using the information in the table, state the order of the electron transport molecules in this chain. Explain your answer.

Summary questions

1 The processes that occur in the electron transport chain are also known as oxidative phosphorylation. Suggest why this term is used.

2 The surface of the inner mitochondrial membrane is highly folded to form cristae. State **one** advantage of this arrangement to the electron transport chain.

3 The oxygen taken up by organisms has an important role in aerobic respiration. Explain this role.

4 As part of which molecule does the oxygen taken in leave an organism after being respired?

AQA Examiner's tip

Oxygen is used as the final acceptor of hydrogen atoms at the end of the electron transport chain. It is therefore used to form water and **not** carbon dioxide, as many candidates wrongly think.

4.4 Anaerobic respiration

Learning objectives:

- How is energy released by respiration in the absence of oxygen?
- How is ethanol produced by anaerobic respiration?
- How is lactate produced by anaerobic respiration?

Specification reference: 3.4.4

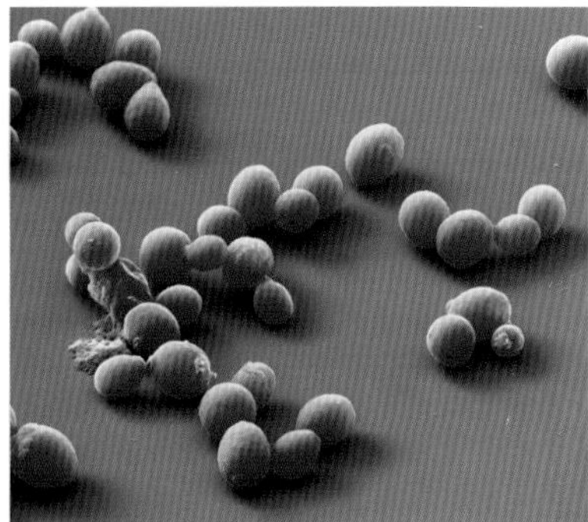

Figure 1 *Coloured SEM of yeast cells. Yeast produces ethanol and carbon dioxide during anaerobic respiration making it useful in brewing*

Hint

During strenuous exercise, muscles carry out aerobic respiration. If this cannot supply ATP fast enough, they also carry out some anaerobic respiration as well. It is a case not of one or the other but of both together.

We saw in Topic 4.3 that oxygen is needed if the hydrogen atoms produced in **glycolysis** and the **Krebs cycle** are to be converted to water and thereby drive the production of ATP. What happens if oxygen is temporarily or permanently unavailable to a tissue or a whole organism?

In the absence of oxygen, neither the Krebs cycle nor the electron transport chain can take place, leaving only the anaerobic process of glycolysis as a potential source of ATP. For glycolysis to continue, its products of pyruvate and hydrogen must be constantly removed. In particular, the hydrogen must be released from the reduced NAD in order to regenerate **NAD.** Without this, the already tiny supply of NAD in cells will be entirely converted to reduced NAD, leaving no NAD to take up the hydrogen newly produced from glycolysis. Glycolysis will then grind to a halt. The replenishment of NAD is achieved by the pyruvate molecule from glycolysis accepting the hydrogen from reduced NAD.

In eukaryotic cells, only two types of anaerobic respiration occur with any regularity:

- In plants, and in microorganisms such as yeast, the pyruvate is converted to ethanol and carbon dioxide.
- In animals, the pyruvate is converted to lactate.

Production of ethanol in plants and some microorganisms

Anaerobic respiration leading to the production of ethanol occurs in organisms such as certain bacteria and fungi (e.g. yeast) as well as in some cells of higher plants, for example, root cells under waterlogged conditions.

The pyruvate molecule formed at the end of glycolysis loses a molecule of carbon dioxide and accepts hydrogen from reduced NAD to produce ethanol. The summary equation for this is:

$$\text{pyruvate + reduced NAD} \longrightarrow \text{ethanol + carbon dioxide + NAD}$$

This form of anaerobic respiration in yeast has been exploited by humans for thousands of years in the brewing industry. In brewing, ethanol is the important product. Yeast is grown in anaerobic conditions in which it ferments natural carbohydrates in plant products, such as grapes (wine production) or barley seeds (beer production) into ethanol.

Production of lactate in animals

Anaerobic respiration leading to the production of lactate occurs in animals as a means of overcoming a temporary shortage of oxygen. Clearly, such a mechanism has considerable survival value, for example, in a baby mammal in the period immediately after birth, and in an animal living in water where the amount of oxygen may sometimes be very low.

However, lactate production occurs most commonly in muscles as a result of strenuous exercise. In these conditions oxygen may be used up more rapidly than it can be supplied and therefore an oxygen debt occurs. It is often essential, however, that the muscles continue to work despite the lack of oxygen, for example, if the organism is fleeing from a predator. In the absence of oxygen, glycolysis would normally cease as reduced NAD accumulates. If glycolysis is to continue and release some energy, the reduced NAD must be removed. To achieve this, each pyruvate molecule produced takes up the two hydrogen atoms from the reduced NAD produced in glycolysis to form lactate as shown below:

pyruvate + reduced NAD ⟶ lactate + NAD

At some point the lactate produced needs to be oxidised back to pyruvate. This can then be either further oxidised to release energy or converted into glycogen. This happens when oxygen is once again available. In any case, lactate will cause cramp and muscle fatigue if it is allowed to accumulate in the muscle tissue. Although muscle has a certain tolerance to lactate, it is nevertheless important that it is removed by the blood and taken to the liver to be converted to glycogen. Figure 3 shows how the NAD needed for glycolysis to continue is regenerated in both common forms of anaerobic respiration.

Figure 2 *During strenuous exercise, muscle may temporarily respire anaerobically*

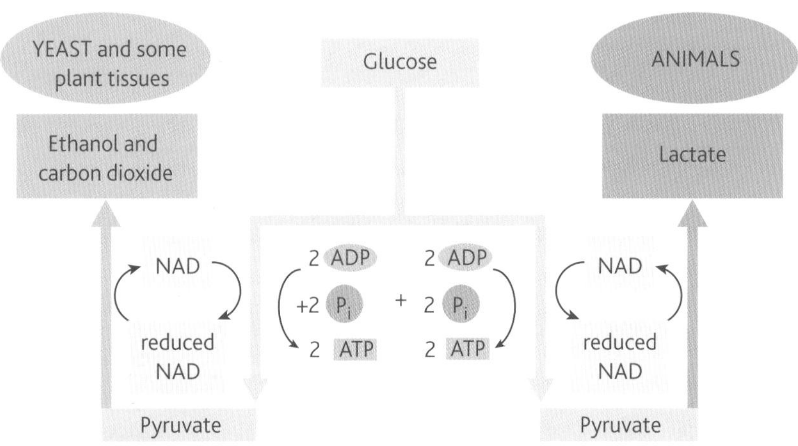

Figure 3 *How the NAD needed for glycolysis is regenerated in various organisms*

Energy yields from anaerobic and aerobic respiration

Energy from cellular respiration is derived in two ways:

- substrate-level phosphorylation in glycolysis and the Krebs cycle. This is the direct linking of inorganic phosphate (P_i) to ADP to produce ATP.
- oxidative phosphorylation in the electron transport chain. This is the indirect linking of inorganic phosphate to ADP to produce ATP using the hydrogen atoms from glycolysis and the Krebs cycle that are carried on NAD and FAD. Cells produce most of their ATP in this way.

In anaerobic respiration, pyruvate is converted to either ethanol or lactate. Consequently it is not available for the Krebs cycle. Therefore in anaerobic respiration neither the Krebs cycle nor the electron transport chain can take place. The only ATP that can be produced by anaerobic respiration is therefore that formed by glycolysis. This amount is very small when compared to the much greater quantity produced during aerobic respiration.

Summary questions

The diagram below shows the relationship between some respiratory pathways.

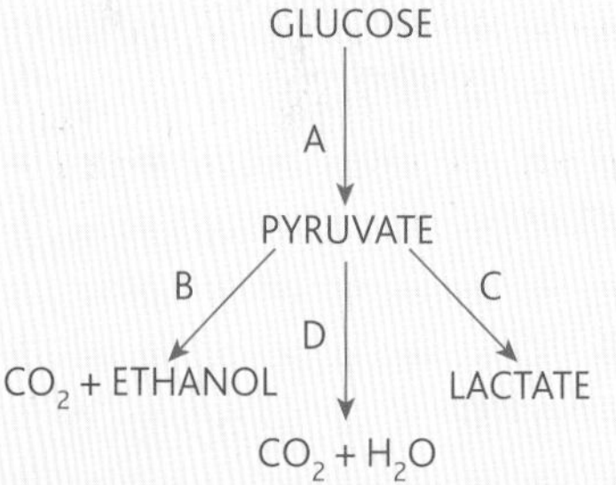

1 State which of the pathways, A, B, C or D, apply to each of the following statements. There may be more than one answer in each case.

a Only occurs in the presence of oxygen.

b Takes place in animals.

c Produces ATP.

d Is carried out by yeast in the absence of oxygen.

e Produces reduced NAD.

f Regenerates NAD from reduced NAD.

g Is known as glycolysis.

Link

To help you follow the experiment described in the application and to answer the questions, it is necessary to understand cell fractionation and enzyme inhibition. It is therefore advisable to review Topics 3.1 and 2.7 of the *AS Biology* book.

Application and How science works

Investigating where certain respiratory pathways take place in cells

Cyanide is a non-competitive inhibitor of an enzyme in the electron transport chain. It therefore prevents the transfer of electrons along this chain.

To determine where in the cell some of the respiratory pathways take place, scientists carried out the following experiment involving cyanide. Given the poisonous nature of cyanide, the scientists carried out risk assessments and imposed appropriate safety precautions to ensure their own safety and that of others. These precautions included the wearing of safety glasses and gloves, working in well-ventilated areas, disposing of the chemical appropriately and having a cyanide antidote kit available at all times. This experiment is an example of How science works (HSW: D).

- Mammalian liver cells were broken up (homogenised) and the resulting homogenate was centrifuged.
- Portions containing only nuclei, ribosomes, mitochondria and the remaining cytoplasm were separated out.
- Samples of each portion, and of the complete homogenate, were incubated as follows:
  - with glucose
  - with glucose and cyanide
  - with pyruvate and cyanide
  - with pyruvate

After incubation the presence or absence of carbon dioxide and lactate in each sample was recorded. The results are shown in Table 1, in which ✓ = present and ✗ = absent.

Table 1

Incubated with	Complete homogenate		Nuclei only		Ribosomes only		Mitochondria only		Remaining cytoplasm only	
	Carbon dioxide	Lactate	Carbon dioxide	Lactate	Carbon dioxide	Lactate	Carbon dioxide	Lactate	Carbon dioxide	Lactate
Glucose	✓	✓	✗	✗	✗	✗	✗	✗	✗	✓
Pyruvate	✓	✓	✗	✗	✗	✗	✓	✗	✗	✓
Glucose + cyanide	✗	✓	✗	✗	✗	✗	✗	✗	✗	✓
Pyruvate + cyanide	✗	✓	✗	✗	✗	✗	✗	✗	✗	✓

1. Briefly describe how the different portions of the homogenate may have been separated out by centrifuging.
2. From the results of this experiment, name **two** organelles that appear not to be involved in respiration. Explain your answer.
3. a In which cell organelle would you expect to find the enzymes of the Krebs cycle?
   b Explain how the results in the table support your answer.
4. Which portion of the homogenate contains the enzymes that convert pyruvate into lactate?
5. Explain why lactate is produced in the presence of cyanide but carbon dioxide is not.
6. Explain why carbon dioxide can be produced by the complete homogenate when none of the separate portions can do so.
7. If glucose were incubated with cytoplasm from yeast cells, which two products would be formed?
8. Which **three** of the following structures might you expect to be rich in mitochondria: xylem vessel, liver cell, red blood cell, epithelial cell, muscle cell?

AQA Examination-style questions

1 **Figure 1** shows the structure of a mitochondrion.

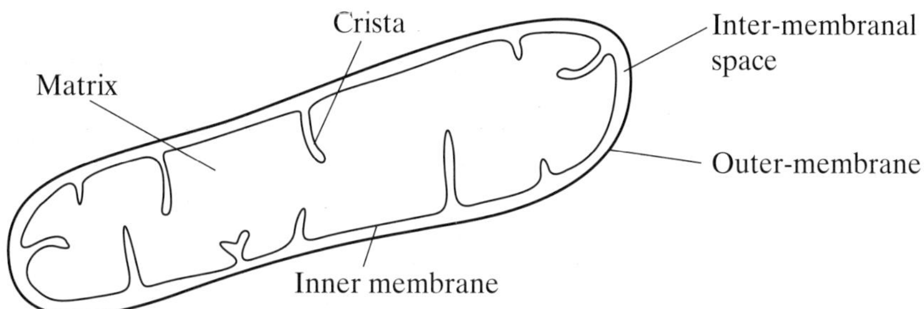

Figure 1

(a) In which part of the mitochondrion does the Krebs cycle take place? *(1 mark)*

(b) Name **two** substances for which there would be net movement into the mitochondrion. *(2 marks)*

(c) Explain how reactions occurring in a mitochondrion generate ATP. *(7 marks)*

AQA, 2006

2 **Figure 2** gives an outline of the process of aerobic respiration.

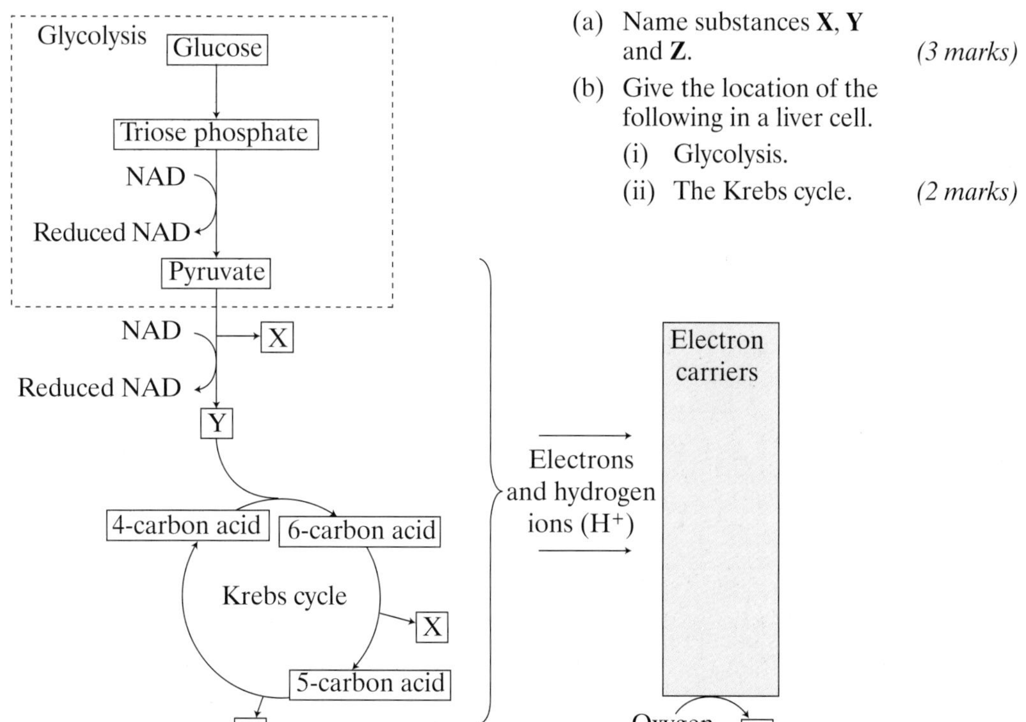

Figure 2

(a) Name substances **X**, **Y** and **Z**. *(3 marks)*

(b) Give the location of the following in a liver cell.

(i) Glycolysis.

(ii) The Krebs cycle. *(2 marks)*

(c) (i) Write the letter **A** on the diagram to show one step where ATP is used.

(ii) Write the letter **B** on the diagram to show one step where ATP is produced. *(2 marks)*

(d) Human skeletal muscle can respire both aerobically and anaerobically. Describe what happens to pyruvate in anaerobic conditions and explain why anaerobic respiration is advantageous to human skeletal muscle. *(4 marks)*

AQA, 2004

3 (a) The main stages in anaerobic respiration in yeast are shown in **Figure 3**.

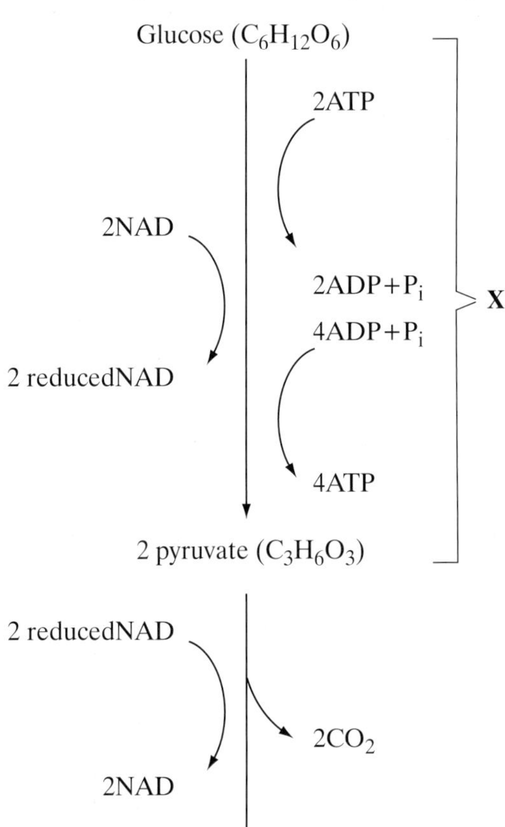

Figure 3

(i) Name process **X**.

(ii) Give **one** piece of evidence from **Figure 3** which suggests that the conversion of pyruvate to ethanol involves reduction.

(iii) Explain why converting pyruvate to ethanol is important in allowing the continued production of ATP in anaerobic respiration. *(4 marks)*

(b) Give **two** ways in which anaerobic respiration of glucose in yeast is:

(i) similar to anaerobic respiration of glucose in a muscle cell;

(ii) different to anaerobic respiration of glucose in a muscle cell. *(4 marks)*

(c) Some students investigated the effect of temperature on the rate of anaerobic respiration in yeast. The apparatus they used is shown in **Figure 4**. The yeast suspension was mixed with glucose solution and the volume of gas collected in five minutes was recorded.

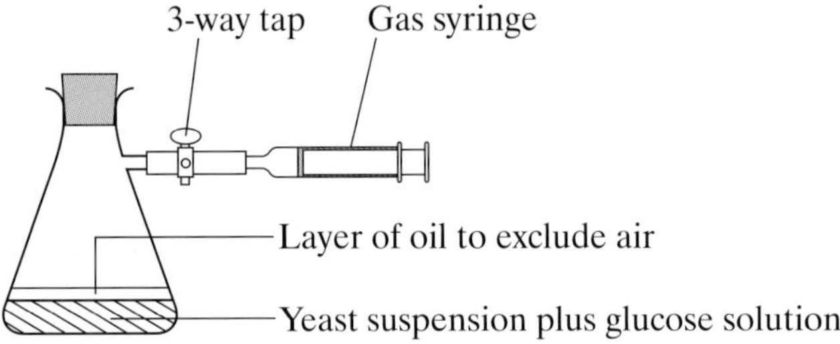

Figure 4

(i) Each student repeated the experiment and the results were pooled. Explain the advantages of collecting a large number of results.

(ii) At 30 °C, one student obtained the following results.

Volume of gas collected in 5 minutes / cm³	**Result 1**	**Result 2**	**Result 3**
	38.3	27.6	29.4

Calculate the mean rate of gas production. Give your answer in $cm^3 s^{-1}$.

(iii) If aerobic respiration had been investigated rather than anaerobic respiration, how would you expect the volumes of gas collected at 30 °C to differ from these results? Explain your answer. *(7 marks)*

AQA, 2005

4 (a) Table 1 contains some statements relating to biochemical processes in an animal cell. Copy and complete the table with a tick (✓) if the statement is true or cross (✗) if it is not true for each biochemical process.

Table 1

Statement	Glycolysis	Krebs Cycle
NAD is reduced		
NADP is reduced		
ATP is produced		
ATP is required		

(2 marks)

(b) An investigation was carried out into the production of ATP by mitochondria. ADP, phosphate, excess substrate and oxygen were added to a suspension of isolated mitochondria.

(i) Suggest the substrate used for this investigation.

(ii) Explain why the concentration of oxygen and amount of ADP fell during the investigation.

(iii) A further investigation was carried out into the effect of three inhibitors, **A**, **B** and **C**, on the electron transport chain in these mitochondria. In each of three experiments, a different inhibitor was added. **Table 2** shows the state of the electron carriers, **W – Z**, after the addition of inhibitor.

Table 2

Inhibitor added	Electron carrier			
	W	**X**	**Y**	**Z**
A	oxidised	reduced	reduced	oxidised
B	oxidised	oxidised	reduced	oxidised
C	reduced	reduced	reduced	oxidised

Give the order of the electron carriers in this electron transport chain. Explain your answer.

(5 marks)

AQA, 2006

5 The boxes in **Figure 5** represent substances in glycolysis, the link reaction and the Krebs cycle.

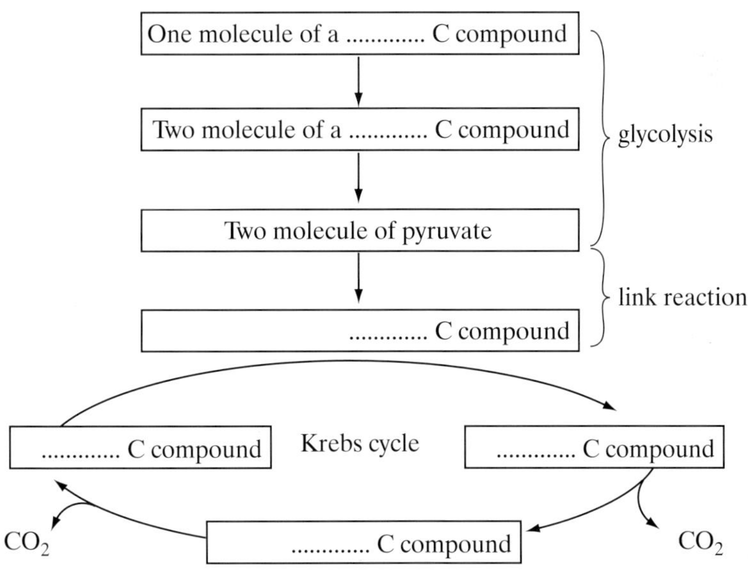

Figure 5

(a) Copy and complete **Figure 5** to show the number of carbon atoms present in **one** molecule of each compound.

(2 marks)

(b) Other substances are produced in the Krebs cycle in addition to the carbon compounds shown in **Figure 5**. Name **three** of these other products.

(3 marks)

AQA, 2005

5 Energy and ecosystems

5.1 Food chains and food webs

Learning objectives:

- How does energy enter an ecosystem?
- How is energy transferred between the organisms in the ecosystem?
- What is meant by the terms: 'trophic level', 'food chain', 'food web', 'producer', 'consumer' and 'decomposer'?
- How is energy lost from the ecosystem?

Specification reference: 3.4.5

The organisms found in any ecosystem rely on a source of energy to carry out all their activities. The ultimate source of this energy is sunlight, which is converted to chemical energy by photosynthesising organisms and is then passed as food between other organisms. In this chapter we shall look at how this energy transfer takes place. We shall also compare natural ecosystems with those based on modern intensive farming and consider how farming practices increase the efficiency of energy conversion.

Organisms can be divided into three groups according to how they obtain their energy and nutrients. These three groups are: producers, consumers and decomposers.

Producers

Producers are photosynthetic organisms that manufacture organic substances using light energy, water and carbon dioxide according to the following equation:

$$\underset{\text{carbon dioxide}}{6CO_2} + \underset{\text{water}}{6H_2O} + \underset{\text{light}}{\text{energy}} \longrightarrow \underset{\text{glucose}}{C_6H_{12}O_6} + \underset{\text{oxygen}}{6O_2}$$

Green plants are producers.

Consumers

Consumers are organisms that obtain their energy by feeding on (consuming) other organisms rather than using the energy of sunlight directly. Animals are consumers. Those that directly eat producers (green plants) are called **primary consumers** because they are the first in the chain of consumers. Those animals eating primary consumers are called **secondary consumers** and those eating secondary consumers are called **tertiary consumers**. Secondary and tertiary consumers are usually predators but they may also be scavengers or parasites.

Decomposers

When producers and consumers die, the energy they contain can be used by a group of organisms that break down these complex materials into simple components again. In doing so, they release valuable minerals and elements in a form that can be absorbed by plants and so contribute to recycling. The majority of this work is carried out by fungi and bacteria, called **decomposers**, and to a lesser extent by certain animals such as earthworms, called **detritivores**.

Hint

You may be familiar with the following terms:

- Herbivore – an animal that eats plants (producers) and is therefore a primary consumer.
- Carnivore – an animal that eats animals and may therefore be a secondary or a tertiary consumer.
- Omnivore – an animal that eats both plants and animals and is therefore a primary consumer and also a secondary or a tertiary consumer.

As you see, these terms are not very precise and you should avoid them when writing about food chains and food webs.

Food chains

The term 'food chain' describes a feeding relationship in which the producers are eaten by primary consumers. These in turn are eaten by secondary consumers, who are then eaten by tertiary consumers. In a long food chain the tertiary consumers may in turn be eaten by further consumers called quaternary consumers. Each stage in this chain is referred to as a **trophic level**. The arrows on food chain diagrams represent the direction of energy flow. Shorter food chains may have three levels:

grass (producer) ⟶ sheep (primary consumer) ⟶ human (secondary consumer)

Longer food chains may have five trophic levels:

nettle (producer) ⟶ aphid (primary consumer) ⟶ ladybird (secondary consumer) ⟶ blue tit (tertiary consumer) ⟶ sparrowhawk (quaternary consumer)

Table 1 describes three further food chains, each from a different habitat.

Table 1 *Examples of food chains*

Trophic level	Grassland	Pond	Seashore
Quaternary consumer	Stoat	Pike	Seagull
Tertiary consumer	Grass snake	Stickleback	Crab
Secondary consumer	Toad	Leech	Whelk
Primary consumer	Caterpillar	Water snail	Limpet
Producer	Grass	Pondweed	Seaweed

Figure 1 *The snake (tertiary consumer) is swallowing an insect-eating frog (secondary consumer) on a plant leaf (producer)*

AQA Examiner's tip

Many students score poorly on ecology questions because they do not use the correct terminology. Make sure you learn simple definitions of ecological terms and use them appropriately.

Food webs

In reality, most animals do not rely upon a single food source and within a single **habitat** many food chains will be linked together to form a food web. For example, on the edge of an oak woodland, the food chain shown above can be combined with others to form the web shown in Figure 2.

The problem with food webs is their complexity. In the example shown in Figure 2, the nettle is shown as being eaten by aphids only. In reality it may be eaten by at least 20 different organisms. The same is true of the oak leaves and some of the consumers in the food web. Indeed, it is likely that all organisms within a habitat, even within an ecosystem, will be linked to others in the food web. Charting the feeding inter-relationships of thousands of species is not feasible. In any case, these relationships are not fixed but change depending on the time of year, age and population size of the organisms. Although not an exact science, describing food webs is valuable in helping us understand populations.

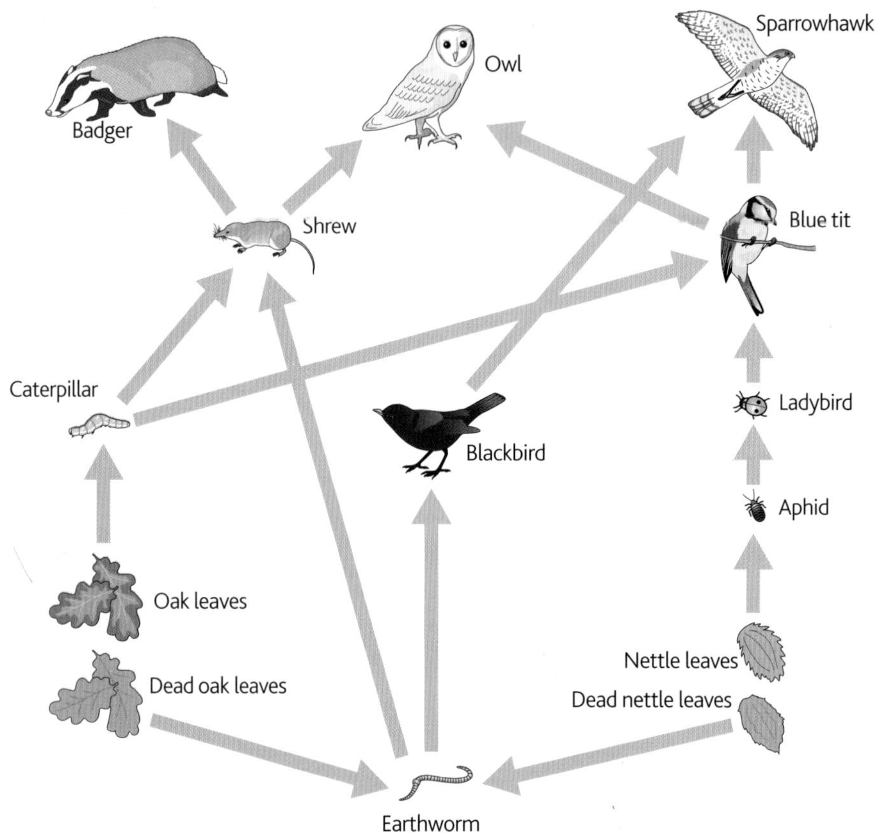

Figure 2 *Part of a woodland food web*

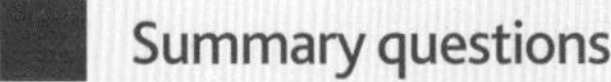

Summary questions

The diagram below shows a simplified food web within an aquatic ecosystem.

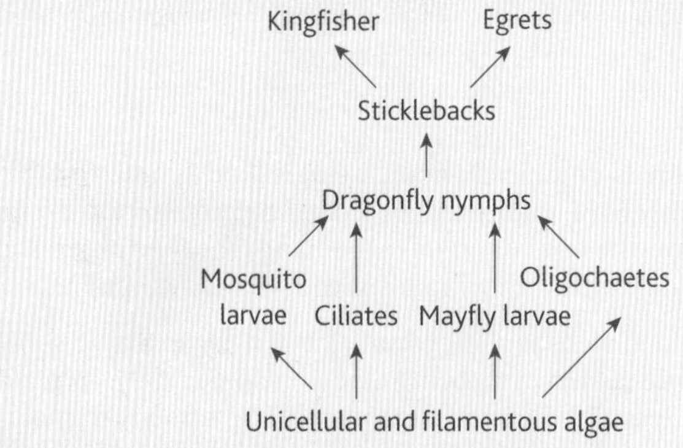

1. Which organisms are secondary consumers?
2. Which organisms carry out photosynthesis?
3. Which organisms are at the fourth trophic level?
4. What do the arrows in the diagram show?
5. When the organisms in this web die they will be broken down by bacteria and fungi. What general term is used to describe these bacteria and fungi?

5.2 Energy transfer between trophic levels

Learning objectives:

- What percentage of energy is transferred from one trophic level to the next?
- How is energy lost along a food chain?
- Why do most food chains have no more than five trophic levels?
- How is the percentage efficiency of energy transfers calculated?

Specification reference: 3.4.5

The Sun is the source of energy for **ecosystems**. However, as little as one per cent of this light energy may be captured by green plants and so made available to organisms in the food chain. These organisms in turn pass on only a small fraction of the energy that they receive to each successive stage in the chain. How then is so much energy lost?

Energy losses in food chains

Plants normally convert between one per cent and three per cent of the Sun's energy available to them into organic matter. Most of the Sun's energy is not converted to organic matter by photosynthesis because:

- over 90 per cent of the Sun's energy is reflected back into space by clouds and dust or absorbed by the atmosphere
- not all wavelengths of light can be absorbed and used for photosynthesis
- light may not fall on a chlorophyll molecule
- a factor, such as low carbon dioxide levels, may limit the rate of photosynthesis (see Topic 3.4).

The total quantity of energy that the plants in a community convert to organic matter is called the **gross production**. However, plants use 20–50 per cent of this energy in respiration, leaving little to be stored. The rate at which they store energy is called the **net production**.

net production = gross production – respiratory losses

Even then, only about ten per cent of this food stored in plants is used by primary consumers for growth. Secondary and tertiary consumers are slightly more efficient, transferring about twenty per cent of the energy available from their prey into their own bodies. The low percentage of energy transferred at each stage is the result of the following:

- Some of the organism is not eaten.
- Some parts are eaten but cannot be digested and are therefore lost in faeces.
- Some of the energy is lost in excretory materials, such as urine.
- Some energy losses occur as heat from respiration and directly from the body to the environment. These losses are high in mammals and birds because of their high body temperature. Much energy is needed to maintain their body temperature when heat is constantly being lost to the environment.

Energy flow along food chains, showing the percentage transferred at each trophic level, is summarised in Figure 1.

It is the relative inefficiency of energy transfer between trophic levels that explains why:

- most food chains have only four or five trophic levels because insufficient energy is available to support a large enough breeding population at trophic levels higher than these
- the total mass of organisms in a particular place (biomass) is less at higher trophic levels
- the total amount of energy stored is less at each level as one moves up a food chain.

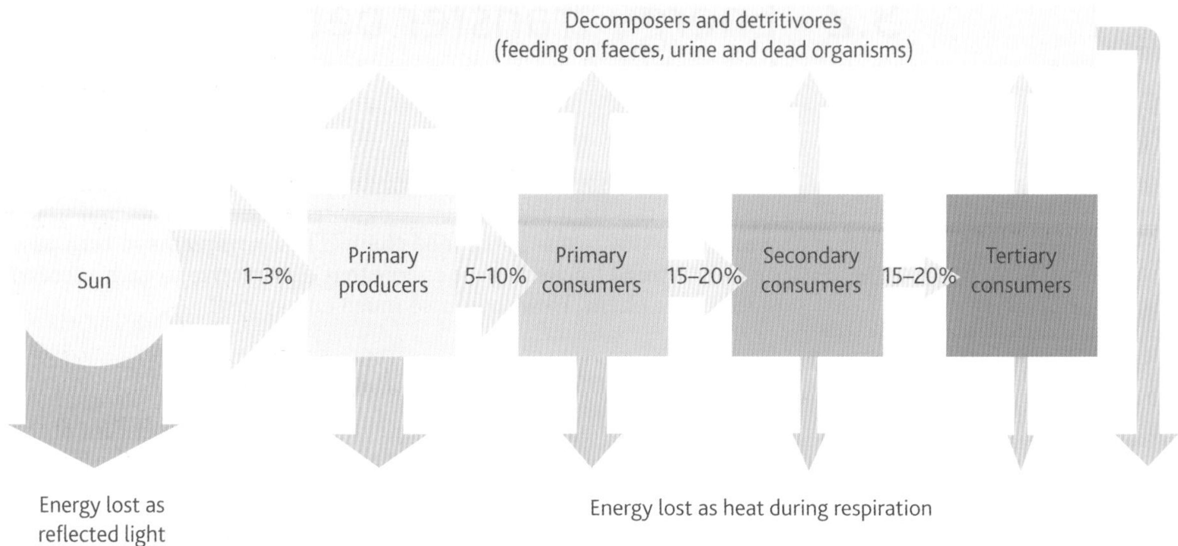

Figure 1 *Energy flow through different trophic levels of a food chain. The arrows are not to scale and give only an idea of the proportion of energy transferred at each stage. Likewise, the figures for % energy transfer between trophic levels are only a rough average as they vary considerably between different plants, animals and habitats*

Hint

If you were ever in any doubt about the considerable loss of energy from organisms, just think about how much food you have eaten in your whole life – and all there is to show for it is what you are now.

Figure 2 *Only about ten per cent of the energy in the plant being eaten by this swallowtail butterfly larva will be used for its growth*

Calculating the efficiency of energy transfers

Data are often presented showing the amount of energy available at each trophic level of a food chain. The energy available is usually measured in kilojoules per square metre per year (kJ m^{-2} $year^{-1}$). It is often useful to calculate the efficiency of the energy transfer between each trophic level of these food chains. This is calculated as follows:

$$\text{energy transfer} = \frac{\text{energy available after the transfer}}{\text{energy available before the transfer}} \times 100$$

Let us take an example. Look at Figure 3, which shows the amount of energy available at different trophic levels in a lake in the USA. Suppose we wanted to calculate the percentage efficiency of the transfer of energy from trout to humans. We would make the calculation as follows.

Energy available after the transfer (i.e. energy available to humans) = 50 kJ m^{-2} $year^{-1}$

Energy available before the transfer (i.e. energy available to trout) = 250 kJ m^{-2} $year^{-1}$

$$\text{percentage efficiency} = \frac{50}{250} \times 100 = \frac{5000}{250} = 20\%$$

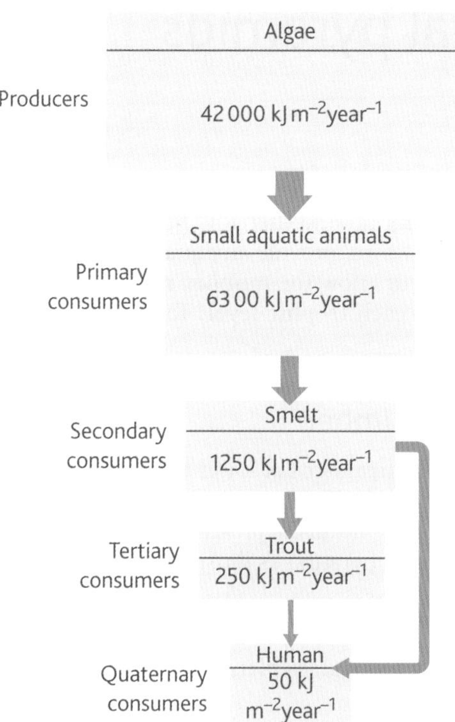

Figure 3 *Food chain in Cayuga Lake, New York State. Figures illustrate the relative amount of energy available at each trophic level in the food chain*

Summary questions

1. State **three** reasons for the small percentage of energy transferred at each trophic level.
2. Explain why most food chains rarely have more than four trophic levels.
3. Using Figure 3, calculate the percentage efficiency of energy transfer between:
   a. primary consumers and secondary consumers
   b. algae and humans.

   Show your working in both cases.

Application

Adding up the totals

Figure 4 shows the flow of energy through a terrestrial ecosystem each year. All the values are in kJ m^{-2}year^{-1}.

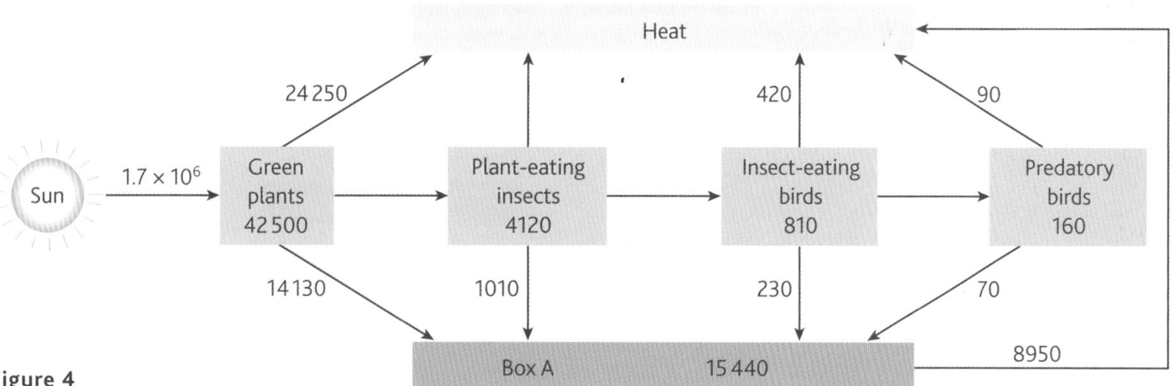

Figure 4

1. Give the name of the group of organisms represented by box A.
2. Which group of organisms are secondary consumers?
3. Calculate the percentage efficiency with which light energy is transferred to energy in green plants. Show your working.
4. State **three** reasons why so little of the solar energy is transferred to energy in green plants.
5. Calculate the amount of energy that is lost as heat from plant-eating insects. Show your working.

Hint

When making calculations involving energy transfers, always remember that energy cannot be created or destroyed. In this type of question, this means that the total amount of energy entering a box must equal the amount of energy in the box plus the amount leaving the box.

5.3 Ecological pyramids

Learning objectives:

- What are the different types of ecological pyramid?
- What are the relative merits and disadvantages of each?

Specification reference: 3.4.5

Diagrams of food chains and food webs are a useful means of showing what different organisms eat and therefore how energy flows between them. They do not, however, provide any quantitative information. Sometimes it is useful to know the number, mass or amount of energy stored by organisms at each **trophic level**. To do this we construct ecological pyramids.

Pyramids of number

Usually the numbers of organisms at lower trophic levels are greater than the numbers at higher levels. This can be shown by drawing bars with lengths proportional to the numbers present at each trophic level. Figure 1a shows the typical number pyramid as illustrated by the food chain:

grass ⟶ rabbits ⟶ foxes

There are considerably more grass plants than rabbits and considerably more rabbits than foxes.

There can be significant drawbacks to using a number pyramid to describe a food chain:

- No account is taken of size – one giant tree is treated the same as one tiny aphid and each parasite has the same numerical value as its larger host. This means that sometimes the pyramid is not a pyramid at all (Figure 1b) or it is inverted (Figure 1c).
- The number of individuals can be so great that it is impossible to represent them accurately on the same scale as other species in the food chain. For example, one tree may have millions of greenfly living off it.

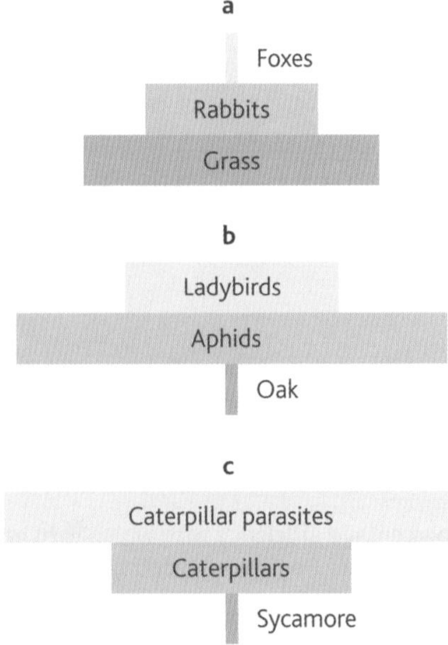

Figure 1 *Pyramids of number*

Pyramids of biomass

A more reliable, quantitative description of a food chain is provided when, instead of counting the organisms at each level, their biomass is measured. **Biomass** is the total mass of the plants and/or animals in a particular place. The fresh mass is quite easy to assess, but the presence of varying amounts of water makes it unreliable. The use of dry mass measurement overcomes this problem but, becausc thc organisms must be killed, it is usually only made on a small sample, and this sample may not be representative. Biomass is measured in grams per square metre ($g\,m^{-2}$) where an area is being sampled, e.g. on grassland or a seashore. Where a volume is being sampled, e.g. in a pond or an ocean, it is measured in grams per cubic metre ($g\,m^{-3}$).

In both pyramids of numbers and pyramids of biomass, only the organisms present at a particular time are shown; seasonal differences are not apparent. This is particularly significant when the biomass of some marine ecosystems is measured. Over the course of a whole year, the mass of phytoplankton (plants) must exceed that of zooplankton (animals), but at certain times of the year this is not seen. For example, in early spring around the British Isles, zooplankton consume phytoplankton so rapidly that the biomass of zooplankton is greater than that of phytoplankton (Figure 2).

Large fish
Small fish
Zooplankton
Phytoplankton

Figure 2 *Pyramid of biomass for a marine ecosystem*

Pyramids of energy

The most accurate representation of the energy flow through a food chain is to measure the energy stored in organisms. However, collecting the data for pyramids of energy (Figure 3) can be difficult and complex. Data are collected in a given area (e.g. one square metre) for a set period of time, usually a year. The results are much more reliable than those for biomass, because two organisms of the same dry mass may store different amounts of energy. For example, one gram of fat stores twice as much energy as one gram of carbohydrate. An organism with more fat will therefore have more stored energy than one with less fat, even though their biomasses are equal. The energy flow in these pyramids is usually measured in $kJ\,m^{-2}\,year^{-1}$.

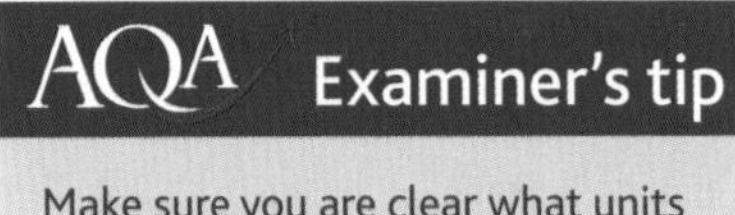

Make sure you are clear what units are used for each type of pyramid.

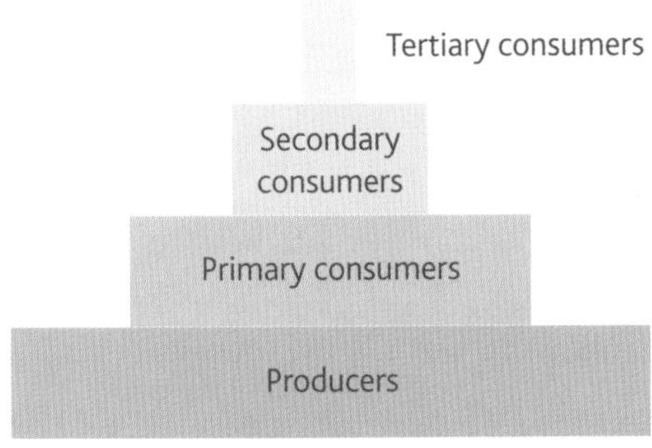

Figure 3 *Pyramid of energy based on oak trees*

Summary questions

1. State **two** advantages of using a pyramid of biomass rather than a pyramid of numbers when representing quantitative information on a food chain.
2. Explain how a pyramid of biomass for a marine ecosystem may sometimes show producers (phytoplankton) with a smaller biomass than primary consumers (zooplankton).
3. Name suitable units for the measurement of biomass.

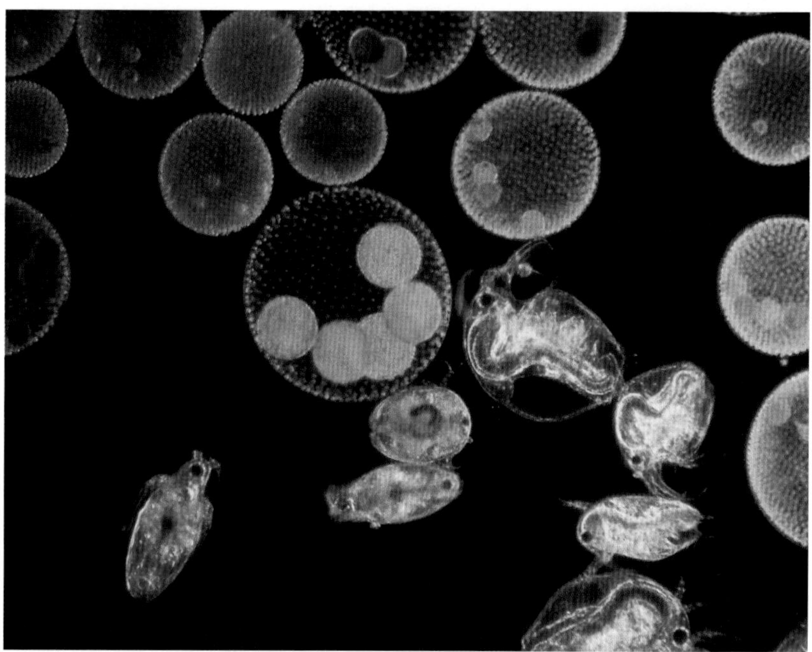

Figure 4 *Dark field photomicrograph of phytoplankton (green algae in upper region) and zooplankton (water fleas in lower region). At certain times of year the biomass of the consumer (zooplankton) may temporarily exceed the biomass of the producer (phytoplankton)*

Application

A woodland food chain

A study of a woodland food chain produced the ecological pyramids shown in Figure 5.

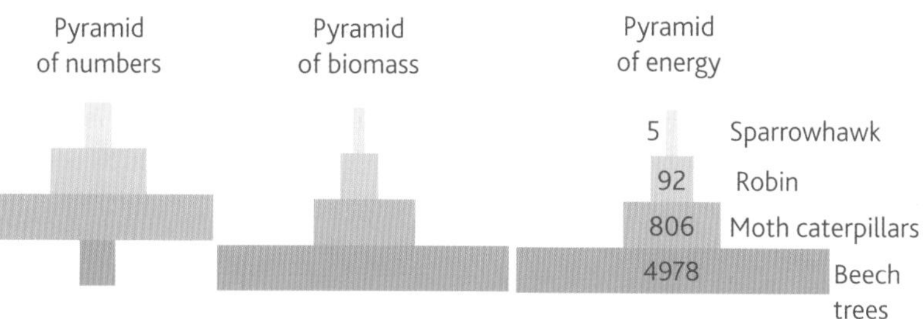

Figure 5

1. Which organisms are the primary consumers?
2. Calculate the percentage efficiency with which energy is transferred from moth caterpillars to robins. Show your working.
3. Suggest suitable units for the figures shown in the pyramid of energy.
4. In the pyramid of numbers, the block representing beech trees is smaller than that of moth caterpillars. In the other pyramids it is larger. Explain this difference.
5. State **two** ways in which energy is lost between the robin and the sparrowhawk.
6. Parasitic fleas obtain their energy from the sparrowhawks on which they live. Redraw each of the pyramids to show how they might appear if parasitic fleas were included.

5.4 Agricultural ecosystems

Learning objectives:

- What is an agricultural ecosystem?
- How do natural and agricultural ecosystems differ?
- What is meant by productivity?
- How is net productivity calculated?

Specification reference: 3.4.5

Hint

Agriculture is about simplifying food webs, allowing more of the available energy to be transferred to humans. This inevitably has a bad effect on the other organisms in the food web!

Figure 1 *Natural ecosystems, such as this oak woodland, have lower productivity*

Figure 2 *Agricultural ecosystems, such as this soya bean field, have higher productivity*

We have looked at how energy is transferred through natural ecosystems. When we look around us, however, much of the landscape that we see is not a natural ecosystem but an agricultural one that has been created by humans. How then do these agricultural ecosystems differ from natural ones?

What is an agricultural ecosystem?

Agricultural ecosystems are made up largely of domesticated animals and plants used to produce food for mankind. We have seen in Topic 5.2 that there are considerable energy losses at each **trophic level** of a food chain. Humans are often at the third, or even fourth, trophic level of a food chain. This means that the energy we receive from the food we eat is often only a tiny proportion of that available from the Sun at the start of the food chain. Agriculture tries to ensure that as much of the available energy from the Sun as possible is transferred to humans. In effect, it is channelling the energy flowing through a food web into the human food chain and away from other food chains. This increases the productivity of the human food chain.

What is productivity?

Productivity is the rate at which something is produced. Plants are called producers because they 'produce' chemical energy (food) by converting light energy into it during photosynthesis. The rate at which plants assimilate this chemical energy is called **gross productivity**. It is measured for a given area over a given period of time, usually in units of $kJ\,m^{-2}\,year^{-1}$.

Some of this chemical energy is utilised by the plant for its respiration, typically 20 per cent. The remainder is known as the **net productivity**. This is available to the next organism in the food chain: the primary consumer. Net productivity can therefore be expressed as:

$$\text{net productivity} = \text{gross productivity} - \text{respiratory losses}$$

Net productivity is important in agricultural ecosystems and is affected by two main factors:

- The efficiency of the crop at carrying out photosynthesis. This is improved if all the necessary conditions for photosynthesis are supplied.
- The area of ground covered by the leaves of the crop.

Comparison of natural and agricultural ecosystems

Although governed by the same basic ecological principles as a natural ecosystem, agricultural ecosystems differ in a number of ways. Some of these differences are shown in Table 1. The two basic differences, energy input and productivity, are also discussed overleaf.

Energy input

In natural ecosystems the only source of energy is the Sun. Most land in Britain would be covered by forest if left to develop naturally. This is known as a climax community (see Topic 7.1). To maintain an agricultural ecosystem we have to prevent this climax community developing. We do this by excluding most of the species in that **community**, leaving only the particular crop that we are trying to grow.

To remove or suppress the unwanted species and to maximise growth requires an additional input of energy. This energy is used to plough fields, sow crops, remove weeds, suppress pests and diseases, feed and house animals, transport materials and the many other tasks carried out by farmers. This additional energy comes in two forms:

- **food**. Farmers and other people that work on farms expend energy as they work. This energy comes from the food they eat.
- **fossil fuels**. As farms have become more mechanised, energy has increasingly come from the fuel used to plough, harvest and transport crops, to produce and apply fertilisers and pesticides and to house, feed and transport livestock.

Hint

Remember that the energy from fossil fuels required in an agricultural ecosystem is **in addition** to solar energy, not instead of it.

Productivity

In natural ecosystems productivity is relatively low. The additional energy input to agricultural ecosystems is used to increase the productivity of a crop by reducing the effect of limiting factors on its growth. The energy used to exclude other species means that the crop has little competition for light, carbon dioxide, water and the minerals needed for photosynthesis. The ground is therefore covered almost exclusively by the crop. Fertilisers are added to provide essential ions, and pesticides are used to destroy pests and prevent disease. Together these factors mean that productivity is much higher in an agricultural ecosystem than in a natural one.

Table 1 *Comparison of natural and agricultural ecosystems*

Natural ecosystem	Agricultural ecosystem
Solar energy only – no additional energy input	Solar energy plus energy from food (labour) and fossil fuels (machinery and transport)
Lower productivity	Higher productivity
More species diversity	Less species diversity
More genetic diversity within a species	Less genetic diversity within a species
Nutrients are recycled naturally within the ecosystem with little addition from outside	Natural recycling is more limited and supplemented by the addition of artificial fertilisers
Populations are controlled by natural means, such as competition and climate	Populations are controlled by both natural means and by use of pesticides and cultivation
Is a natural climax community	Is an artificial community prevented from reaching its natural climax

Summary questions

1. What is meant by the term 'net productivity'?
2. In what units is net productivity usually measured?
3. Explain why the productivity of an agricultural ecosystem is greater than that of a natural ecosystem.
4. What are the differences between the ways that energy is provided in a natural ecosystem and in an agricultural ecosystem?

Application

Increasing productivity

Food production depends upon photosynthesis. As the rate of photosynthesis is determined by the factor that is in shortest supply (the limiting factor) it follows that there is a commercial value in determining which factor is limiting photosynthesis at any one time. By supplying more of this factor, photosynthesis, and hence food production, can be increased.

It is not feasible to control the environment of crops in natural conditions. Plants grown in greenhouses are a different matter however. In the enclosed environment of a greenhouse it is possible to regulate temperature, humidity, light intensity and carbon dioxide concentration. Scientists are able to predict the effects of changing these factors on the rate of photosynthesis. They can then advise commercial growers on the beneficial applications of their findings in order to increase the rate of photosynthesis and hence the growth of their crops. This is an example of How science works (HSW: I).

Figure 3 *Commercial greenhouse*

It may seem logical to simply increase the level of all factors, to ensure maximum yield from photosynthesis. In practice different plants have different optimum conditions and too high a level of a particular factor may reduce yield or kill the plant altogether. For example, high temperatures may increase the yield of one species but **denature** the enzymes of another, thereby killing it. It is also uneconomic and wasteful to expend energy raising temperature and other levels beyond what is necessary. Precise control of the environment is therefore essential. This can be brought about in ways ranging from totally manual control to the use of sophisticated computerised systems.

Let us take the example of carbon dioxide concentration. The average concentration in the atmosphere is around 400 parts per million (ppm). It has been shown that by raising this level to 1000 ppm the yields from tomato plants can be increased by 20 per cent or more. The actual level to which carbon dioxide concentration should be enhanced depends on many factors. Table 2 shows some of the recommended levels of carbon dioxide for maximum yield depending on certain factors.

Table 2 *Recommended levels of carbon dioxide for maximum yield*

Conditions	CO_2/ppm
Bright sunny weather (short duration)	5000
Bright sunny weather (long duration)	1000
Cloudy weather	750
Young plants	700
Greenhouse open for ventilation	400

1 Why is the suggested level of carbon dioxide set lower on a cloudy day than a sunny one?

2 Suggest a reason for the level of 400 ppm when the greenhouse is open to the outside air.

3 Prolonged exposure to high levels of carbon dioxide causes stomata to close. Suggest how this fact may have influenced the recommended levels of carbon dioxide on bright, sunny days.

5.5 Chemical and biological control of agricultural pests

Learning objectives:

- What are pests and pesticides?
- What are the features of an efficient pesticide?
- How are biological agents used to control pests?
- What is integrated pest management?

Specification reference: 3.4.5

We saw in Topic 5.4 that agricultural ecosystems attempt to channel as much available energy from the Sun as possible along human food chains. We also know that each food chain is part of a much more complex food web. This means that many other organisms are competing for the energy in our food chains. To us, these competing organisms are pests. They may be controlled using chemicals, biological agents or a combination of both (integrated system).

What are pests and pesticides?

Although it is difficult to define what is meant by a **pest**, it is generally taken to be an organism that competes with humans for food or space, or it could be a danger to health.

Pesticides are poisonous chemicals that kill pests. They are named after the pests they destroy: herbicides kill plants (herbs), fungicides kill fungi and insecticides kill insects.

An effective pesticide should:

- **be specific**, so that it is only toxic to the organisms at which it is directed. It should be harmless to humans and other organisms, especially the natural predators of the pest, to earthworms, and to pollinating insects such as bees
- **biodegrade**, so that, once applied, it will break down into harmless substances in the soil. At the same time, it needs to be chemically stable, so that it has a long shelf-life
- **be cost-effective**, because development costs are high and new pesticides remain useful only for a limited time. This is because pests can develop genetic resistance, making the pesticide useless
- **not accumulate**, so that it does not build up, either in specific parts of an organism or as it passes along food chains.

Hint

A pest is just an organism growing or living where we do not want it.

Biological control

It is possible to control pests by using organisms that are either predators or parasites of the pest organism. The aim is to control the pest, not to eradicate it, which might be counterproductive. If the pest was reduced to such an extent that there was insufficient food for its predators, the predators would die. The surviving pests would then be able to multiply unchecked. Ideally, the control agent and the pest should exist in balance with one another, at a level where the pest has little, or no, adverse effect.

Link

Details of the predator–prey relationship and the relative sizes of their populations are covered in Topic 1.5.

Using biological control instead of chemical pesticides can have some advantages. These are shown in Table 1. Biological control methods also have a number of disadvantages:

- They do not act as quickly, so there is often some interval of time between introducing the control organism and a significant reduction in the pest population.
- A control organism may itself become a pest. For example, its population may increase, especially where there are few natural

predators. As the pest population is reduced, the control organism may use alternative sources of food such as crops.

Figure 1 *Cane toad eating a pygmy possum. Pest controller turned pest*

Table 1 *Comparison of biological and chemical control of organisms*

Biological control	Chemical pesticides
Very specific	Always have some effect on non-target species
Once introduced, the control organism reproduces itself	Must be reapplied at intervals, making them very expensive
Pests do not become resistant	Pests develop genetic resistance, and new pesticides have to be developed

Integrated pest-control systems

Integrated pest-control systems aim to integrate all forms of pest control rather than being reliant on one type. The emphasis is on deciding the acceptable level of the pest rather than trying to eradicate it altogether. Eradication is, in any case, costly to carry out, counterproductive and almost impossible to achieve. Integrated control involves:

- choosing animal or plant varieties that suit the local area and are as pest-resistant as possible
- managing the environment to provide suitable habitats, close to the crops, for natural predators
- regularly monitoring the crop for signs of pests so that early action can be taken
- removing the pests mechanically (hand-picking, vacuuming, erecting barriers) if the pest exceeds an acceptable population level
- using biological agents if necessary and available
- using pesticides as a last resort if pest populations start to get out of control

Such systems can be effective with minimum impact on the environment.

Hint

Remember that pests are part of food webs. They provide food for other organisms. Their removal can disrupt these food webs and have long-term consequences for us all.

How controlling pests affects productivity

Pests reduce productivity in agricultural ecosystems. Weeds compete with crop plants for water, mineral ions, carbon dioxide, space and light. As these are often in limited supply, any amount taken by the pest means less is available for the crop plant. One or more of them may become the limiting factor in photosynthesis, thus reducing the rate of photosynthesis, and hence productivity. Insect pests may damage the leaves of crops, limiting their ability to photosynthesise and thus reducing their productivity. Alternatively, they may be in direct competition with humans, eating the crop itself. Many crops are now grown in **monoculture**, and this enables insect and fungal pests to spread rapidly. Pests of domesticated animals may cause disease. The animals may not grow as rapidly, be unfit for human consumption or die – all of which lead to reduced productivity.

Figure 2 *Monocultures, such as this area of rape seed, enable insect and fungal pests to spread rapidly*

The aim of pest control is to limit the effect of pests on productivity to a commercially acceptable level. In other words, to balance the cost of pest control with the benefits it brings. The problem is that at least two different interests are involved: the farmer who has to satisfy our demand for cheap food while still making a living, and the **conservation**

Summary questions

1 Pesticides are used to increase productivity. Explain how their use might sometimes reduce productivity.

2 State **two** advantages and **two** disadvantages of biological pest control.

3 Weeds are growing amongst wheat in a field. Explain how these weeds might reduce the productivity of a crop.

of natural resources, which will enable us to continue to have food in the future. The trick is to balance these two, often conflicting, interests.

Application

To weed or not to weed?

The graph in Figure 3 shows the effects of weeds on the productivity of two crops: wheat and soya bean.

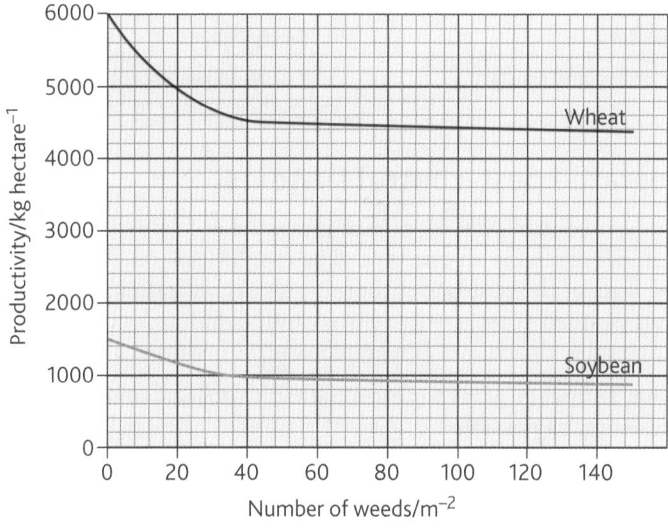

Figure 3

1 Describe the effects of weeds on the productivity of wheat.

2 A herbicide that reduces the number of weeds from 40 m^{-2} to 0 m^{-2} is applied to both crops. Which crop would show the greatest percentage change in productivity?

3 It will cost a farmer £100 to treat each hectare of his wheat crop with a herbicide. The herbicide will reduce the number of weeds from 40 m^{-2} to 20 m^{-2}. He can sell his wheat at £150 a tonne (one tonne = 1000 kg). Is it economically worthwhile for the farmer to apply the herbicide to his crop? Explain your reasoning.

Application

A mighty problem

The two-spotted spider mite, *Tetranychus urticae*, is an important pest of crops, especially those in greenhouses. Control is mostly achieved using chemicals. However, the spider mite has increasingly developed resistance to these chemicals and they are therefore less effective in controlling its populations.

Studies have been carried out to investigate the use of biological control to combat spider mites. In one such study the predatory mite *Phytoseiulus persimilis* was used to test its effectiveness against the two-spotted spider mite. This predatory mite feeds on the spider mite. Mites were introduced into two separate groups of 100 bean plants as follows:

Experiment 1 – spider mites only

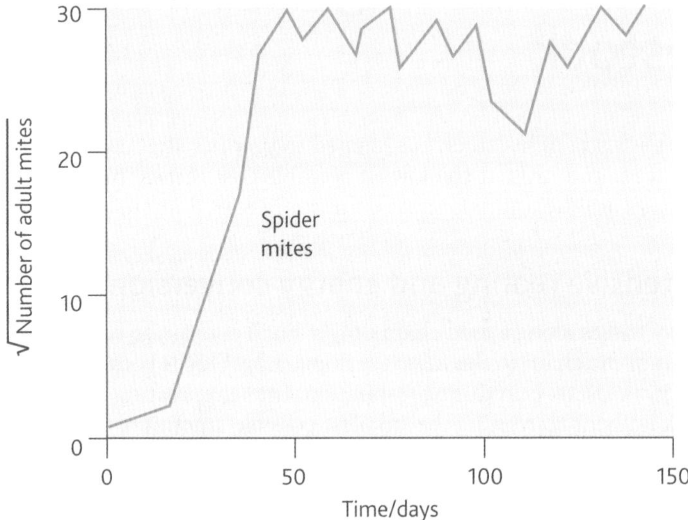

Experiment 2 – spider mites and predatory mites

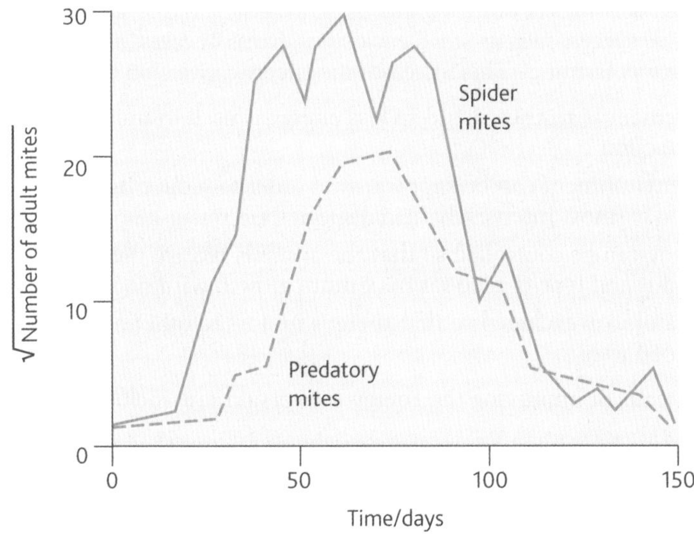

Figure 4

Figure 5 *Biological pest control: orange predatory mite (*Phytoseiulus persimilis*) attacking the red spider mite (*Tetranychus urticae*)*

1 Describe and explain the differences between the spider mite populations in experiment 1 and experiment 2.

2 Comment on the effectiveness of predatory mites in controlling populations of spider mites.

3 Suggest what the levels of the two populations in experiment 2 might be over a period of 150–300 days if the experiment was continued. Explain the reasons for the levels you suggest.

5.6 Intensive rearing of domestic livestock

Learning objectives:

- How does rearing animals intensively increase the efficiency of energy conversion?

Specification reference: 3.4.5

Intensive rearing of livestock is designed to produce the maximum yield of meat, eggs and milk at the lowest possible cost. Cows, pigs, chickens and turkeys are the animals most commonly reared intensively.

Intensive rearing and energy conversion

As energy passes along a food chain only a small percentage passes from one organism in the chain to the next (see Topic 5.2). This is because much of the energy is lost as heat during respiration. Intensive rearing of domestic livestock is about converting the smallest possible amount of food energy into the greatest quantity of animal mass. One way to achieve this is to minimise the energy losses from domestic animals during their lifetime. This means that more of the food energy taken in by the animals will be converted into body mass, ready to be passed on to the next link in the food chain, namely us. Energy conversion can be made more efficient by ensuring that as much energy from respiration as possible goes into growth rather than other activities or other organisms. This is achieved by keeping animals in confined spaces, such as small enclosures, barns or cages, a practice often called 'factory farming'. This increases the energy-conversion rate because:

- movement is restricted and so less energy is used in muscle contraction
- the environment can be kept warm in order to reduce heat loss from the body (most intensively reared species are warm-blooded)
- feeding can be controlled so that the animals receive the optimum amount and type of food for maximum growth with no wastage
- predators are excluded so that there is no loss to other organisms in the food web.

Other means of improving the energy-conversion rate include:

- **selective breeding** of animals to produce varieties that are more efficient at converting the food they eat into body mass
- using hormones to increase growth rates.

Hint

It is worth remembering that intensively reared animals still require large amounts of plant material (producers) as a source of food and this needs to be grown somewhere.

Summary questions

1. How does rearing animals intensively in small covered enclosures increase their energy-conversion rate?
2. Suggest a reason why keeping animals in the dark for longer periods might improve the energy-conversion rate.

Application and How science works

Features of intensive rearing of livestock

Food is essential for life. With an ever-expanding human population, there is pressure to produce more and more food intensively. What then are the main features of intensive rearing? These include:

- **efficient energy conversion**. By restricting wasteful loss of energy, more energy is passed to humans along the food chain.
- **low cost**. Foods such as meat, eggs and milk can be produced more cheaply than by other methods.
- **quality of food**. It is often argued that the taste of the foods produced by intensive rearing is inferior to foods produced less intensively.

- **use of space**. Intensive rearing uses less land while efficient production means that less of the countryside is required for agriculture, leaving more as natural **habitats.**
- **safety**. Smaller, concentrated units are easier to control and regulate. The high density animal **populations** are more vulnerable to the rapid spread of disease but it is easier to prevent infections being introduced from the outside and to isolate the animals if this happens.
- **disease**. Large numbers of animals living in close proximity means that infections can spread easily amongst them. To control this, the animals are regularly given antibiotics.
- **use of drugs**. Over-use of antibiotics to prevent disease in animals has lead to the evolution of **antibiotic resistance**. This resistance can be transferred to bacteria that cause human diseases, making their treatment with certain antibiotics ineffective. Other drugs may be given to animals to improve their growth or reduce aggressive behaviour. These may alter the flavour of the food or pass into the foods and then into humans, affecting their health.
- **animal welfare**. The larger intensive farms have the resources to maintain a high level of animal welfare and are more easily regulated. However, animals are kept unnaturally and this may cause stress, resulting in aggressive behaviour. This may cause them to harm each other or themselves, which is why battery chickens are de-beaked. Restricted movement can lead to osteoporosis and joint pain. The well-being of animals may be sacrificed for financial gain.
- **pollution**. Intensively reared animals produce large concentrations of waste in a small area. Rivers and ground water may become polluted. Pollutant gases may be dangerous and smell. Large intensive farms may have their own disposal facilities that enable them to treat waste more effectively than smaller non-intensive farms.
- **reduced genetic diversity**. Selective breeding is used to develop animals with high energy-conversion rates and a tolerance of confined conditions. This reduces the genetic diversity of domestic animals, resulting in the loss of genes that might later prove to have been beneficial.
- **use of fossil fuels**. High energy-conversion rates are possible because fossil fuels are used to heat the buildings that house the animals, in the production of the materials in the buildings (especially cement) and in the production and transportation of animal feeds. The carbon dioxide emitted increases global warming.

This is an example of How science works (HSW: I & J).

Figure 1 *Rows of intensively reared battery hens*

Figure 2 *Intensively reared domestic pigs inside a shed on a pig farm*

1. Explain how antibiotic resistance in bacteria causing animal diseases can be transferred to bacteria causing human disease.
2. Egg production often involves hens being kept in battery cages. Give reasons for and against this form of egg production.
3. Discuss some of the ethical issues that arise from the features of intensive farming listed above.

Figure 3 *Farmland with small fields and many hedgerows*

Application and How science works

Economic and environmental issues concerned with intensive food production

Economic issues

As consumers we want a reliable supply of a wide range of foods at minimum cost. We therefore create a highly competitive market, which puts farmers under pressure to cut costs and supply cheap food in order to stay in business. As a result they have had to resort to using intensive methods of food production, with its consequent impact on the environment. There is clearly a conflict between our desire for cheap and plentiful food and our desire to conserve the environment. The question for the future is: how skilfully can we balance these two conflicting needs? Are we prepared to pay a little more for our food to ensure a sustainable agriculture that has a reduced impact on wildlife?

Environmental issues

In the UK, food production has doubled over the past 40 years. This has been achieved by the use of improved genetic varieties of plant and animal species, greater use of chemical fertilisers and pesticides, greater use of biotechnology and changes in farm practices that have led to larger farms. These changes have had many ecological impacts. The main effect of intensive food production has been to diminish the variety of habitats within ecosystems. This has led to a reduction in **species diversity**.

Certain practices have directly removed habitats and reduced species diversity. Examples of these practices include:

- removal of hedgerows and woodland
- creation of monocultures, for example, replacing natural meadows with cereal crops or grass for silage
- filling in ponds and draining marshes and other wetlands
- over-grazing of land, for example, upland areas by sheep, thereby preventing regeneration of woodland.

Other practices have had a more indirect effect:

- use of pesticides and inorganic fertilisers, thus reducing species diversity and polluting watercourses
- escape of farm wastes into water courses, thus killing fish and other organisms
- absence of crop rotation, leading to poor soil structure.

Figure 4 *Large area of arable land without hedgerows*

Despite the obvious conflicts between intensive food production and **conservation**, there are a number of management techniques that can be used to increase species and habitat diversity without unduly raising food costs or lowering yields. Examples of these conservation techniques include:

- maintaining existing hedgerows at the most beneficial height and shape to provide better habitats
- planting hedges rather than erecting fences as field boundaries
- maintaining existing ponds and, where possible, creating new ones

- leaving wet corners of fields rather than draining them
- planting native trees on land with a low species diversity rather than in species-rich areas
- reducing the use of pesticides – using biological control where possible or genetically modified organisms that are resistant to pests
- using organic rather than inorganic fertilisers
- using crop rotation that includes a **nitrogen-fixing** crop, rather than fertilisers, to improve soil fertility
- creating natural meadows and using hay rather than grasses for silage
- leaving the cutting of verges and field edges until after flowering and after the seeds have dispersed.

To encourage farmers to adopt some of these conservation measures, a number of financial incentives are offered by the Department for Environment and Rural Affairs (Defra) in the UK and by the European Union.

It is possible that intensive farming methods may lead to some areas being unable to support any farming. For example, in the late 1930s, intensive farming and the lack of crop rotation in South Dakota led to degradation of the soil structure. A period of drought followed by strong winds led to the top soil being blown away. The 'dust bowls' that resulted were unfit to grow crops of any kind.

Examples such as the American dust bowl suggest that conservation in some form is essential to ensure the long-term prospects for food production. It must, however, be remembered that produce from farms carrying out conservation is likely to be more expensive. This might be acceptable to the populations of relatively affluent countries, such as the UK, but might be unpopular in less developed countries, where more expensive food would cause social and economic hardship.

This is an example of How science works (HSW: I & L).

1. Suggest **two** reasons why removing hedgerows may increase the productivity of a farmer's land.
2. Hedges provide a habitat for many different species, especially insects and birds. Suggest **one** possible way in which these organisms might actually improve the yield of nearby crops.
3. Without financial incentives it would not be economic for most farmers to carry out conservation techniques. Suggest why this is so.
4. Suggest **one** advantage and **one** disadvantage to consumers of buying food that has been produced by a farm that practises all the conservation techniques listed above.

AQA Examination-style questions

1 **Figure 1** shows a pyramid of energy for an ecosystem.

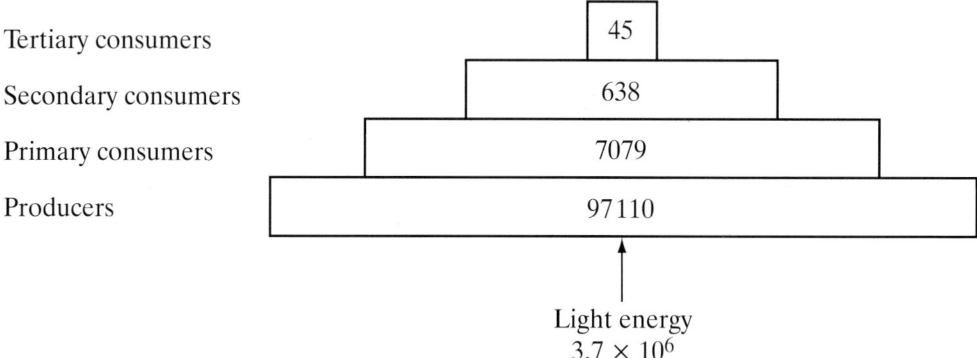

Figure 1

(a) Suggest suitable units for the measurement of energy transfer in this pyramid of energy. *(1 mark)*

(b) (i) Calculate the percentage of energy transferred from primary consumers to tertiary consumers.

(ii) Give **two** reasons why the percentage of energy transferred between consumers is generally low. *(3 marks)*

(c) Give **two** reasons why all the light energy reaching the producers cannot be used in photosynthesis. *(2 marks)*

AQA, 2007

2 **Figure 2** shows a simplified food web in an aquatic ecosystem.

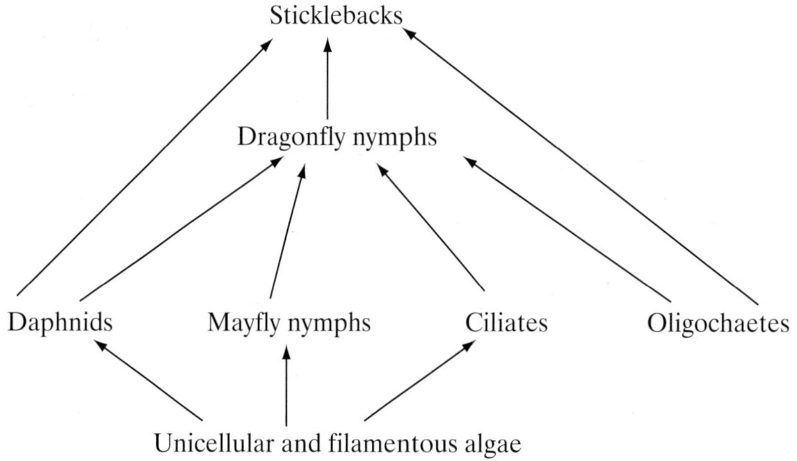

Figure 2

(a) In this food web, which organisms feed as tertiary consumers? *(1 mark)*

(b) The biomass of organisms in an ecosystem can change during the year. In this aquatic ecosystem, the biomass of primary consumers is temporarily greater than that of the producers during the early summer.

(i) Sketch the pyramids of biomass in early summer and autumn for this ecosystem. Name the trophic levels.

(ii) Suggest suitable units to represent biomass in these pyramids. *(3 marks)*

(c) Explain why food chains rarely have more than five trophic levels. *(2 marks)*

AQA, 2003

3 Potato plants originate from the Andes mountains in South America. They are adapted for survival in a cool climate. The potatoes we eat are food storage organs, called tubers, and are produced on underground stems.

Figure 3 shows the rate of photosynthesis and respiration for one variety of potato plant.

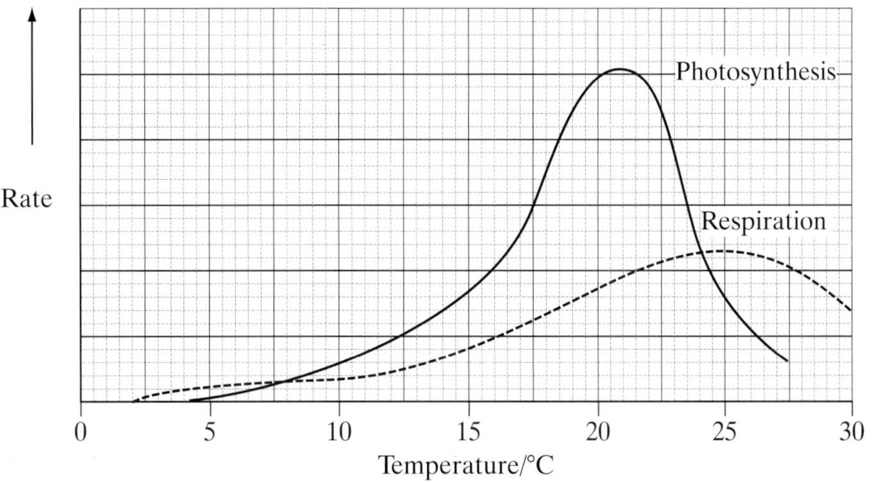

Figure 3

(a) Between which temperatures is there a net gain in energy by the potato plant? *(1 mark)*

(b) When this variety was grown in a hot climate, with a mean daytime temperature of 23.5 °C, it failed to produce tubers. Use information in **Figure 3** to explain why no tubers were produced. *(2 marks)*

(c) Suggest what causes the rate of photosynthesis to decrease at temperatures above 21 °C. *(2 marks)*

AQA, 2003

4 Purple loosestrife is a plant which grows in Europe. It was introduced into the USA where it became a pest.

(a) Suggest why purple loosestrife became a pest when it was introduced into the USA, but is not a pest in Europe. *(2 marks)*

(b) A European beetle was tested to see whether it could be used for the biological control of purple loosestrife in the USA. In an investigation beetles were released in an area where purple loosestrife was a pest. The table shows some of the results.

Time after releasing beetles / years	Mean number of purple loosestrife stems per square metre	Mean number of beetles per square metre
1	22	5
2	8	40
3	6	68
4	7	62

Are the beetles effective in controlling purple loosestrife? Give evidence from the table to support your answer. *(2 marks)*

(c) Fire-ants are a serious pest in parts of the USA. An investigation was carried out to find the best way to control the fire-ant population. **Figure 4** shows the results of this investigation.

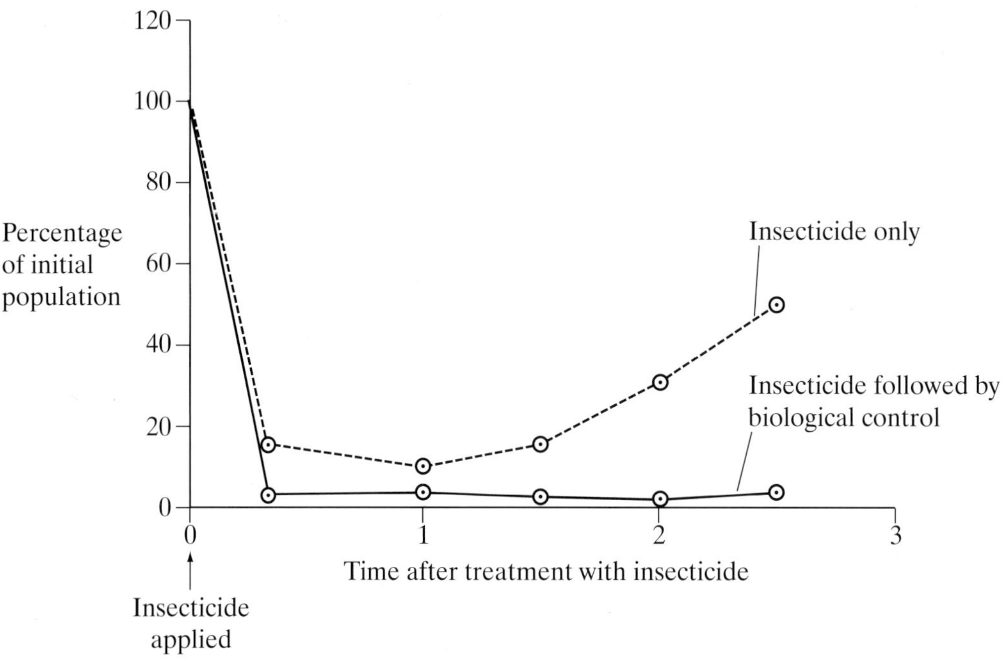

Figure 4

(i) Describe the effect of using insecticide followed by biological control.

(ii) Explain the change in fire-ant population over the period when they were treated with insecticide alone. *(5 marks)*

(d) Give the advantages and disadvantages of using biological control. *(6 marks)*

AQA, 2006

5 (a) Insecticides are pesticides which kill insects. A low concentration of insecticide was sprayed on the leaves of rose plants to kill greenfly which were feeding on the plants. Ladybirds eat greenfly. One month after spraying, the concentration of insecticide in the tissues of ladybirds was found to be higher than the concentration sprayed on the rose plants. Suggest why. *(3 marks)*

(b) Spotted knapweed is a common weed in the USA. Two methods, chemical control and biological control, have been used to reduce the numbers of spotted knapweed plants. The table shows the results of an investigation comparing the effectiveness of these two methods.

Month	**Mean number of spotted knapweed plants per m²**	
	Chemical control	**Biological control**
February	2	2
March	15	3
April	3	3
May	20	5
June	3	4
July	16	3
August	2	2

(i) Describe the pattern of plant numbers resulting from the use of:
chemical control;
biological control.

(ii) Explain how chemical control leads to the changes in the number of spotted knapweed plants from March to June. *(3 marks)*

(c) Explain why the spotted knapweed plants were never completely eliminated when using:
 (i) chemical control;
 (ii) biological control. *(4 marks)*

AQA, 2004

6 **Figure 5** shows the transfer of energy through a cow. The figures are in kJ × 10^6 year^{-1}.

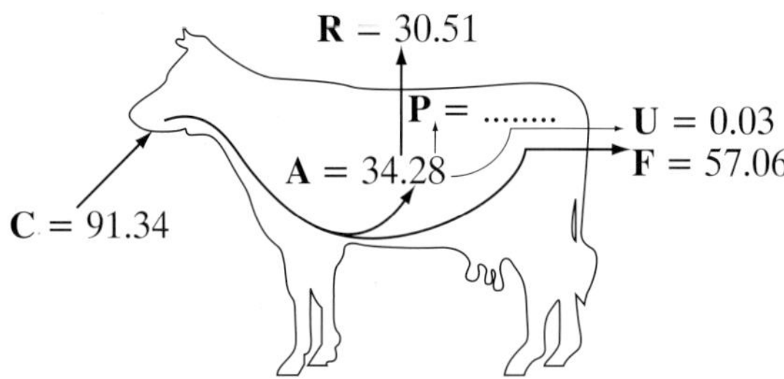

Key: **A** = energy absorbed from the gut
C = energy consumed in food
F = energy lost in faeces
P = energy used in production of new tissue
R = energy lost by respiration
U = energy lost in urine

Figure 5

(a) (i) Complete the following equation for the energy used in the production of new tissue. Use only the letters **C**, **F**, **R** and **U**.

P =

(ii) Calculate the value of **P** in kJ x 10^6 year^{-1}. *(2 marks)*

(b) It has been estimated that an area of 8100 m^2 of grassland is needed to keep one cow. The productivity of grass is 21 135 kJ m^{-2} year^{-1}. What percentage of the energy in the grass is used in the production of new tissue in one cow? Show your working. *(2 marks)*

(c) Keeping cattle indoors, in barns, leads to a higher efficiency of energy transfer. Explain why. *(1 mark)*

AQA, 2004

7 A food chain found in an oak woodland is shown below.

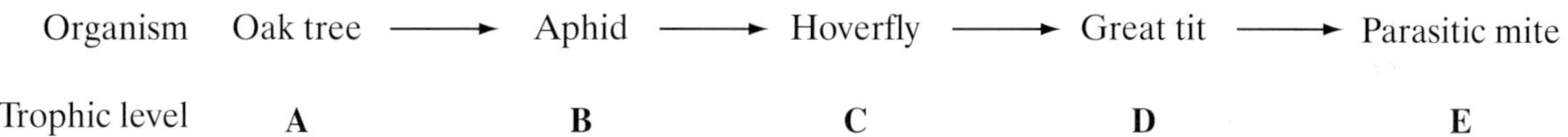

The pyramid of numbers and pyramid of biomass representing this food chain are shown in **Figure 6**.

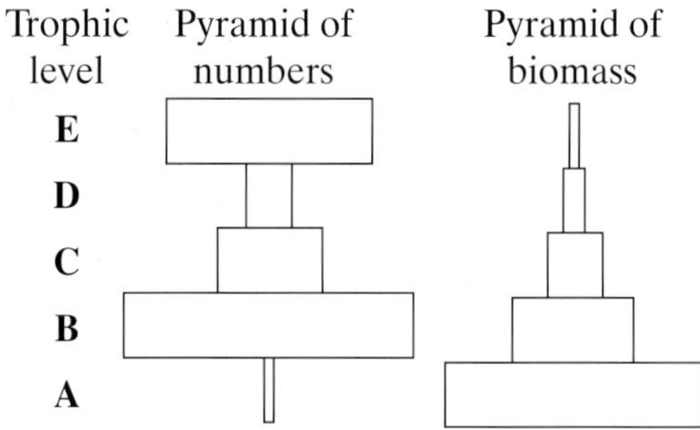

Figure 6

(a) Not all the light energy entering the leaves of the oak tree is used in photosynthesis. Give **one** reason for this. *(1 mark)*

(b) Give **two** ways in which energy is lost between trophic levels **A** and **B**. *(2 marks)*

(c) Explain the difference between the shapes of the two pyramids at trophic levels **D** and **E**. *(2 marks)*

AQA, 2005

6 Nutrient cycles

6.1 The carbon cycle

Learning objectives:

- Where does carbon enter the living component of an ecosystem?
- Where does carbon enter the non-living component of an ecosystem?
- What role is played by saprobiotic organisms in the carbon cycle?

Specification reference: 3.4.6

We saw in Topic 5.2 that energy enters an **ecosystem** as sunlight and is lost as heat. This heat cannot be recycled. The flow of energy through an ecosystem is therefore in one direction, that is, it is linear. Provided the Sun continues to supply energy to Earth, this is not a problem. Nutrients, by contrast, do not have an extraterrestrial source. There is only a certain quantity of them on Earth. It is essential therefore that elements such as carbon and nitrogen are recycled. The flow of nutrients within an ecosystem is not linear, but cyclic.

All nutrient cycles have one simple sequence at their heart.

- The nutrient is taken up by **producers** (plants) as simple, inorganic molecules.
- The producer incorporates the nutrient into complex organic molecules.
- When the producer is eaten, the nutrient passes into **consumers** (animals).
- It then passes along the food chain when these animals are eaten by other consumers.
- When the producers and consumers die, their complex molecules are broken down by **saprobiotic microorganisms** (decomposers) that release the nutrient in its original simple form. The cycle is then complete.

Although other processes and non-living sources are also involved, it is this sequence, illustrated in Figure 1, that forms the basis of all nutrient cycles.

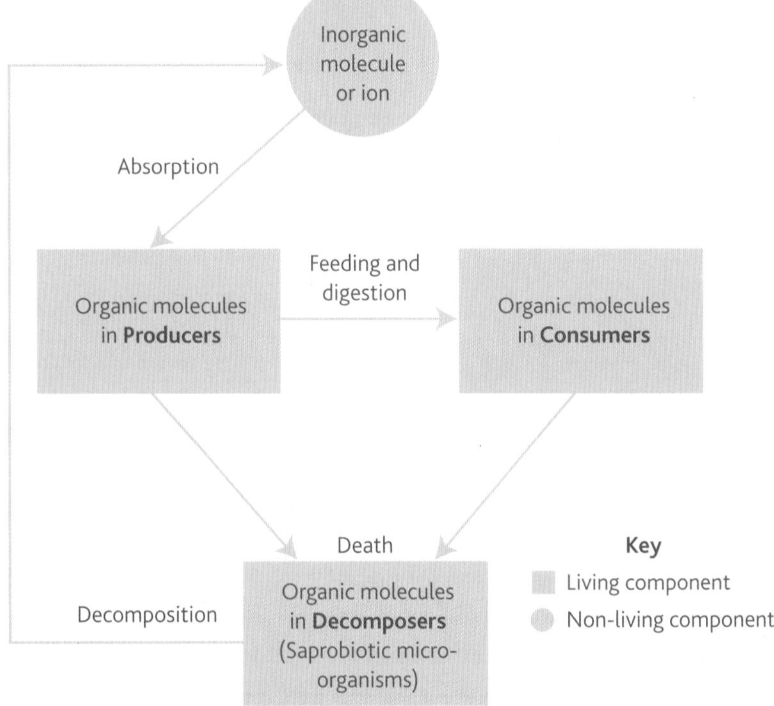

Figure 1 *Basic sequence of all nutrient cycles*

Carbon is a component of all the major macromolecules in living organisms. It is the basic building block for life itself. The main source of carbon for terrestrial organisms is the carbon dioxide in the atmosphere and yet it makes up a mere 0.04 per cent of the air around us. It is not surprising therefore that the turnover of atmospheric carbon dioxide is considerable. Photosynthetic organisms remove it from the air to build it up into macromolecules such as carbohydrates, fats and proteins. All organisms return it to the air through the process of respiration.

Variations in the rates of respiration and photosynthesis give rise to short-term fluctuations in the proportions of oxygen and carbon dioxide in the atmosphere. For example, the concentration of carbon dioxide at night is greater than during the day. This is because the absence of light means that no photosynthesis can take place. Respiration is still occurring, however, although its rate may be slightly reduced as the temperature is normally lower at night. Similarly, the daytime concentration of carbon dioxide on a summer's day, when photosynthesis is at its greatest, is lower than on a winter's day.

Globally the level of carbon dioxide in the atmosphere has increased over the past few hundred years. The main reasons for this are two human activities:

- The **combustion of fossil fuels**, such as coal, oil and peat, has released carbon dioxide that was previously locked up within these fuels.
- **Deforestation**, especially of the rain forests, has removed enormous amounts of photosynthesising **biomass** and so less carbon dioxide is being removed from the atmosphere.

The additional production of carbon dioxide through these human activities is threatening to upset the delicate balance of the carbon cycle. As carbon dioxide is a **greenhouse gas** this additional production is contributing to global warming. This issue is discussed in Topic 6.2.

The oceans contain a massive reserve of carbon dioxide. This store is some 50 times greater than that in the atmosphere. It helps to keep the level of atmospheric carbon dioxide more or less constant. Some of any excess carbon dioxide in the atmosphere dissolves in the waters of the oceans. When atmospheric levels are low, the reverse occurs. Aquatic photosynthetic organisms (phytoplankton) use this dissolved carbon dioxide to form the macromolecules that make up their bodies.

The carbon in photosynthetic organisms passes along food chains to animals. On their death, both plants and animals are usually broken down by saprobiotic microorganisms known collectively as **decomposers**. Saprobiotic microorganisms secrete enzymes on to the dead organisms. These enzymes break down complex molecules into smaller, soluble molecules that the saprobiotic microorganisms absorb by diffusion. The carbon in the dead organisms is then released as carbon dioxide during respiration by the decomposer.

If decay is prevented for any reason, then the organisms may become fossilised into coal, oil or peat. Not all parts of organisms decompose. The shells and bones of aquatic organisms sink to the bottom of the oceans and, over millions of years, form carbon-containing sedimentary rocks such as chalk and limestone. This carbon eventually returns to the atmosphere as these rocks are weathered.

AQA Examiner's tip

It is important to remember the key reactions in photosynthesis, in which carbon dioxide is fixed into biological molecules, and in respiration, in which carbon dioxide is released from these molecules.

Link

The processes of photosynthesis and respiration are dealt with in Chapters 3 and 4 respectively.

Figure 2 *Burning rain forest contributes to global warming not only by releasing carbon dioxide but also by removing enormous amounts of photosynthesising biomass that would have absorbed carbon dioxide from the atmosphere*

Hint

All organisms die and so decomposers feed on all trophic levels in food chains and food webs.

Figure 3 *Fungi such as this bracket fungus are decomposers that help to recycle carbon*

The carbon cycle is summarised in Figure 4.

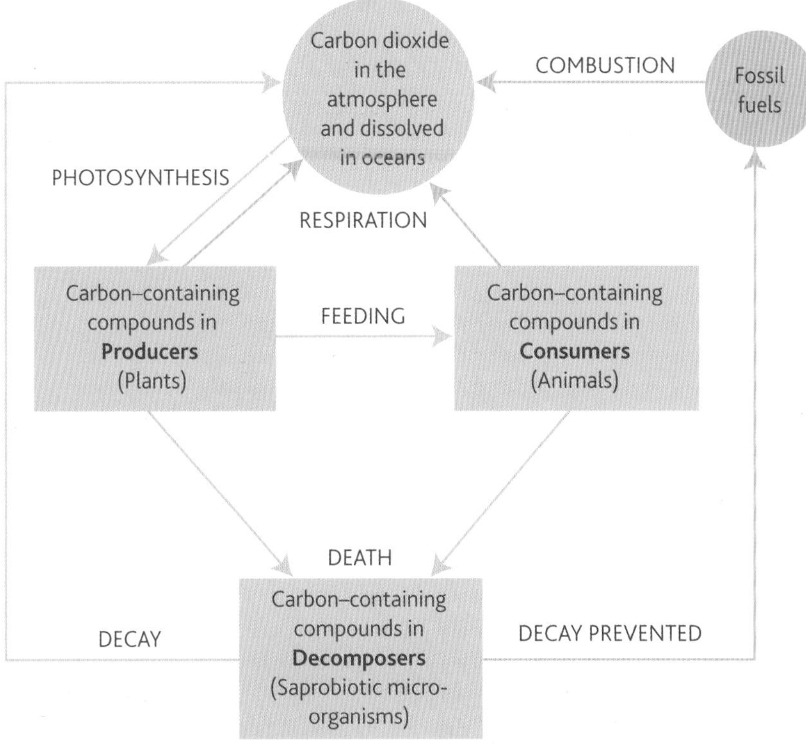

Figure 4 *The carbon cycle*

Summary questions

1. Explain why the carbon dioxide concentration of the atmosphere is often less on a summer's day than on a winter's day.
2. Figure 5 is an illustration of the carbon cycle. Each box represents a process. Name the process in each of the boxes A, B, C and D.

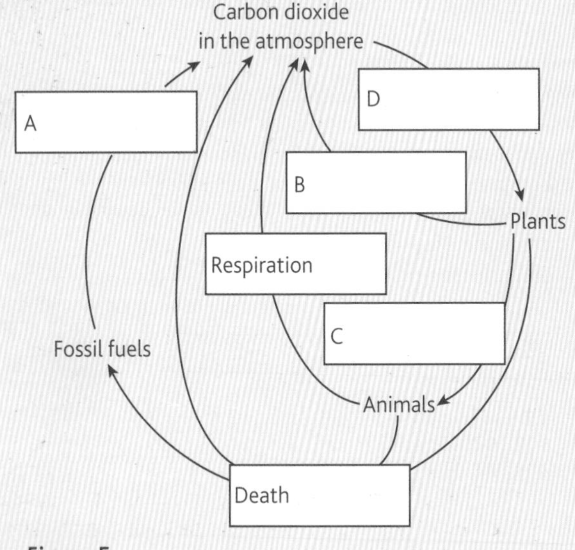

Figure 5

3. Suggest which of the sequences below best represents the flow of carbon (C) through the carbon cycle:
   a. atmospheric C ⟶ respiration ⟶ animal C ⟶ plant C ⟶ decay ⟶ atmospheric C
   b. atmospheric C ⟶ plant C ⟶ photosynthesis ⟶ atmospheric C
   c. atmospheric C ⟶ animal C ⟶ respiration and decay ⟶ atmospheric C
   d. atmospheric C ⟶ plant C ⟶ animal C ⟶ respiration and decay ⟶ atmospheric C
   e. atmospheric C ⟶ plant C ⟶ photosynthesis ⟶ animal C ⟶ atmospheric C

6.2 The greenhouse effect and global warming

Learning objectives:

- What is the greenhouse effect?
- Which are the major greenhouse gases and where do they come from?
- Why is the production of greenhouse gases increasing?
- How do greenhouse gases contribute to global warming?
- What are the consequences of global warming?

Specification reference: 3.4.6

It was mentioned in Topic 6.1 that carbon dioxide is a greenhouse gas. A greenhouse gas is any gas that has a greenhouse effect.

The greenhouse effect

The greenhouse effect is a natural process that occurs all the time and keeps average global temperatures at around 17 °C. Without it, the average temperature at the surface of the Earth would be about minus 18 °C. This effect is the result of the heat and light of the Sun (solar radiation) that reaches our planet. Some solar radiation is reflected back into space, some is absorbed by the atmosphere and, fortunately, some reaches the Earth's surface. Some of this radiation reaching the Earth's surface is reflected back as heat and is lost into space. However, some is radiated back to Earth by clouds and the 'greenhouse gases' that form part of the atmosphere. The gases trap this heat close to the Earth's surface, keeping it warm. The greenhouse gases act like the glass in a greenhouse by trapping heat beneath them, hence the name 'greenhouse effect'. The process is illustrated in Figure 1.

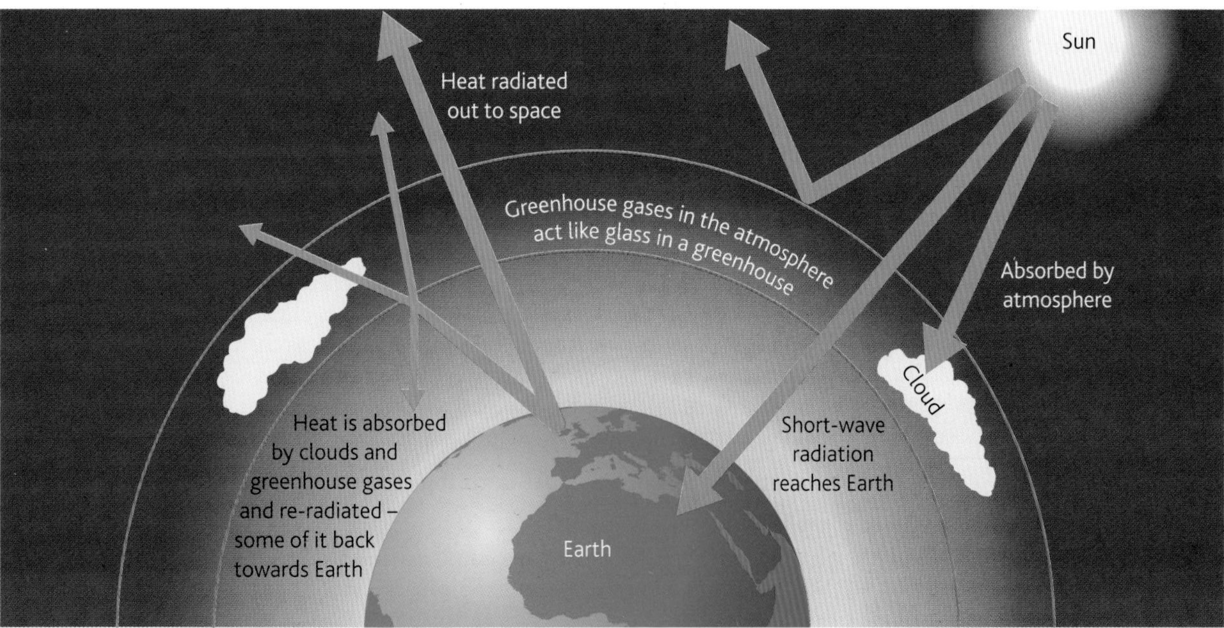

Figure 1 *The greenhouse effect*

Greenhouse gases

The most important greenhouse gas is carbon dioxide, partly because there is so much of it and partly because it remains in the atmosphere for so much longer than other greenhouse gases (100 years, compared with 10 years for methane). It has been estimated that 50–70 per cent of global warming is due to carbon dioxide in the atmosphere. It is mainly as a result of human activities that the concentration of carbon dioxide is increasing, enhancing the greenhouse effect and causing environmental concerns. Another natural greenhouse gas is methane. Methane is

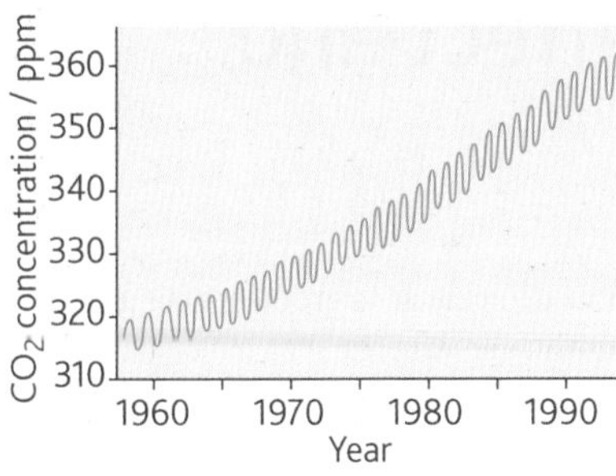

Figure 2 *Trends in atmospheric carbon dioxide concentration since 1958 (recorded at Mauna Loa, Hawaii)*

produced when microorganisms break down the organic molecules of which organisms are made. This occurs mostly in two situations:

- when decomposers break down the dead remains of organisms
- when microorganisms in the intestines of primary consumers, such as cattle, digest the food that has been eaten.

Global warming

The mean global temperature has increased by around 0.6 °C since 1900, a change known as global warming. The Earth has always shown periodic fluctuations in temperature. We cannot therefore say for certain that this recent temperature increase is due to human activities, such as the burning of fossil fuels and deforestation, generating additional carbon dioxide. What we can say is that the concentration of carbon dioxide has risen from 270 parts per million (ppm) before the industrial revolution to 370 ppm today. This rise in carbon dioxide concentration over recent years is shown in Figure 2. We also know that carbon dioxide is a greenhouse gas. Most scientists therefore think that these human activities have contributed to global warming.

Link

To help you understand the terms in this section, it is worth revising Topic 1.1 'Populations and ecosystems'.

Hint

Global warming changes the niches of many organisms and so affects the composition of communities.

Consequences of global warming

Even if humans continue to release greenhouse gases at the present rate, it is by no means certain what the effects would be. Carbon dioxide levels could even be reduced naturally, as phytoplankton and other plants increase their rates of photosynthesis.

Global warming is expected to bring about changes in temperature and precipitation, the timing of the seasons and the frequency of extreme events such as storms, floods and drought. Climate change will affect the **niches** that are available in a community. As each organism is adapted to a particular niche the distribution of species will alter. If the rate of climate change is slow, species may have time to gradually migrate to new areas, where they will compete for the available niches. This could lead to the loss of native species that occupy those niches. It is already being noted that many species are moving in the same direction as the climate. For example, in the northern hemisphere, many species of butterflies, lichens and birds are moving northwards. Alternatively, some species may be able to survive in their current locations by adapting to different food sources.

If present trends in global warming continue, the following changes are possible:

- Melting of polar ice caps could cause the extinction of some wild plants and animals, for example, the polar bear, and cause sea levels to rise.
- A rise in sea level due to the thermal expansion of oceans could flood low-lying land, including much of Bangladesh. It could also flood many major cities, as well as fertile land such as the Nile delta. Salt water would extend further up rivers and make cultivation of crop plants difficult.
- Higher temperatures and less rainfall could lead to the failure of the present crops in some areas. More drought-resistant species would have to be grown and, in severe cases, the land might become desert and so not sustain any crops at all. The distribution of wild plants in these areas would naturally change, with only **xerophytes** being able to survive. As animals are ultimately dependent on plants for food, their distribution would also be affected. Species that can feed on xerophytes and can withstand hot dry conditions would move in.

Figure 3 *Carbon dioxide from domestic and industrial burning of fossil fuels may contribute to global warming*

- Greater rainfall and intense storms would occur in some areas, due to the disturbance of climate patterns. Again the distribution of plants and animals would change in favour of those adapted to withstand such conditions.
- The life-cycles and populations of insect pests would alter as they adapt to the changed conditions. As insects carry many human and crop **pathogens**, tropical diseases could spread towards the poles. Species that damage crops could move towards the poles into areas that they have not previously inhabited. For example, the green shield bug, a common pest of vegetable crops in Mediterranean countries, has recently been found in southern England for the first time.

However, it should be remembered that there could also be benefits to some parts of the world. The increased rainfall would fill reservoirs, the warmer temperatures would allow crops to be grown where it is presently too cold, the rate of photosynthesis (and hence productivity) could increase and it might be possible to harvest twice a year instead of once.

Figure 4 *Global warming is contributing to the melting of polar ice*

Application and How science works

Digging into the past

To determine whether there is a correlation between the concentration of carbon dioxide in the atmosphere and global temperature requires measurements of both these factors over many thousands of years. But no measurements were taken thousands of years ago. How then can we look into the past? The answer lies in the ingenuity of scientists.

We can determine the age of fossils by measuring the amount of radioactive decay of an **isotope** of carbon (radiocarbon dating). Fossil shells from beneath the ocean bed contain oxygen isotopes as well as carbon ones. The proportion of these oxygen isotopes reflects the proportion of them in the sea water at the time the organism took them into its shell when it was alive. The proportion of each oxygen isotope in the sea water varied according to the coming and going of each ice age, which in turn reflected the global temperature. By measuring the isotopes of carbon and oxygen in fossil shells we get both an estimate of its age and the temperature of the Earth at that time.

What about the carbon dioxide concentration? When ice forms, air bubbles become trapped in it. If we drill down into the Antarctic glaciers, the deeper we go the older is the ice. The air bubbles in the ice will reflect the composition of the atmosphere at the time when the ice was formed. Scientists have drilled 3.6 km down through an Antarctic glacier. At this depth the ice is 420 000 years old. By measuring the carbon dioxide concentration of air bubbles in the ice at different depths, we can find out how its concentration in the atmosphere has changed over the past 420 000 years.

The results of these two different studies are shown in the graphs in Figure 5. This is an example of How science works (HSW: E).

1 Giving reasons from the information in Figure 5, state whether you think there is a correlation between the concentration of carbon dioxide in the atmosphere and global temperature.

2 Do the data indicate that changes in carbon dioxide concentration cause changes in global temperature. Explain your answer.

Summary questions

1 In Figure 2 the line on the graph fluctuates. Suggest an explanation for this regular up-and-down pattern.

2 Tropical rain forest around the world is being burnt so that the land can be used for other purposes. Explain all the possible effects on carbon dioxide levels in the atmosphere and global warming if the cleared land was used to produce palm oil for the manufacture of biofuels for use in vehicles.

3 State how the effect on global warming might be different if the land was used to raise beef cattle.

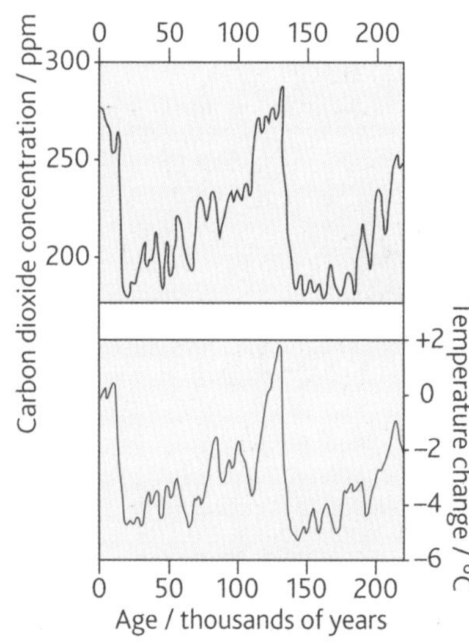

Figure 5 *Comparison of changes in atmospheric carbon dioxide (from analysis of bubbles in glacial ice) and temperature changes (from oxygen isotope studies)*

Application

Global warming and crop yields

Most studies of the influence of global warming on crop yields have concentrated on the effect of an increase in daytime temperature. A study carried out on the yield of rice, however, looked at the effects of an increase in both the maximum (day) temperature and the minimum (night) temperature.

The study found that there was no correlation between the maximum (day) temperature and crop yields. The results for the effects of the minimum (night) temperature on the yield of rice are shown in the graphs in Figure 6.

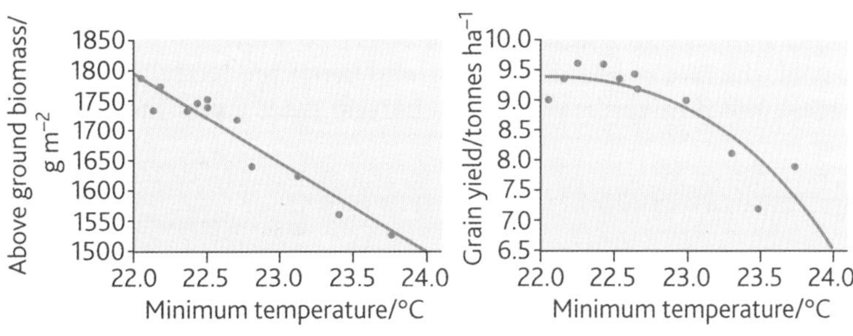

Figure 6 *Source: Peug* et al, *PNAS, vol. 101, July 2004 (Copyright 2004, National Academy of Science, USA)*

1. Describe the effect of minimum (night) temperature on the grain yield of rice.
2. Calculate the percentage decrease in above-ground biomass that occurs when the minimum temperature is increased from 22.0 °C to 24.0 °C. Show your working.
3. Suggest a possible explanation for some of the reduction in yield when the minimum temperature increases.

Application

Global warming and insect pests

Global warming can affect the numbers and life-cycles of insect pests. Rice crops can be infected by the rice stem-borer, *Chilo suppressalis*, a moth whose larvae (caterpillars) over-winter in the paddy fields. They emerge in spring and enter the stems of rice plants, limiting growth and sometimes resulting in the death of the plant.

A study in Japan looked at the mean winter (November to April) temperature over the past 50 years, how this temperature affected the number of larvae in rice plants and how, in turn, this affected the yield of rice. The results are summarised in the graphs in Figure 7.

1. Describe and explain the relationship shown in graph B.
2. From the graphs, suggest an explanation of how global warming can lead to a loss of yield in rice crops.
3. From the graphs, calculate the likely % loss in rice yield in 1980 as a result of rice stem-borer larvae. Show your working.
4. Suggest an explanation why infection with stem-borer larvae might kill rice plants.

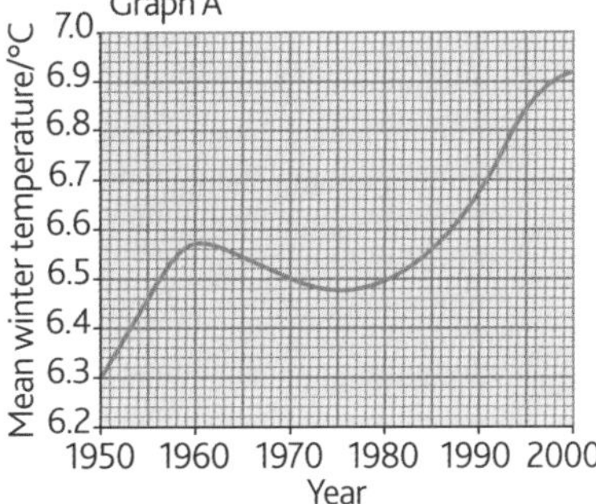

Graph B

Number of larvae per plant

40
30
20
10
0

5.75 6.00 6.25 6.50 6.75 7.00

Mean water temperature/°C

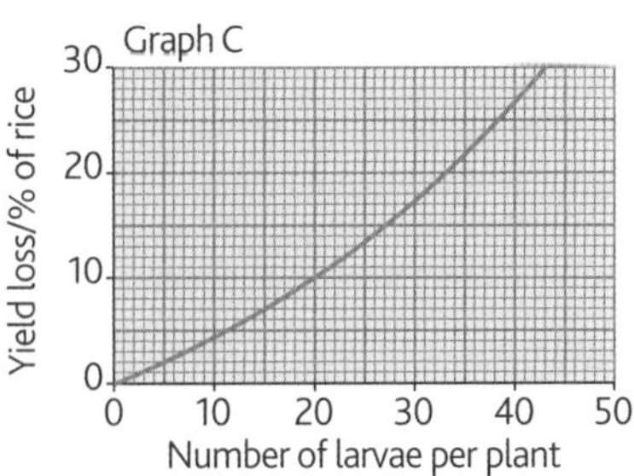

Figure 7

6.3 The nitrogen cycle

Learning objectives:

- How is nitrogen recycled in ecosystems?
- What is the role of saprobiotic microorganisms in this recycling?
- What do you understand by the terms 'ammonification', 'nitrification', 'nitrogen fixation' and 'denitrification'?
- Where does nitrogen enter the living component of an ecosystem?
- Where does nitrogen enter the non-living component of an ecosystem?

Specification reference: 3.4.6

All living organisms require a source of nitrogen from which to manufacture proteins, nucleic acids and other nitrogen-containing compounds. Although 78 per cent of the atmosphere is nitrogen, there are very few organisms that can use nitrogen gas directly. Plants take up most of the nitrogen they need in the form of nitrate ions (NO_3^-), from the soil. These ions are absorbed, using **active transport**, by the root hairs. This is where nitrogen enters the living component of the ecosystem. Animals obtain nitrogen-containing compounds by eating and digesting plants.

Nitrate ions are very soluble and easily leach through the soil, beyond the reach of plant roots. In natural ecosystems, the nitrate levels are restored through the recycling of nitrogen-containing compounds. In agricultural ecosystems, the level of soil nitrate can be further increased by the addition of fertilisers. When plants and animals die, the process of decomposition begins, in a series of steps by which microorganisms replenish the nitrate levels in the soil. This release from decomposition is most important because, in natural ecosystems, there is very little nitrate available from other sources.

There are four main stages in the nitrogen cycle (Figure 1), **ammonification**, **nitrification**, **nitrogen fixation** and **denitrification**, each of which involves **saprobiotic** microorganisms.

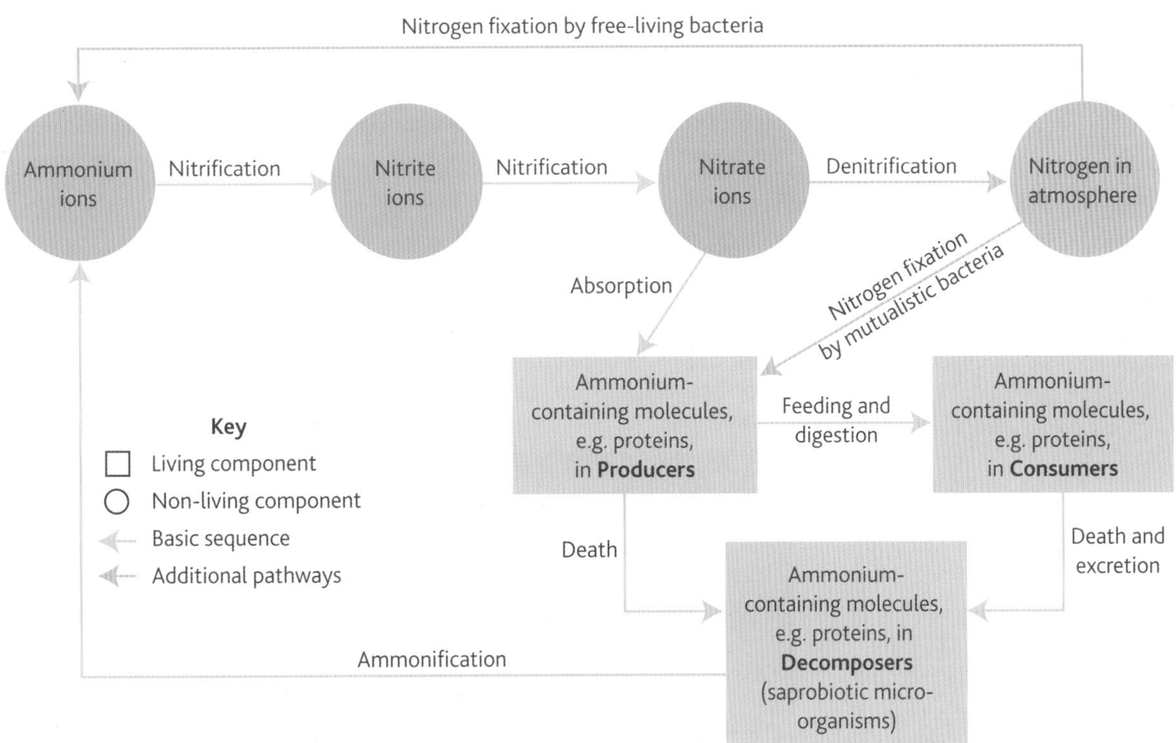

Figure 1 *The nitrogen cycle*

Link

The processes of absorption by root hairs and active transport formed part of your AS Biology course and are dealt with in Topics 13.7 and 3.8 respectively of the *AS Biology* book.

Hint

In many ecosystems, the availability of nitrates is the factor that limits plant growth. As plants are the primary producers, this means that nitrate availability affects the whole ecosystem.

AQA Examiner's tip

The word 'nitrogen' is often misused by candidates. Nitrogen is an element which forms a part of ions, such as nitrites and nitrates, as well as part of complex molecules, such as proteins and nucleic acids. Do not use the term 'nitrogen' when referring to these substances. Instead, use 'nitrogen-containing ions' or 'nitrogen-containing molecules'.

Ammonification

Ammonification is the production of ammonia from organic ammonium-containing compounds. In nature, these compounds include urea (from the breakdown of excess amino acids) and proteins, nucleic acids and vitamins (found in faeces and dead organisms). Saprobiotic microorganisms, mainly fungi and bacteria, feed on these materials, releasing ammonia, which then forms ammonium ions in the soil. This is where nitrogen returns to the non-living component of the ecosystem.

Nitrification

Plants use light energy to produce organic compounds. Some bacteria, however, obtain their energy from chemical reactions involving inorganic ions. One such reaction is the conversion of ammonium ions to nitrate ions. This is an **oxidation** reaction and so releases energy. It is carried out by free-living soil microorganisms called nitrifying bacteria. This conversion occurs in two stages:

1 oxidation of ammonium ions to nitrite ions (NO_2^-)
2 oxidation of nitrite ions to nitrate ions (NO_3^-).

Nitrifying bacteria require oxygen to carry out these conversions and so they need a soil that has many air spaces. To raise productivity, it is important for farmers to keep soil structure light and well aerated by ploughing. Good drainage also prevents the air spaces from being filled with water and so prevents air being forced out of the soil.

Nitrogen fixation

This is a process by which nitrogen gas is converted into nitrogen-containing compounds. It can be carried out industrially and also occurs naturally when lightning passes through the atmosphere. By far the most important form of nitrogen fixation is carried out by microorganisms, of which there are two main types:

- **free-living nitrogen-fixing bacteria**. These bacteria reduce gaseous nitrogen to ammonia, which they then use to manufacture amino acids. Nitrogen-rich compounds are released from them when they die and decay.
- **mutualistic nitrogen-fixing bacteria**. These bacteria live in nodules on the roots of plants such as peas and beans (Figure 2). They obtain carbohydrates from the plant and the plant acquires amino acids from the bacteria.

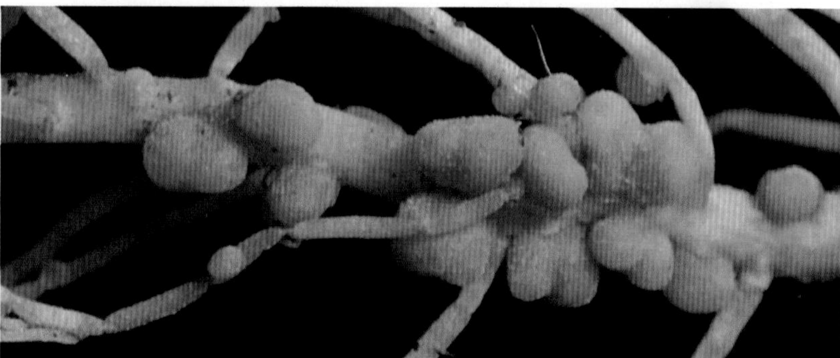

Figure 2 *Nitrogen-fixing nodules on the roots of a pea plant allow the plant to use free nitrogen in the atmosphere and soil. Mutualistic bacteria in the nodules fix the nitrogen, transforming it into a form useable by the plant*

Denitrification

When soils become waterlogged, and therefore short of oxygen, the type of microorganism present changes. Fewer **aerobic** nitrifying and nitrogen-fixing bacteria are found, and there is an increase in **anaerobic denitrifying bacteria**. These convert soil nitrates into gaseous nitrogen. This reduces the availability of nitrogen-containing compounds for plants. For land to be productive, the soils on which crops grow must therefore be kept well aerated to prevent the build-up of denitrifying bacteria.

As with any nutrient cycle, the delicate balance can be easily upset by human activities. Some of the effects of these activities are considered in Topic 6.4.

Figure 3 *Ploughing helps to aerate the soil and so prevents the build-up of denitrifying bacteria that can reduce the level of soil nitrates*

Summary questions

In the following passage, suggest the most appropriate word to replace each of the numbers in brackets.

A few organisms can convert nitrogen gas into compounds useful to other organisms in a process known as (1). These organisms can be free-living or live in a relationship with certain (2). Most plants obtain their nitrogen by absorbing (3) from the soil through their (4) by active transport. They then convert this to (5), which is passed to animals when they eat the plants. On death, (6) break down these organisms, releasing (7), which can then be oxidised to form nitrites by (8) bacteria. Further oxidation by the same type of bacteria forms (9) ions. These ions may be converted back to atmospheric nitrogen by the activities of (10) bacteria.

6.4 Use of natural and artificial fertilisers

Learning objectives:

- Why are fertilisers needed in agricultural ecosystems?
- How do natural and artificial fertilisers differ?
- How do fertilisers increase productivity?

Specification reference: 3.4.6

In Topic 5.4, we saw how agricultural ecosystems increase the efficiency of energy conversion along human food chains. They do so by improving productivity. One farming practice that contributes to this improved productivity is the use of fertilisers. Let us see how this is achieved.

The need for fertilisers

All plants need mineral ions, especially nitrogen, from the soil. Much food production in the developed world is intensive, that is, it is concentrated on specific areas of land that are used repeatedly to achieve maximum yield from the crops and animals grown on them. Intensive food production makes large demands on the soil because mineral ions are continually taken up by the crops being grown on it. These crops are either used directly as food or as fodder for animals that are then eaten. Either way, the mineral ions that the crops have absorbed from the soil are removed.

In natural ecosystems the minerals that are removed from the soil by plants are returned when the plant is broken down by microorganisms on its death. In agricultural systems the crop is harvested and then transported from its point of origin for consumption. The urine, faeces and dead remains of the consumer are rarely returned to the same area of land. Under these conditions the levels of the mineral ions in agricultural land will fall. It is therefore necessary to replenish these mineral ions because, otherwise, their reduced levels will become the main limiting factor to plant growth. Productivity will consequently be reduced. To offset this loss of mineral ions, fertilisers need to be added to the soil. These fertilisers are of two types:

- **natural (organic) fertilisers**, which consist of the dead and decaying remains of plants and animals as well as animal wastes such as manure and bone meal
- **artificial (inorganic) fertilisers**, which are mined from rocks and deposits and then converted into different forms and blended together to give the appropriate balance of minerals for a particular crop. Compounds containing the three elements, nitrogen, phosphorus and potassium, are almost always present.

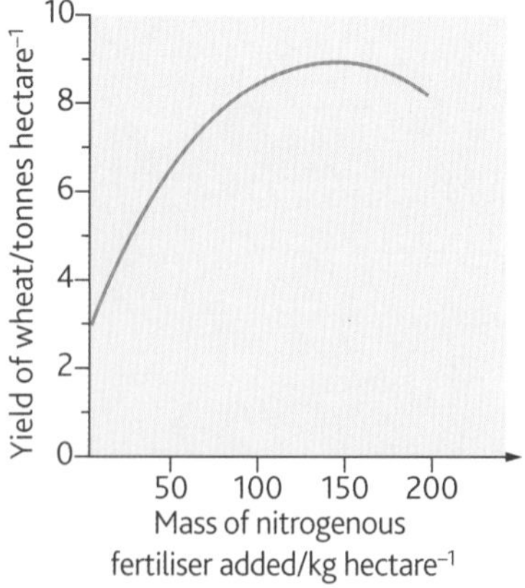

Figure 1 *The effect of different quantities of nitrogenous fertiliser on the yield of wheat*

Research suggests that a combination of natural and artificial fertilisers gives the greatest long-term increase in productivity. However, it is important that minerals are added in appropriate quantities as there is a point at which further increases in the quantity of fertiliser no longer results in increased productivity. This is illustrated in Figure 1.

Figure 2 *Cattle slurry, a natural fertiliser, being spread on to a crop of wheat*

How fertilisers increase productivity

Plants require minerals for their growth. Let us look at nitrogen as an example. Nitrogen is an essential component of proteins and DNA. Both are needed for plant growth. Where nitrates are readily available, plants are likely to develop earlier, grow taller and have a greater leaf area. This increases the rate of photosynthesis and improves crop productivity. There can be no doubt that nitrogen fertilisers have been of considerable benefit in providing us with cheaper food. It is estimated that the use of fertilisers has increased agricultural food production in the UK by around 100 per cent since 1955.

Application

Different forms of nitrogen fertiliser

Nitrogen fertiliser can be applied to crops in a number of different forms. These include ammonium salts, animal manure, the ground-up bones of animals (bone meal) and urea (a waste product found in the urine of mammals). An investigation was carried out in which the same crop was grown on six separate plots of land each of the same area. No nitrogen fertiliser was added to the first plot. To each of the remaining five plots, a different form of nitrogen fertiliser was added at the rate of 140 kg total nitrogen per hectare. The graph in Figure 3 shows the results of the investigation.

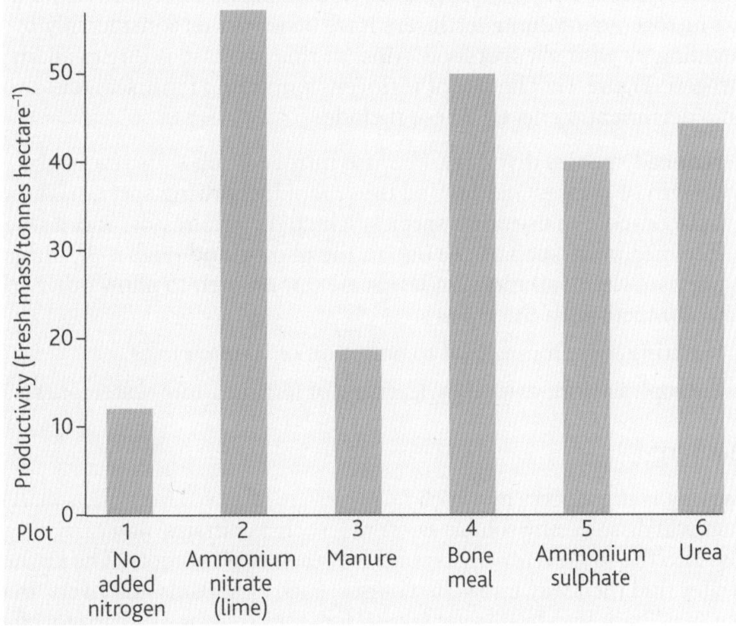

Figure 3

1. Which forms of nitrogen used in the investigation are natural fertilisers?
2. Why did the investigation include a plot to which no nitrogen fertiliser was added?
3. Suggest how the addition of nitrogen fertiliser, in whatever form, increased productivity.
4. The mass of each fertiliser used was different in each case. Suggest why this was necessary.
5. It is sometimes claimed that nitrogen fertilisers in the form of ammonium salts increase productivity of crops better than other forms of nitrogen fertilisers. State, with your reasons, whether or not you think the results of this experiment support this view.
6. The increase in productivity when manure was applied was lower than for other forms of nitrogen fertiliser. This is because the manure has to break down before its nitrogen is released and this process takes a few months. How might a farmer who spreads manure on his/her crops, use this information in order to improve productivity?

Summary questions

1. Explain why fertilisers are needed in an agricultural ecosystem.
2. Using Figure 1, state what concentration of fertiliser you would advise a farmer to apply to a field of wheat.
3. Suggest a reason why, after a certain point, the addition of more fertiliser no longer improves the productivity of a crop.
4. Distinguish between natural and artificial fertilisers.

6.5 Environmental consequences of using nitrogen fertilisers

Learning objectives:

- What are the main environmental effects of using nitrogen fertilisers?
- What is meant by 'leaching' and 'eutrophication'?
- How do these processes affect the environment?

Specification reference: 3.4.6

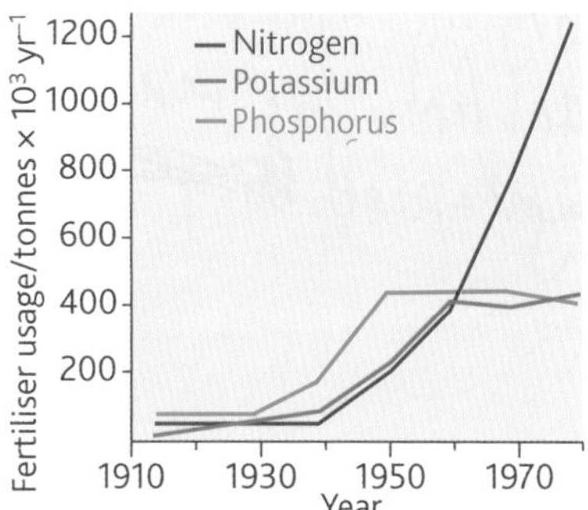

Figure 1 *Use of fertilisers in the UK*

Figure 2 *Low species diversity in a field grown for silage that has had nitrogen-containing fertiliser added*

Figure 3 *High species diversity in a meadow grown for hay without the addition of nitrogen-containing fertiliser*

In natural ecosystems minerals such as nitrate, which are removed from the soil by plants, are returned when the plant is broken down. However, in agricultural systems, the crop is removed and so the nitrate is not returned and has to be replaced. This is done by the addition of natural or artificial fertilisers.

Effects of nitrogen fertilisers

Nitrogen is an essential component of proteins and is needed for growth and, therefore, an increase in the area of leaves. This increases the rate of photosynthesis and improves crop productivity. There can be no doubt that nitrogen-containing fertilisers have benefited us considerably by providing us with cheaper food. Most of this increase is due to additional nitrogen (Figure 1). The use of nitrogen-containing fertilisers has also had some detrimental effects. These include:

- **reduced species diversity**, because nitrogen-rich soils favour the growth of grasses, nettles and other rapidly growing species. These out-compete many other species, which die as a result. Species-rich hay meadows, such as the one in the photograph (Figure 3), only survive when soil nitrogen levels are low enough to allow other species to compete with the grasses
- **leaching**, which may lead to pollution of watercourses
- **eutrophication**, caused by leaching of fertiliser into watercourses.

Leaching

Leaching is the process by which nutrients are removed from the soil. Rain water will dissolve any soluble nutrients, such as nitrates, and carry them deep into the soil, eventually beyond the reach of plant roots. The leached nitrates find their way into watercourses, such as streams and rivers, that in turn may drain into freshwater lakes. Here they may have a harmful effect on humans if the river or lake is a source of drinking water. Very high nitrate levels in drinking water can prevent efficient oxygen transport in babies and a link to stomach cancer in humans has been suggested. The leached nitrates are also harmful to the environment as they can cause eutrophication.

Eutrophication

Eutrophication is the process by which nutrients build up in bodies of water. It is a natural process that occurs mostly in freshwater lakes and the lower reaches of rivers. Eutrophication consists of the following sequence of events:

1. In most lakes and rivers there is naturally very little nitrate and so nitrate is a limiting factor for plant and algal growth.
2. As the nitrate concentration increases as a result of leaching, it ceases to be a limiting factor for the growth of plants and algae and both grow exponentially.
3. As algae mostly grow at the surface, the upper layers of water become densely populated with algae. This is called an 'algal bloom'.
4. This dense surface layer of algae absorbs light and prevents it from penetrating to lower depths.

5 Light then becomes the limiting factor for the growth of plants and algae at lower depths and so they eventually die.

6 The lack of dead plants and algae is no longer a limiting factor for the growth of saprobiotic algae and so these too grow exponentially, using the dead organisms as food.

7 The saprobiotic bacteria require oxygen for their respiration, creating an increased demand for oxygen.

8 The concentration of oxygen in the water is reduced and nitrates are released from the decaying organisms.

9 Oxygen then becomes the limiting factor for the population of **aerobic** organisms, such as fish. These organisms ultimately die as the oxygen is used up altogether.

10 Without the aerobic organisms, there is less competition for the **anaerobic** organisms, whose populations now rise exponentially.

11 The anaerobic organisms further decompose dead material, releasing more nitrates and some toxic wastes, such as hydrogen sulphide, which make the water putrid.

Figure 4 *Algal bloom in a canal as a result of eutrophication caused by nitrogen fertiliser run-off*

Organic manures, animal slurry, human sewage, ploughing old grassland and natural leaching can all contribute to eutrophication, but the leaching of artificial fertilisers is the main cause.

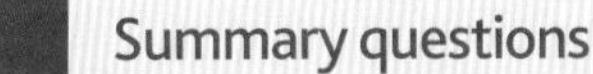

Summary questions

1 What is eutrophication?

2 How may an increase in algal growth at the surface lead to the death of plants growing beneath them?

3 Explain how the death of these plants can result in the death of animals such as fish.

Application

Troubled waters

A farmer applied a large quantity of fertiliser to fields next to a small lake. A period of heavy rain followed. After 10 days, scientists monitoring the lake noticed changes to the algal population, the clarity of the water and the levels of dissolved oxygen. These changes are shown in the three graphs in Figure 5. Secchi depth is a measure of the clarity of water. Measurements are taken by lowering a black-and-white disc (called a Secchi disc) into the water and recording the depth at which it is no longer visible.

1 Suggest a reason why the changes in the lake do not occur until 10 days after the application of the fertiliser to the fields.

2 Explain a possible cause of the increase in the density of algae after 10 days.

3 Describe and explain the relationship between the density of algae and water clarity in the lake.

4 Describe and explain changes to the levels of dissolved oxygen over the 100-day period.

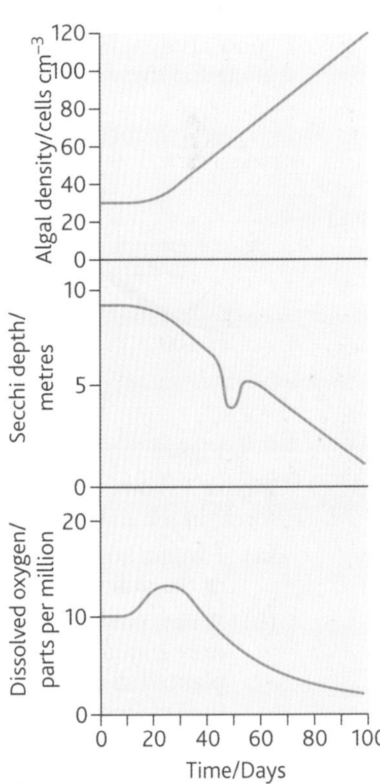

Figure 5

AQA Examination-style questions

1 Arctic tundra is an ecosystem found in very cold climates. **Figure 1** shows some parts of the carbon and nitrogen cycles in arctic tundra.

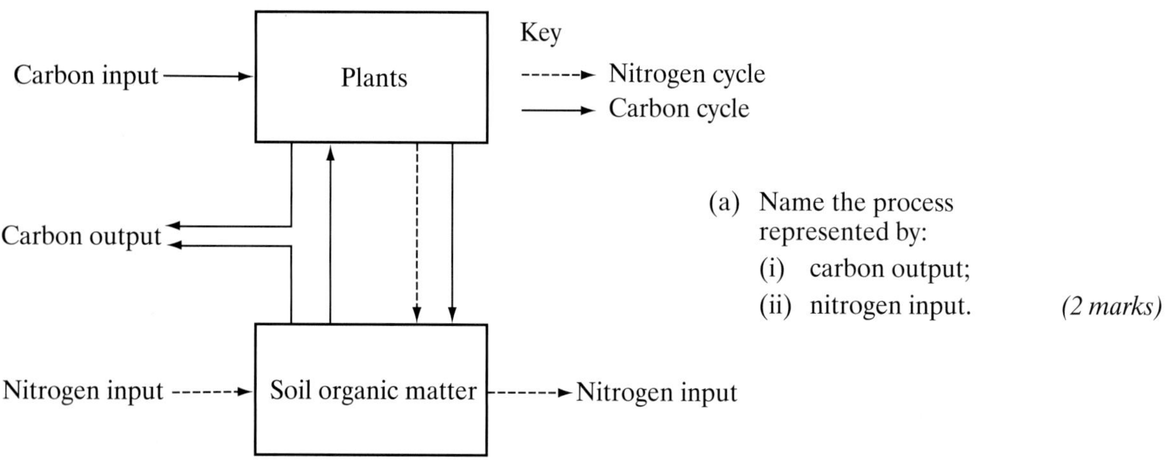

Figure 1

(a) Name the process represented by:
   (i) carbon output;
   (ii) nitrogen input. *(2 marks)*

(b) An increase in temperature causes an increase in carbon input. Explain why. *(2 marks)*

(c) The nitrogen compounds in the organic matter in soil are converted to nitrates. Explain how nitrogen output can occur from these nitrates. *(3 marks)*

AQA, 2007

2 **Figure 2** shows a river system in an area of farmland. The numbers show the nitrate concentration in parts per million (ppm) in water samples taken at various locations along the river. Concentrations above 250 ppm encourage eutrophication in the river.

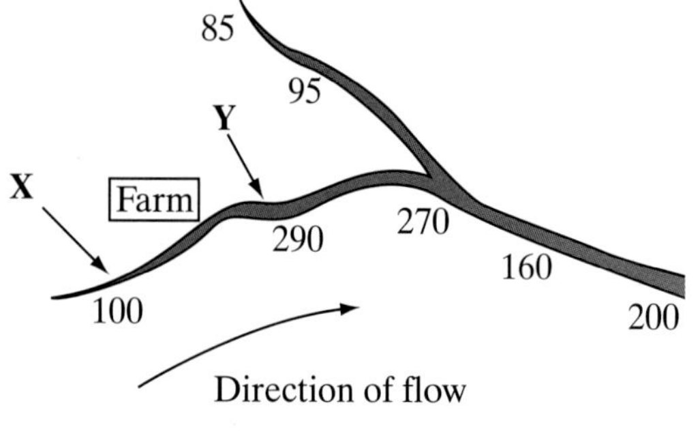

Figure 2

(a) Explain how farming practices might be responsible for the change in nitrate concentration in the water between point **X** and point **Y**. *(2 marks)*

(b) Describe the effect the nitrate concentration may have in the river at point **Y**. *(5 marks)*

AQA, 2005

3 **Figure 3** shows the cumulative mass of carbon removed from the atmosphere by a pine forest in the 20 years after planting.

(a) Explain how the growth of the forest results in a decrease in the carbon content of the atmosphere. *(2 marks)*

(b) A new power station is to be built which will emit a total of 3800 tonnes of carbon over 20 years. In order to balance the carbon emissions a pine forest will be planted to remove an equivalent amount over 20 years. Use the graph to work out the smallest area of forest that would be needed. Show your working. *(2 marks)*

(c) Explain how carbon-containing compounds present in the pine leaves that fall from the trees are used for growth by microorganisms that live in the soil. *(3 marks)*

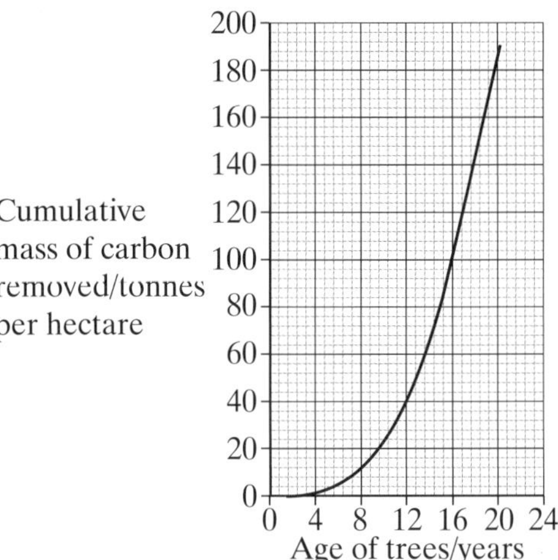

Figure 3

(d) Give **one** reason to explain why the rate of recycling of carbon would be greater in summer than in winter. *(1 mark)*

AQA, 2004

4 (a) Wet moorland soils often contain low concentrations of nitrogen compounds, as a result of denitrification. Sundew is a plant which lives in wet moorlands. Its leaves have sticky hairs which can trap small insects that are then digested.

(i) Describe the process of denitrification.

(ii) Explain how digestion of insects helps the sundew to obtain additional nitrogen compounds. *(4 marks)*

(b) Samples of plant and animal tissue were analysed to determine the proportions of the elements, carbon and nitrogen. In the plant tissue the ratio of carbon to nitrogen was 40:1. In the animal tissue the ratio was 8:1.

Explain why the ratio is much higher in the plant tissue than in the animal tissue. *(2 marks)*

(c) Describe how nitrogen in compounds in a dead plant is made available for use by other plants. *(6 marks)*

AQA, 2003

5 Two fields, **A** and **B**, were used to grow the same crop. In the previous year field **A** was used for grazing cattle and field **B** was used for the same crop. The fields were divided into plots. Different masses of fertiliser containing sodium nitrate were applied to these plots. After six weeks, samples of crop plants from each plot were collected and their mass determined. The results are shown in the table.

Mass of fertiliser added / kg ha^{-1}	**Mass of crop / kg m^{-2}**	
	Field A	Field B
0	14.5	6.4
10	16.7	9.8
20	17.4	12.9
30	17.5	16.2
40	17.5	17.1
50	17.5	17.1
60	17.5	17.1

(a) (i) Describe the pattern shown by the data for field **B**.

(ii) Explain the change in the mass of crop produced from field **B** when the mass of fertiliser added increases from 0 to 20 kg ha^{-1}.

(iii) Explain why the mass of crop produced stays the same in both fields when more than 40 kg of fertiliser is added. *(5 marks)*

(b) When no fertiliser was added, the mass of crop from field **A** was higher than from field **B**. Explain this difference. *(2 marks)*

(c) Explain **two** advantages and **one** disadvantage of an inorganic fertiliser such as sodium nitrate compared with an organic fertiliser such as manure. *(3 marks)*

AQA, 2004

7 Ecological succession

7.1 Succession

Learning objectives:

- What changes occur in the variety of species that occupy an area over time?
- What are meant by the terms succession and climax community?
- How can managing succession help to conserve habitats?

Specification reference: 3.4.7

We have seen that **ecosystems** are made up of all the interacting **biotic** and **abiotic** factors in a particular area within which there are a number of **communities** of organisms. As we look around at natural ecosystems, such as moorland or forest, we may get the impression that they have been there forever. This is far from the case. Ecosystems constantly change, sometimes slowly and sometimes very rapidly. **Succession** is the term used to describe these changes, over time, in the species that occupy a particular area.

One example of succession is when bare rock or other barren land is first colonised. This may occur as a result of:

- a glacier retreating and depositing rock
- sand being piled into dunes by wind or sea
- volcanoes erupting and depositing lava
- lakes or ponds being created by land subsiding
- silt and mud being deposited at river estuaries.

The first stage of this type of succession is the colonisation of an inhospitable environment by organisms called **pioneer species**. Pioneer species often have features that suit them to colonisation. These may include:

- the production of vast quantities of wind-dispersed seeds or spores, so they can easily reach isolated situations such as volcanic islands
- rapid germination of seeds on arrival as they do not require a period of dormancy
- the ability to photosynthesise, as light is normally available but other 'food' is not. They are therefore not dependent on animal species
- the ability to fix nitrogen from the atmosphere because, even if there is soil, it has few or no nutrients
- tolerance to extreme conditions.

Hint

Pioneer communities put some organic material into the soil when they die. This allows recycling to start and increases mineral ions in the soil.

Succession takes place in a series of stages. At each stage, certain species can be identified which change the environment, especially the soil, so that it becomes more suitable for other species. These other species may then out-compete the species in the existing community and so a new community is formed.

Imagine an area of bare rock. One of the few kinds of organism capable of surviving on such an inhospitable area is lichens. Lichens are therefore pioneer species. Lichens can survive considerable drying out.

Figure 1 *Lichens, with their ability to withstand dry conditions and to colonise bare rock, are frequently the first pioneer species on barren terrain*

In time, weathering of the base rock produces sand or soil, although this in itself cannot support other plants. However, as the lichens die and decompose they release sufficient nutrients to support a community of small plants. In this way the lichens change the abiotic environment by creating soil and nutrients for the organisms that follow. Mosses are typically the next stage in succession, followed by ferns. With the continuing erosion of the rock and the increasing amount of organic matter available from the death of these plants, a thicker layer of soil is built up. Again these species change the abiotic environment, making it more suitable for the organisms that follow, for example, small flowering plants such as grasses and, in turn, shrubs and trees. In the UK the ultimate community is most likely to be **deciduous** oak woodland. This stable state comprises a balanced equilibrium of species with few, if any, new species replacing those that have become established. In this state, many species flourish. This is called the **climax community**. This community consists of animals as well as plants.

The animals have undergone a similar series of successional changes, which have been largely determined by the plant types available for food and as **habitats**. The dead lichens provide food for animals such as detritus-feeding mites. The growth of mosses and grasses provides food and habitats for insects, millipedes and worms. These are followed in turn by secondary consumers, such as centipedes, which feed on these organisms. The development of flowering plants, including trees, helps to support communities of butterflies and moths as well as larger organisms, such as reptiles, mammals and birds.

Hint

The climax community is determined by the main abiotic factor. For example, trees may not develop on very high mountains because it is too windy or the soil layer is too thin.

Within the climax community there is normally a dominant plant species and a dominant animal species.

During any succession there are a number of common features that emerge:

- **the non-living environment becomes less hostile**, for example, soil forms, nutrients are more plentiful and plants provide shelter from the wind. This leads to:
- **a greater number and variety of habitats** that in turn produce:
- **increased biodiversity** as different species occupy these habitats. This is especially evident in the early stages, reaching a peak in mid-succession, but decreasing as the climax community is reached. The decrease is due to dominant species out-competing pioneer and other species, leading to their elimination from the community. With increased biodiversity comes:
- **more complex food webs**, leading to:
- **increased biomass**, especially during mid-succession.

Figure 2 *Deciduous woodland is normally the climax community in lowland Britain*

Climax communities are in a stable equilibrium with the prevailing climate. It is abiotic factors such as climate that determine the dominant species of the community. In the lowlands of the UK, the climax community is deciduous woodland. In other climates of the world it may be tundra, steppe or rain forest.

Another type of succession occurs when land that has already sustained life is suddenly altered. This may be the result of land clearance for agriculture or a forest fire. The process by which the ecosystem returns to its climax community is the same as described above, except that it normally occurs more rapidly. This is because spores and seeds often remain alive in the soil, and there is an influx of animals and plants through dispersal and migration from the surrounding area. This type of succession therefore does not begin with pioneer species, but with organisms from subsequent successional stages. Because the land has been altered in some way, for example, by fire, some of the species in the climax community will be different.

Figure 3 summarises the events of ecological succession on land.

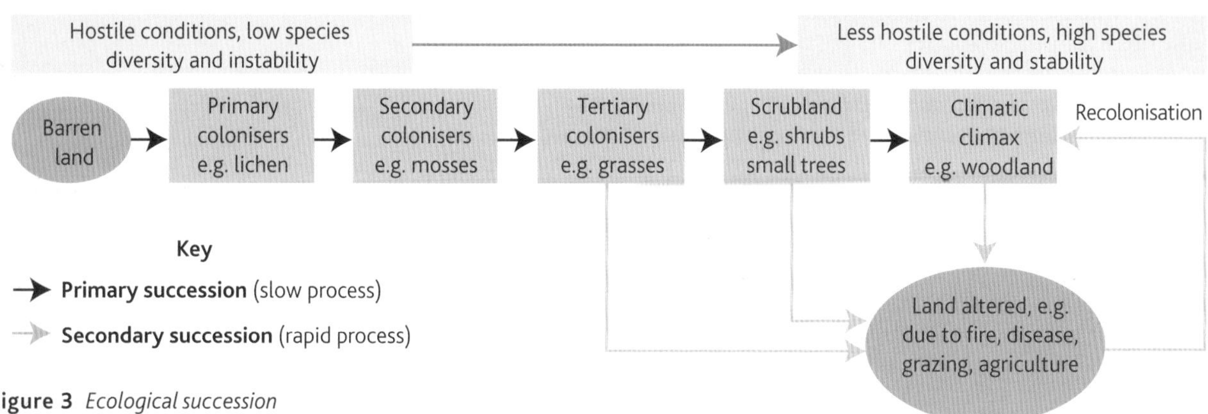

Figure 3 *Ecological succession*

Summary questions

1. What name is given to the first organisms to colonise bare land?
2. Describe how changes in the environment lead to increased biodiversity during succession.
3. What name is given to the stable, final stage of any succession?

Figure 4 *The grassland in the foreground is grazed by sheep and so is prevented from reaching its natural climax. The land behind the fence has not been grazed for many years and has reverted to the climax community of woodland. This is therefore an example of secondary succession*

Application

Warming to succession

Many glaciers in the northern hemisphere have been melting over the past 200 years. This retreat is, in part, the result of the additional global warming that has taken place since the industrial revolution and the burning of fossil fuels that has accompanied it. When glaciers melt and retreat they leave behind gravel deposits known as moraines. The retreat of the glaciers in Glacier Bay, Alaska, has been measured since 1794 and so the age of the moraines in this region is recorded.

Although no ecologist has been present to watch the succession that has taken place on these moraines, they can infer the changes that have occurred by examining the plant and animal communities on the moraines of different ages. The 'youngest' moraines (those nearest the retreating glacier) have the earliest colonisers (pioneer species), whereas those successively further away from the glacier show a time sequence of later communities.

Each stage of a succession has its own distinctive community of plants and animals that alters the environment in a way that allows the next stage and its community to develop. The stages that follow the retreat of an Arctic glacier are:

- **pioneer stage**. In the early years after the ice has retreated, photosynthetic bacteria and lichens colonise patches of land. Both of these pioneers fix nitrogen This is essential because nitrogen is virtually absent from glacial moraines. They also form tough mats that help to stabilise the loose surface of the moraines. When these pioneer species die, they decompose to form humus. Humus provides the nutrients that enable mosses to colonise. The pioneer stage occurs when the land has been ice free for 10–20 years.
- ***Dryas* stage**. Some 30 years after the ice has retreated, the ground is an almost continuous mat of the herbaceous plant *Dryas*. Its roots stabilise the thin and fragile soil layer formed from the erosion of the rocks that make up the moraine. *Dryas* also fixes nitrogen, further adding nitrogenous nutrients to this poor-quality soil. Other plants found at this stage are the Arctic poppy and moss campion.
- **alder stage**. This arises about 60 years after the ice has retreated. Alder is a small shrub-like tree that has nitrogen-fixing nodules on its roots, enabling it to grow on nitrogen-poor soil. Alder

sheds its leaves, which decompose into nitrogen-rich humus that further enriches the soil. The alder stage occurs some 50–70 years after the retreat of the glacial ice.

- **spruce stage**. About 100 years after the ice has retreated, spruce trees develop amongst the alder. A period of transition takes place and during the next 50 years or so the taller spruce out-competes the alder and ultimately displaces it altogether.

Figure 5 summarises changes in soil nitrogen, plant diversity and biomass following the retreat of an Arctic glacier.

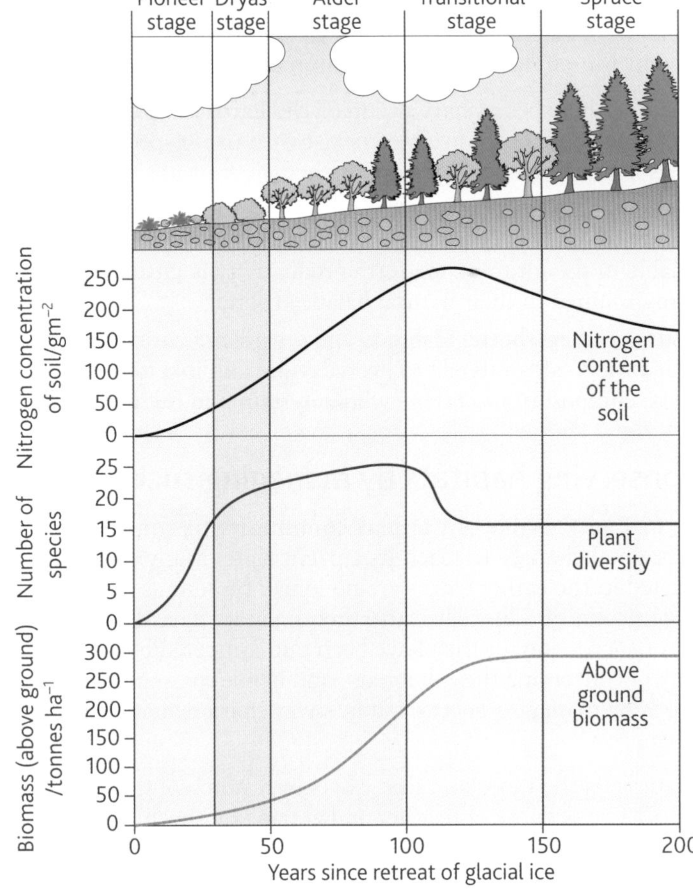

Figure 5

Figure 6 *Dryas (mountain avens) is the most common pioneer species in Glacier Bay, Alaska. It is able to fix nitrogen and forms dense mats and therefore enriches and stabilises the thin fragile soil*

Figure 7 *Arctic poppy (yellow flower) and moss campion (pink flowers) are early flowering pioneer species on Arctic moraines*

Figure 8 *Spruce trees are the final succession stage following the retreat of glacial ice in the Arctic. They begin to grow around 100 years after the ice has retreated and persist as the dominant vegetation for centuries*

1 Using the information on the graphs, describe and explain the changes in above-ground biomass over the 200-year period.

2 a Using your knowledge of the nitrogen cycle, explain how nitrogen from the atmosphere becomes incorporated into the soil, causing its level to increase during the first 100 years after the glacier retreats.

b Suggest two reasons for the fall in soil nitrogen levels after 150 years.

3 Suggest a reason for:

a the rapid increase in plant species during the first 30 years after the retreat of the glacier.

b the fall in the number of plant species 100 years after the retreat of the glacier.

4 Explain why it would be more appropriate to use a transect rather than random quadrats when investigating this succession.

7.2 Conservation of habitats

Learning objectives:

- What is conservation?
- How can managing succession help to conserve habitats?

Specification reference: 3.4.7

Figure 1 *Moorland is an example of the conservation of a habitat by managing succession. Burning of heather and grazing by sheep has prevented shrubs and trees from developing*

Summary questions

Fenland is an area of waterlogged marsh and peat land. It supports a rich and unique community of plants and animals. If left alone, reeds initially dominate and the area gradually dries out as dead vegetation accumulates. Grasses, shrubs and trees in turn replace the fenland species.

1. Give reasons for conserving habitats such as fenland.
2. Suggest practical measures that may be taken to prevent succession by grasses, shrubs and trees in fenland.

What is conservation?

Conservation is the management of the Earth's natural resources in such a way that maximum use of them can be made in the future. This involves active intervention by humans to maintain **ecosystems** and **biodiversity**. It is therefore a dynamic process that entails careful management of existing resources and reclamation of those already damaged by human activities. The main reasons for conservation are:

- **ethical**. Other species have occupied the Earth far longer than we have and should be allowed to coexist with us. Respect for living things is preferable to disregard for them.
- **economic**. Living organisms contain a gigantic pool of genes with the capacity to make millions of substances, many of which may prove valuable in the future. Long-term productivity is greater if ecosystems are maintained in their natural balanced state.
- **cultural and aesthetic**. Habitats and organisms enrich our lives. Their variety adds interest to everyday life and inspires writers, poets, artists, composers and others who entertain and fulfill us.

Conserving habitats by managing succession

We saw in Topic 7.1 that any **climax community** has undergone a series of successional changes to reach its current state. Many of the species that existed in the earlier stages are no longer present as part of the climax community. This is because their habitats have disappeared as a result of succession, or they have been out-competed by other species. One way of conserving these habitats, and hence the species they contain, is by managing succession in a way that prevents a change to the next stage.

One example is the moorland that exists over much of the higher ground in the UK. The burning of heather and grazing by sheep has prevented this land from reaching its climax community. The burning and grazing destroy the young tree saplings and so prevent the natural succession into deciduous woodland.

Around 4000 years ago, much of lowland UK was a climax community of oak woodland, but most of this forest was cleared to allow grazing and cultivation. The many heaths and grasslands that we now refer to as 'natural' are the result of this clearance and subsequent grazing by animals.

If the factor that is preventing further succession is removed, then the ecosystem develops naturally into its climatic climax (secondary succession). For example, if grasslands are no longer grazed or mowed, or if farmland is abandoned, shrubs initially take over, followed by deciduous woodland.

Application and How science works

Conflicting interests

One challenging conservation issue in the UK is the conflict between the conservation of hen harriers and the commercial hunting of red grouse.

One scientific survey investigated the effect of predation by hen harriers on the breeding success of red grouse on managed moorland in Scotland. Some of the results included:

- On moorland where hen harriers were present there were, on average, 17 per cent fewer young grouse than on moorlands without hen harriers.
- Over a 3-year period grouse nests were intensively observed during the 6 weeks following the hatching of chicks. In this period, predation by harriers accounted for 91 per cent of grouse chick losses.
- Prey remains found around harrier nests were examined. Of the 300 items identified, 32 per cent were grouse chicks.

1 How many of the items of prey identified around harrier nests were grouse chicks?

2 Harriers also feed on voles and meadow pipits. Explain how a rise in the population of these organisms might affect the population of grouse.

Moorland is considered one of the most attractive landscapes in the UK. Many of the national parks are made up of moorland and are visited by millions of people each year. To rear grouse, moorland has to be carefully managed. Controlled grazing by sheep and the periodic burning of vegetation are used to maintain low-growing plant populations of heather, bilberry and crowberry that grouse feed on and nest within. The money to support this management comes largely from charges made to those who shoot grouse.

3 Explain what might happen to moorland if sheep-grazing and burning of the vegetation ceased.

The population of grouse in the UK is in decline due mainly to disease. Currently there are around 250 000 breeding pairs. The hen harrier was persecuted to such an extent that, by 1900, it was only found on a few Scottish islands. It recolonised the UK mainland in the 1970s and there are now around 750 breeding pairs. Both harriers and grouse normally produce one clutch of eggs each year. Hen harriers are protected by law and it is illegal to kill them, collect their eggs or destroy their nests. Conservationists want to retain this protection so that the population of hen harriers can increase. Grouse managers want to be allowed to control hen harrier populations to prevent them threatening the declining grouse populations.

4 Outline the arguments for and against continued protection of hen harrier populations.

To try to help resolve this conflict, scientists are currently conducting experiments to test whether hen harrier populations can be increased at the same time as reducing their negative impact on grouse populations. The information can then be used to inform decisions about how best to conserve grouse, harriers and moorland habitats. This is an example of How science works (HSW: F & L).

The experiment will be carried out in two large areas where harriers are currently rare. Within these areas, the results of two strategies on the size of harrier and grouse populations will be measured:

- Killing hen harrier chicks, or moving them to a different location, when the harrier population size reaches an agreed ceiling.
- Providing alternative sources of food for hen harriers.

Figure 2 *Hen harrier*

Figure 3 *Red grouse*

4 In each of the following, suggest a reason why:

a The experiment will take at least 5 years to produce any findings.

b An independent body, acceptable to both conservation groups and grouse managers will be needed to oversee the experiment.

c The sites chosen for the experiment are ones where harriers can be expected to colonise relatively quickly.

d Each experimental area will contain a number of different moorland sites managed by different individuals.

e Some people are concerned about the long-term implications of a suspension, however temporary, to the legal protection of harriers that would be required during the experiment.

5 How do scientific experiments such as this one help to inform decision-making?

AQA Examination-style questions

1 Heather plants are small shrubs. Heather plants are the dominant species in the climax community of some moorlands. The structure and shape of a heather plant changes as it ages. This results in changes in the species composition of the community. A large area of moorland was burnt leaving bare ground. The table shows four stages of succession in this area.

Time after burning/ years	Appearance of heather plant	Mean percentage cover of heather	Other plant species present
4		10	Many
12		90	Few
19		75	Several
24		30	Many

(a) Explain what is meant by:

(i) succession;

(ii) a climax community. *(3 marks)*

(b) Explain why the number of other plant species decreases between 4 and 12 years after burning. *(2 marks)*

(c) The rate at which a heather plant produced new biomass was measured in g per kg of heather plant per year. This rate decreased as the plant aged. Use the information in the table to explain why. *(3 marks)*

AQA, 2006

2 Attempts are being made to conserve the natterjack toad, which breeds in ponds. A number of recommendations have been made about how to do this. Some of these are shown in the table, together with a reason for each one.

Recommendation	**Reason**
Keep, or create, shallow ponds.	Deep ponds contain insect predators that eat natterjack tadpoles.
Keep pH of ponds between 5 and 7.	Optimum pH for natterjack tadpoles.
Keep vegetation content of ponds low.	Large amounts of vegetation attract insect predators.
Remove shrubs from around ponds.	Shrubs provide cover for common toads, whose tadpoles kill natterjack tadpoles.
Keep grass around ponds short by mowing, or allowing rabbits to graze.	Short grass suits beetles that natterjack toads eat.

(a) Explain two ways in which these recommendations could reduce populations of animals other than common toads. *(2 marks)*

(b) Most of the UK populations of natterjack toads are found on sand dune systems. The communities of plants and animals on such systems usually change over time as a result of succession. Explain how the natterjack conservation recommendations could affect succession in a sand dune system. *(3 marks)*

AQA, 2005

8 Inheritance and selection

8.1 Studying inheritance

Learning objectives:

- What are meant by the terms 'genotype' and 'phenotype'?
- What are dominant, recessive and co-dominant alleles?
- What are multiple alleles?

Specification reference: 3.4.8

In this chapter we shall look at the way in which characters are passed from one generation to the next and how this can produce genetic variety within a population. We shall see that, when **populations** are geographically isolated, there is no interbreeding between members of each population and so their genes are also isolated. In time, the genes within each population will change and this can lead to the formation of new species. Let us begin by looking at some of the terms and conventions that are used in studying inheritance.

The fact that children resemble both their parents to a greater or lesser degree and yet are identical to neither has long been recognised. However, it took the re-discovery, at the beginning of the last century, of the work of a scientist and monk, called Gregor Mendel, to establish the basic laws by which characteristics are inherited.

Link

An understanding of inheritance depends upon an understanding of the way chromosomes behave during meiosis. It is therefore advisable to study Topic 8.4 in the *AS Biology* book before starting this chapter.

Genotype and phenotype

Genotype is the genetic constitution (make-up) of an organism. It describes all the **alleles** that an organism contains. The genotype sets the limits within which the characteristics of an individual may vary. It may determine that a human baby could grow to be 1.8 m tall, but the actual height that this individual reaches is affected by other factors, such as diet. A lack of an element such as calcium (for the growth of bone) at a particular stage of development could mean that the individual never reaches his/her potential maximum height. Any change to the genotype as a result of a change to the DNA is called a mutation and it may be inherited if it occurs in the formation of **gametes**.

Phenotype is the observable characteristics of an organism. It is the result of the interaction between the expression of the genotype and the environment. The environment can alter an organism's appearance. Any change to the phenotype that does not affect the genotype is not inherited and is called a modification.

Genes and alleles

A gene is a section of DNA, that is, a sequence of **nucleotide** bases, that usually determines a single characteristic of an organism (for example, eye colour). It does this by coding for particular polypeptides. These make up the enzymes that are needed in the biochemical pathway that leads to the production of the characteristic (for example, a gene could code for a brown pigment in the iris of the eye). Genes exist in two, or occasionally more, different forms called alleles. The position of a gene on a chromosome is known as the locus.

An allele is one of the different forms of a gene. In pea plants, for example, there is a gene for the colour of the seed pod. This gene has two different forms, or alleles: an allele for a green pod and another allele for a yellow pod.

Hint

All individuals of the same species have the same genes, but not necessarily the same alleles of these genes.

Only one allele of a gene can occur at the locus of any one chromosome. However, in sexually reproducing organisms the chromosomes occur in pairs called **homologous chromosomes**. There are therefore two loci that can each carry one allele of a gene. If the allele on each of the chromosomes is the same (for example, both alleles for green pods are present) then the organism is said to be **homozygous** for the character. If the two alleles are different (for example, one chromosome has an allele for green pods and the other chromosome has an allele for yellow pods) then the organism is said to be **heterozygous** for the characteristic.

In most cases where two different alleles are present in the genotype (heterozygous state) only one of them shows itself in the phenotype.

For instance, in our example where the alleles for green pods and yellow pods are present in the genotype, the phenotype is always green pods. The allele of the heterozygote that expresses itself in the phenotype is said to be **dominant**, while the one that is not expressed is said to be **recessive**. A homozygous organism with two dominant alleles is called **homozygous dominant**, whereas one with two recessive alleles is called **homozygous recessive**. The effect of a recessive allele is apparent in the phenotype of a **diploid** organism only when it occurs in the presence of another identical allele, that is, when it is in the homozygous state.

These different genetic types are shown in Figure 1.

AQA Examiner's tip

A surprisingly large number of candidates do not know the difference between an allele and a gene. Make certain you are not one of them.

AQA Examiner's tip

Many candidates fail to appreciate that, in diploid cells, there are **two** copies of each allele: one copy inherited from the mother and the other copy from the father.

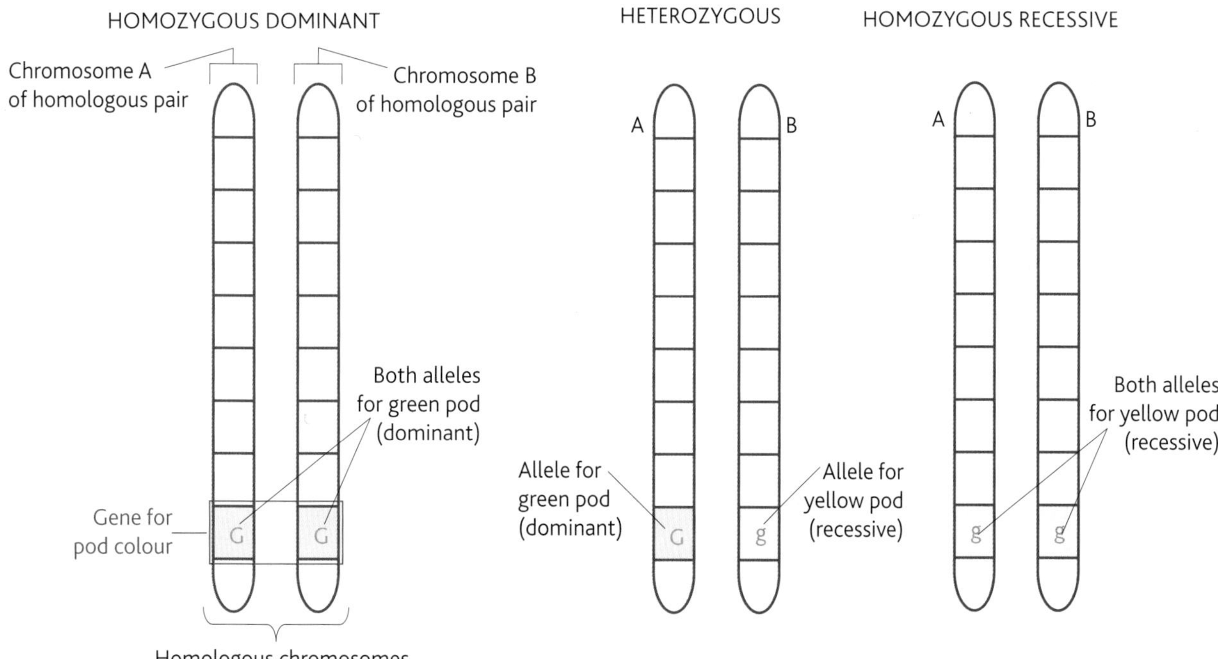

Figure 1 *Pair of homologous chromosomes showing different possible pairings of dominant and recessive alleles*

In some cases, two alleles both contribute to the phenotype, in which case they are referred to as **co-dominant**. In this situation when both alleles occur together, the phenotype is either a blend of both features (for example, snapdragons with pink flowers resulting from an allele for red-coloured flowers and an allele for white-coloured flowers) or both features are represented (for example, the presence of both A and B antigens in blood group AB). We will learn more about co-dominance in Topic 8.4.

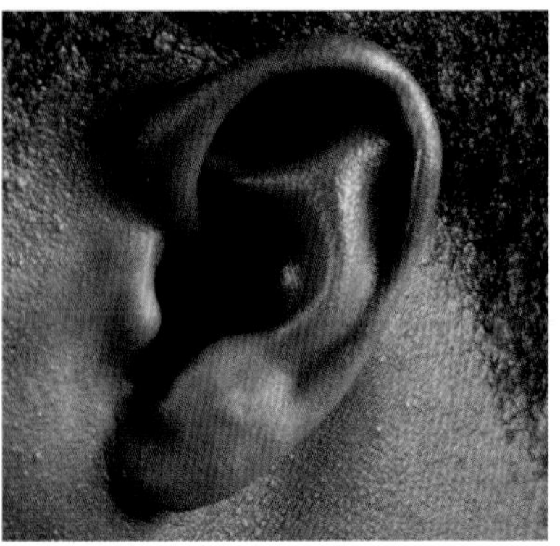

Figure 2 *'Attached' earlobe. As with other inherited physical body characteristics, the earlobe is subject to genetic differences. Here, the earlobe is firmly attached to the facial skin rather than hanging freely*

Sometimes a gene has more than two allelic forms. In this case, the organism is said to have **multiple alleles** for the character. However, as there are always only two chromosomes in a homologous pair, it follows that only two of the three or more alleles in existence can be present in a single organism. Multiple alleles occur in the human ABO blood grouping system. Again we shall learn more about multiple alleles in Topic 8.4.

Figure 3 summarises the different terms used in genetics.

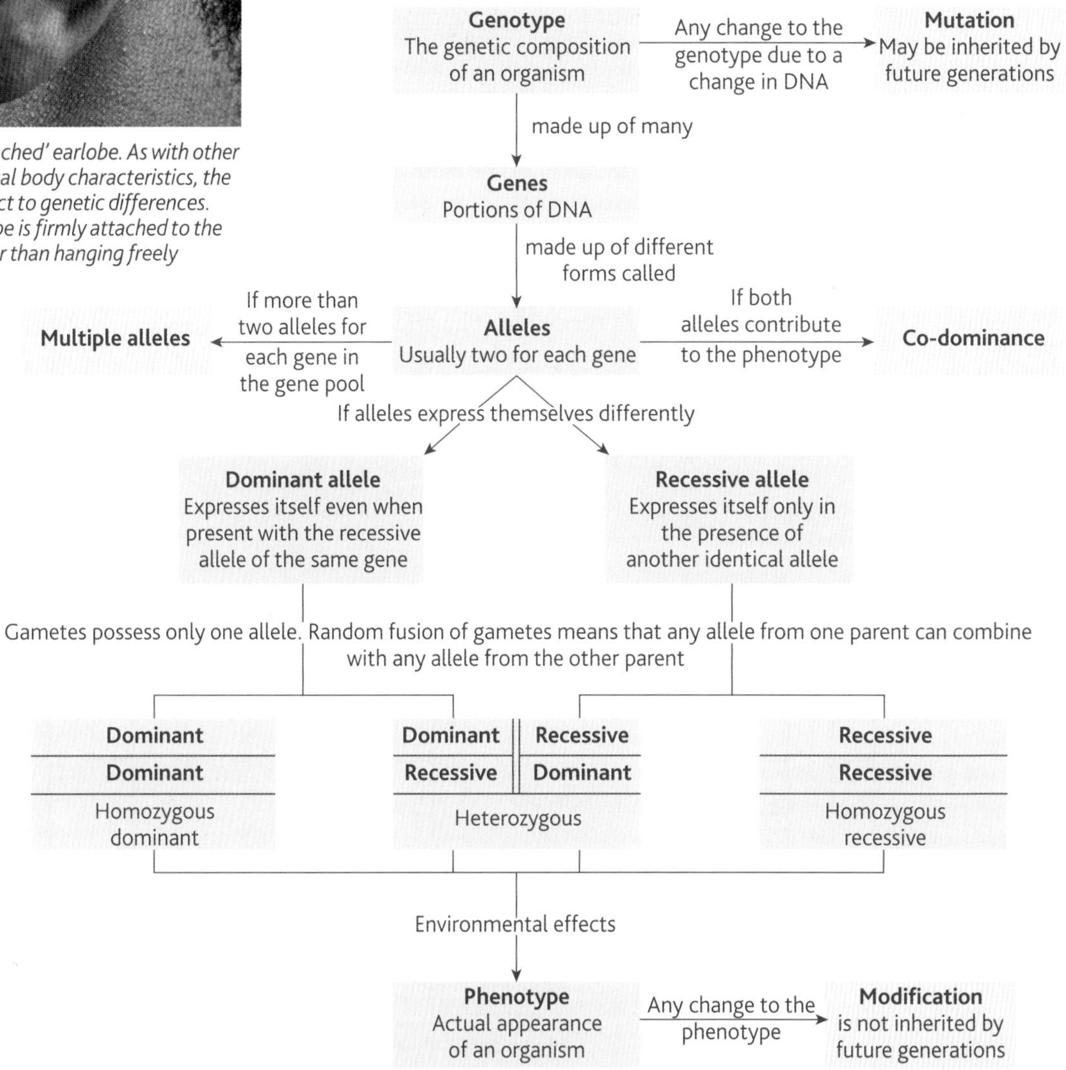

Genotype
The genetic composition of an organism

Any change to the genotype due to a change in DNA →

Mutation
May be inherited by future generations

made up of many ↓

Genes
Portions of DNA

made up of different forms called ↓

Alleles
Usually two for each gene

If more than two alleles for each gene in the gene pool →

Multiple alleles

If both alleles contribute to the phenotype →

Co-dominance

If alleles express themselves differently

Dominant allele
Expresses itself even when present with the recessive allele of the same gene

Recessive allele
Expresses itself only in the presence of another identical allele

Gametes possess only one allele. Random fusion of gametes means that any allele from one parent can combine with any allele from the other parent

Dominant	**Dominant** | **Recessive**	**Recessive**
Dominant	**Recessive** | **Dominant**	**Recessive**
Homozygous dominant	Heterozygous	Homozygous recessive

Environmental effects ↓

Phenotype
Actual appearance of an organism

Any change to the phenotype →

Modification
is not inherited by future generations

Figure 3 *Summary of genetic terms*

Summary questions

In the following passage, give the word that best replaces the number in brackets.

The genetic composition of an organism is called the (1) and any change to it is called a (2) and may be inherited by future generations. The actual appearance of an organism is called the (3). A gene is a sequence of (4) along a piece of DNA that determines a single characteristic of an organism. It does this by coding for particular (5) that make up the enzymes needed in a biochemical pathway. The position of a gene on a chromosome is called the (6). Each gene has two or more different forms called alleles. If the two alleles on a homologous pair of chromosomes are the same they are said to be (7), but if they are different, they are said to be (8). An allele that is not apparent in the phenotype when paired with a dominant allele is said to be (9). Two alleles are called (10) where they contribute equally to the appearance of a characteristic.

8.2 Monohybrid inheritance

Learning objectives:

- How are genetic crosses represented?
- How is a single gene inherited?

Specification reference: 3.4.8

Representing genetic crosses

Genetic crosses are usually represented in a standard form of shorthand. This shorthand form is described in Table 1. Although you may occasionally come across variations to this scheme, that outlined in Table 1 is the one normally used. Once you have practised a number of crosses, you may be tempted to miss out stages or explanations. Not only is this likely to lead to errors, it often makes your explanations difficult for others to follow.

Table 1 *Representing genetic crosses*

Instruction	Reason/notes	Example [green pod and yellow pod]
Questions in examinations usually give the symbols to be used in which case always use the ones provided. Choose a single letter to represent each characteristic.	An easy form of shorthand.	–
Choose the first letter of one of the contrasting features.	When more than one character is considered at one time such a logical choice means it is easy to identify which letter refers to which character.	Choose **G** (green) or **Y** (yellow).
If possible, choose the letter in which the higher and lower case forms differ in shape as well as size.	If the higher and lower case forms differ it is almost impossible to confuse them, regardless of their size.	Choose **G** because the higher case form (**G**) differs in shape from the lower case from (**g**) whereas **Y** and **y** are very similar and are likely to be confused.
Let the higher case letter represent the dominant feature and the lower case letter the recessive one. Never use two different letters where one character is dominant.	The dominant and recessive feature can easily be identified. Do not use two different letters as this indicates co-dominance.	Let **G** = green and **g** = yellow. Do not use **G** for green and **Y** for yellow.
Represent the parents with the appropriate pairs of letters. Label them clearly as 'parents' and state their phenotypes.	This makes it clear to the reader what the symbols refer to.	Green pod × Yellow pod Parents **GG** × **gg**
State the gametes produced by each parent. Label them clearly, and encircle them. Indicate that meiosis has occurred.	This explains why the gametes only possess one of the two parental factors. Encircling them reinforces the idea that they are separate.	Meiosis Meiosis Gametes (G) (G) (g) (g)
Use a type of chequerboard or matrix, called a Punnett square, to show the results of the random crossing of the gametes. Label male and female gametes even though this may not affect the results.	This method is less liable to error than drawing lines between the gametes and the offspring. Labelling the sexes is a good habit to acquire – it has considerable relevance in certain types of crosses, e.g. sex-linked crosses.	♂ GAMETES: (G) (G) ♀ GAMETES (g): Gg Gg ♀ GAMETES (g): Gg Gg
State the phenotypes of each different genotype and indicate the numbers of each type. Always put the higher case (dominant) letter first when writing out the genotype.	Always putting the dominant feature first can reduce errors in cases where it is not possible to avoid using symbols with the higher and lower case letters of the same shape.	All offspring are plants producing green pods (Gg).

Inheritance of pod colour in peas

Monohybrid inheritance is the inheritance of a single gene. To take a simple example we will look at one of the features Gregor Mendel studied: the colour of the pods of pea plants. Pea pods come in two basic colours: green and yellow.

If pea plants with green pods are bred repeatedly with each other so that they consistently give rise to plants with green pods, they are said to be **pure breeding** for the character of green pods. Pure-breeding strains can be bred for almost any character. This means that the organisms are homozygous (that is, they have two **alleles** that are the same) for that particular gene.

If these pure-breeding green-pod plants are then crossed with pure-breeding yellow-pod plants, all the offspring, known as the **first filial**, or **F_1, generation**, produce green pods. This means that the allele for green pods is **dominant** to the allele for yellow pods, which is therefore **recessive**. This cross is shown in Figure 1.

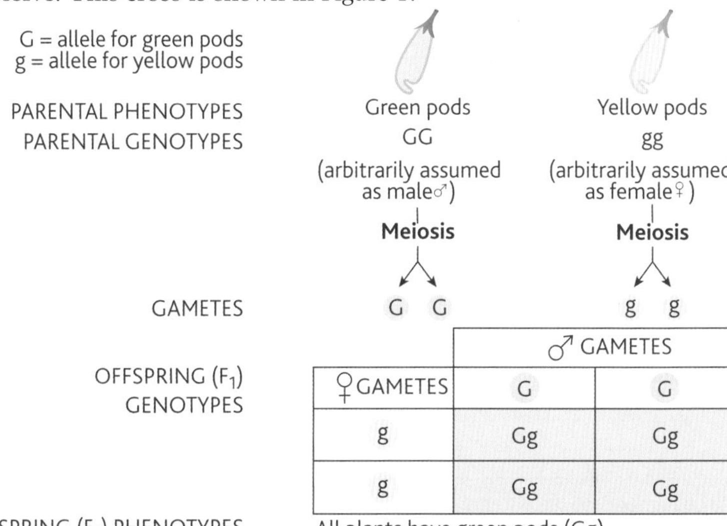

Figure 1 *Cross between a pea plant that is pure breeding for green pods and one that is pure breeding for yellow pods*

When the heterozygous plants (Gg) of the F_1 generation are crossed with one another (= F_1 intercross), the offspring (known as the second filial, or F_2, generation) are always in an approximate ratio of three plants with green pods to each one plant with yellow pods. This cross is shown in Figure 2.

These observed facts led to the formation of a basic law of genetics (the law of segregation). This states:

In diploid organisms, characteristics are determined by alleles that occur in pairs. Only one of each pair of alleles can be present in a single gamete.

Application

Determining genotypes

One common problem that arises when studying inheritance is that an organism whose **phenotype** displays a dominant characteristic may possess either of two **genotypes**:

- two dominant alleles (homozygous dominant)
- one dominant allele and one recessive allele (heterozygous).

AQA Examiner's tip

It is important to always use the accepted method of representing genetic crosses and to do so in full. You may understand what you are doing, but if your teacher or examiner cannot follow it, you are most unlikely to get full credit for your efforts.

Hint

The term 'F_1' should be used only for the offspring of crosses between the original parents, whereas the term 'F_2' should be used only for the offspring resulting from crossing the F_1 individuals.

Summary questions

1 In humans, Huntington's disease is caused by a dominant, mutant gene. Draw a genetic diagram to show the possible genotypes and phenotypes of the offspring produced by a man with one allele for the disease and a woman who does not suffer from the disease.

2 In cocker spaniels, black coat colour is the result of a dominant allele and red coat colour is the result of a corresponding recessive allele.

a Draw a genetic diagram to show a cross between a pure-breeding bitch with a black coat and a pure-breeding dog with a red coat.

b If a dog and a bitch from this first cross are mated, what is the probability that any one of the offspring will have a red coat? Use a genetic diagram to show your working.

It is not possible to tell which genotype an organism has from outward appearances. It is, however, possible to determine the actual genotype by carrying out a specific genetic cross. It is a very common type of cross known as a test cross because it tests whether an unknown genotype is homozygous dominant or heterozygous.

To see how we carry out this cross, let us use our example of pea plants with different seed-pod colours. Suppose we have a plant that produces green seed pods. This plant has two possible genotypes with respect to pod colour:

- homozygous dominant (GG)
- heterozygous (Gg).

To discover its actual genotype, we cross the plant with an organism displaying the recessive phenotype of the same character, i.e. in our case with a pea plant producing yellow pods (gg).

1 Draw genetic diagrams to show the results of a cross between a pea plant with yellow pods and:

a a pea plant that has a homozygous dominant genotype for green pods;

b a pea plant that has a heterozygous genotype for green pods.

2 A cross was carried out between a pea plant producing green pods and one producing yellow pods. The seeds from this cross were germinated and, of the 63 plants grown, all produced green pods.

a What is the probable genotype of the parent plant with green pods?

b Explain why we cannot be absolutely certain of the parent plant's genotype.

3 In a cross between a different pea plant with green pods and a pea plant with yellow pods, 96 plants were produced. 89 of these had green pods and 7 had yellow pods.

a What is the probable genotype of the parent plant with green pods?

b How certain can we be of the genotype of the parent plant with green pods?

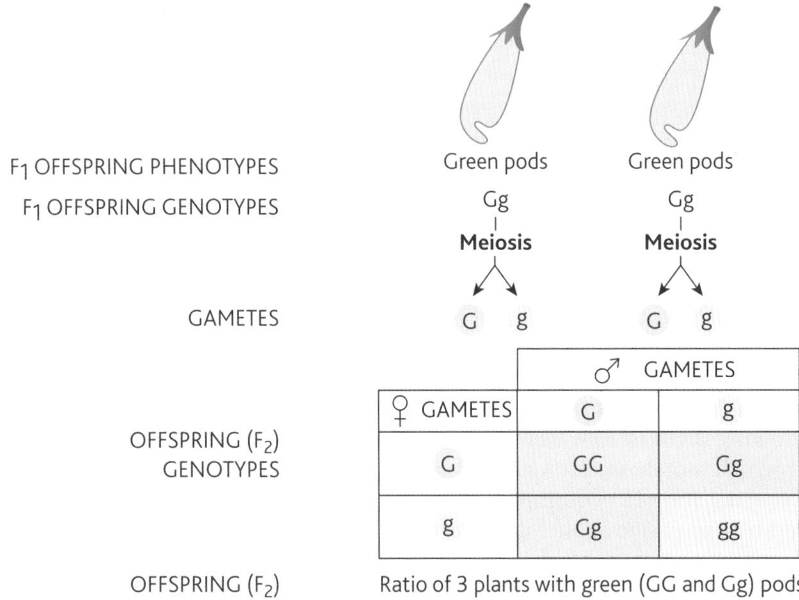

Figure 2 *F_1 intercross between pea plants that are heterozygous for green pods*

8.3 Sex inheritance and sex linkage

Learning objectives:

- How is sex determined genetically?
- What is sex linkage?
- How are sex linked diseases such as haemophilia inherited?

Specification reference: 3.4.8

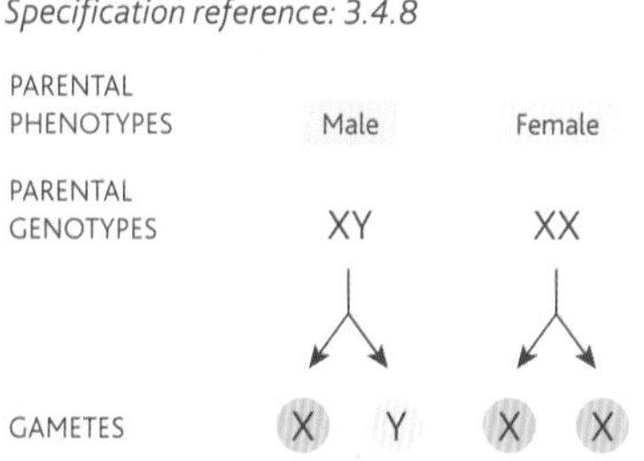

OFFSPRING GENOTYPES

	♂ GAMETES	
♀ GAMETES	X	Y
X	XX	XY
X	XX	XY

OFFSPRING PHENOTYPES

50% Male (XY)
50% Female (XX)

Figure 1 *Sex inheritance in humans*

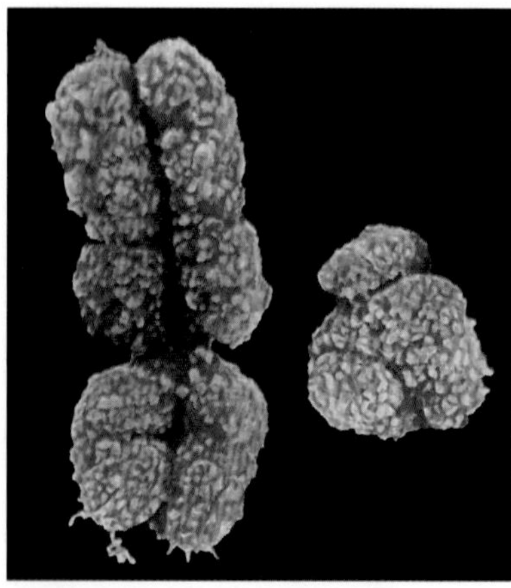

Figure 2 *Scanning electron micrograph (SEM) of human X (left) and Y chromosomes as found in a male*

Humans have 23 pairs of chromosomes. 22 of these pairs have partners that are identical in appearance, whether in a male or a female. The remaining pair are the sex chromosomes. In human females, the two sex chromosomes appear the same and are called the **X chromosomes**. In the human male there is a single X chromosome like that in the female, but the second one of the pair is smaller in size and shaped differently. This is the **Y chromosome**.

Sex inheritance in humans

Unlike other features of an organism, sex is determined by chromosomes rather than **genes**. In humans:

- as females have two X chromosomes, all the gametes are the same in that they contain a single X chromosome
- as males have one X chromosome and one Y chromosome, they produce two different types of gamete – half have an X chromosome and half have a Y chromosome.

The inheritance of sex is shown in Figure 1.

Sex linkage – haemophilia

Any gene that is carried on either the X or Y chromosome is said to be sex linked. However, the X chromosome is much longer than the Y chromosome. This means that, for most of the length of the X chromosome, there is no equivalent homologous portion of the Y chromosome. Those characteristics that are controlled by **recessive alleles** on this non-homologous portion of the X chromosome will appear more frequently in the male. This is because there is no homologous portion on the Y chromosome that might have the **dominant allele**, in the presence of which the recessive allele does not express itself.

By contrast the X chromosome carries many genes. One example in humans is the condition called haemophilia, in which the blood clots only slowly and there may be slow and persistent internal bleeding, especially in the joints. As such it is potentially lethal if not treated. This has resulted in some selective removal of the gene from the population, making its occurrence relatively rare (about 1 person in 20 000 in Europe). Although haemophiliac females are known, the condition is almost entirely confined to males.

One of a number of causes of haemophilia is a recessive allele with altered DNA nucleotides that therefore do not code for the required protein. This results in an individual being unable to produce a protein that is required in the clotting process. The extraction of this protein from donated blood means that it can now be given to haemophiliacs, allowing them to lead near-normal lives. Figure 3 shows the usual way in which a male inherits haemophilia. Note that the alleles are shown in the usual way (H = dominant allele for production of the clotting protein, and h = recessive allele for the non-production of clotting protein). However, as they are linked to the X chromosome, they are not shown separately, but always attached to the X chromosome, that is, as X^H and X^h respectively. There is no equivalent allele on the Y chromosome as it does not carry the gene for producing clotting protein.

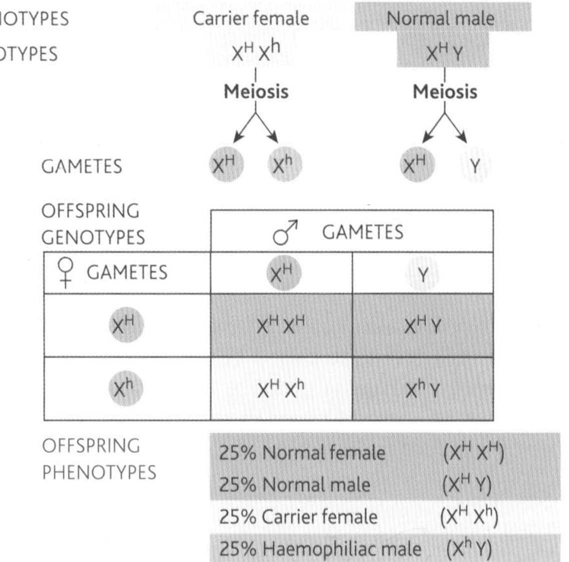

Figure 3 *Inheritance of haemophilia from a carrier female*

As males can *only* obtain their Y chromosome from their father, it follows that their X chromosome comes from their mother. As the defective allele that does not code for the clotting protein is linked to the X chromosome, males always inherit the disease from their mother. If their mother does not suffer from the disease, she may be **heterozygous** for the character (X^HX^h). Such females are called carriers because they carry the allele without showing any signs of the character in their phenotype.

As males pass the Y chromosome on to their sons, they cannot pass haemophilia to them. However, they can pass the allele to their daughters, via the X chromosome, who would then become carriers of the disease (Figure 4).

Hint

Remember that males have a Y chromosome and this could only have come from their fathers. Their X chromosome must therefore have come from their mothers.

AQA Examiner's tip

Sometimes candidates show the X chromosome passing from father to son. If this were so, the mother must have provided the son's Y chromosome, in which case she would be male! Clearly this is nonsense.

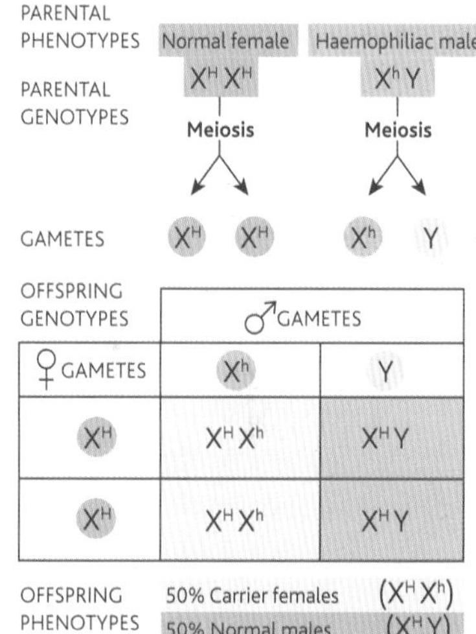

Figure 4 *Inheritance of the haemophiliac allele from a haemophiliac male*

Pedigree charts

One useful way to trace the inheritance of sex-linked characters such as haemophilia is to use a pedigree chart. In these:

- a male is represented by a square
- a female is represented by a circle
- shading within either shape indicates the presence of a character, such as haemophilia, in the phenotype
- a dot within a circle signifies a woman with a normal phenotype but who carries the defective allele.

Summary questions

Red-green colour blindness is linked to the X chromosome. The allele (r) for red-green colour blindness is recessive to the normal allele (R). Figure 5 on the next page shows the inheritance of this characteristic in a family.

1. What sex chromosomes are present in individuals labelled E and F?
2. In terms of colour blindness, what are the phenotypes of each of the individuals labelled A, B and D?
3. In terms of colour blindness, what are the genotypes of each of the individuals labelled G, H, I and J?
4. If individual C were to have children with a normal female (one who does not have any r alleles), what would the probability be of any sons having colour blindness?
5. Individual J is colour blind. From the family tree, suggest how this might have occurred.

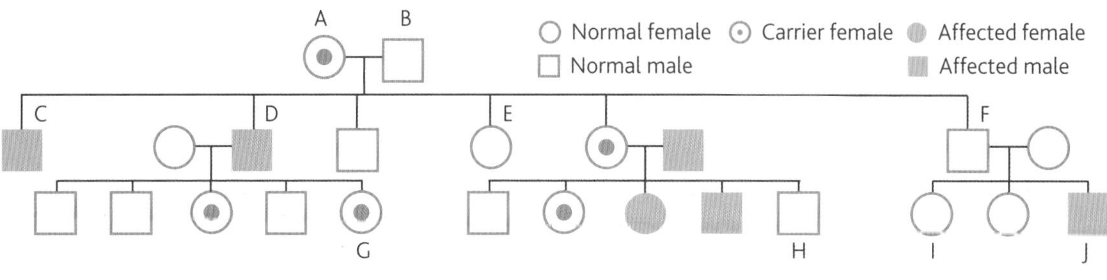

Figure 5

Application

A right royal disease

The royal families of Europe have been affected by haemophilia for the last few centuries. The origins of the disease stretch back to Queen Victoria. A pedigree chart showing the inheritance of haemophilia from Queen Victoria in members of various European royal families is shown in Figure 6.

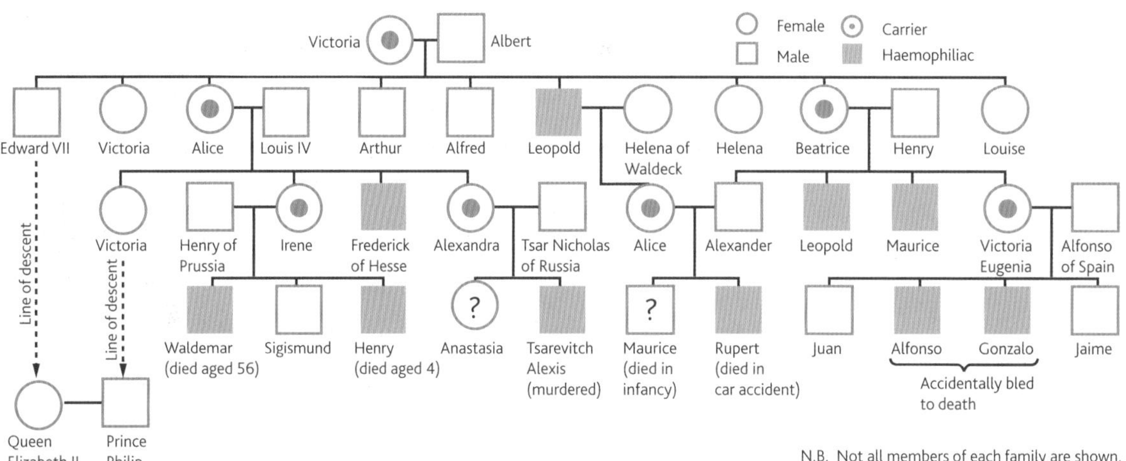

Figure 6 *Pedigree chart showing the transmission of haemophilia from Queen Victoria*

1. Explain why haemophilia is not present in the current British royal family of Queen Elizabeth II and Prince Philip, and their children.
2. Give evidence from the chart which shows that haemophilia is:
   a. sex-linked
   b. recessive.
3. Using the symbols X^H for the chromosome carrying an allele that produces a clotting protein and X^h for a chromosome carrying an allele that does not produce a clotting protein, list the possible genotypes of the following people:
   a. Queen Elizabeth II
   b. Gonzalo
   c. Irene
4. Suppose Waldemar and Anastasia had married and produced children. Using the same symbols, list all the possible genotypes of their:
   a. sons
   b. daughters.

   Explain your answers.

8.4 Co-dominance and multiple alleles

Learning objectives:

- How does co-dominance affect the inheritance of characteristics?
- How do multiple alleles affect inheritance?
- How are blood groups in humans inherited?

Specification reference: 3.4.8

In Topics 8.2 and 8.3 we dealt with straightforward situations in which there were two possible alleles at each locus on a chromosome, one of which was dominant and the other recessive. We shall now look at two different situations.

- **co-dominance**, in which both alleles are equally dominant
- **multiple alleles**, where there are more than two alleles, of which only two may be present at the loci of an individual's homologous chromosomes.

Co-dominance

Co-dominance occurs where, instead of one allele being dominant and the other recessive, both alleles are dominant to some extent. This means that both alleles of a gene are expressed in the **phenotype**.

Figure 1 *Snapdragons*

One example occurs in the snapdragon plant, in which one allele codes for an enzyme that catalyses the formation of a red pigment in flowers. The other allele codes for an altered enzyme that lacks this catalytic activity and so does not produce the pigment. If these alleles showed the usual pattern of one dominant and one recessive, the flowers would have just two colours: red and white. As they are co-dominant, however, three colours of flower are found:

- In plants that are **homozygous** for the first allele, both alleles code for the enzyme, and hence pigment, production. These plants have red flowers.
- In plants that are **homozygous** for the other allele, no enzyme and hence no pigment is produced. These plants have white flowers.
- **Heterozygous** plants, with their single allele for the functional enzyme, produce just sufficient red pigment to produce pink flowers.

Hint

Remember to use different letters, such as R and W, when answering questions on co-dominance and to put these as superscripts attached to the letter of the gene (C), e.g. C^R and C^W.

If a snapdragon with red flowers is crossed with one with white flowers, the resulting seeds give rise to plants with pink flowers. Note that we cannot use upper and lower case letters for the alleles, as this would imply that one (the upper case) was dominant to the other (the lower case). We therefore use different letters – in this case R for red and W for white – and put them as superscripts on a letter that represents the gene, in this case C for colour. Hence the allele for red pigment is written as C^R and the allele for no pigment as C^W. Figure 2 shows a cross between a red and a white snapdragon while Figure 3 shows a cross between the resultant pink-flowered plants.

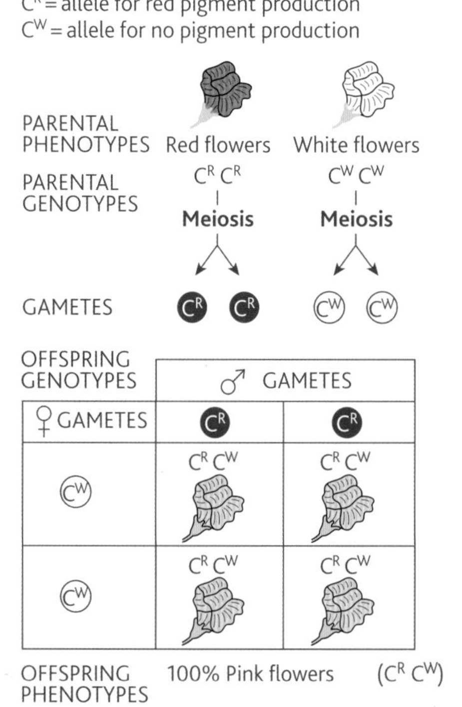

Figure 2 *Cross between a snapdragon with red flowers and one with white flowers*

PARENTAL PHENOTYPES: Pink flowers, Pink flowers

PARENTAL GENOTYPES: $C^R C^W$, $C^R C^W$

Meiosis / Meiosis

GAMETES: C^R C^W / C^R C^W

OFFSPRING GENOTYPES

♀ GAMETES	♂ GAMETES C^R	♂ GAMETES C^W
C^R	$C^R C^R$	$C^R C^W$
C^W	$C^R C^W$	$C^W C^W$

OFFSPRING PHENOTYPES: 50% Pink flowers ($C^R C^W$); 25% Red flowers ($C^R C^R$); 25% White flowers ($C^W C^W$)

Figure 3 *Cross between two snapdragons with pink flowers*

Multiple alleles

Sometimes a gene has more than two alleles, that is, it has multiple alleles. The inheritance of the human ABO blood groups is an example. There are three alleles associated with the gene I (immunoglobulin gene), which lead to the production of different **antigens** on the surface membrane of red blood cells:

- allele I^A, which leads to the production of antigen A
- allele I^B which leads to the production of antigen B
- allele I^O, which does not lead to the production of either antigen.

Although there are three alleles, only two can be present in an individual at any one time, as there are only two homologous chromosomes and therefore only two gene loci. The alleles I^A and I^B are co-dominant, whereas the allele I^O is recessive to both. The possible genotypes for the four blood groups are shown in Table 1. There are obviously many different possible crosses between different blood groups, but two of the most interesting are:

Table 1 *Possible genotypes of blood groups in the ABO system*

Blood group	Possible genotypes
A	I^AI^A or I^AI^O
B	I^BI^B or I^BI^O
AB	I^AI^B
O	I^OI^O

1 A cross between an individual of blood group O and one of blood group AB, rather than producing individuals of either of the parental blood groups, produces only individuals of the other two groups, A and B (see Figure 4).

2 When certain individuals of blood group A are crossed with certain individuals of blood group B, their children may have any of the four blood groups (see Figure 5).

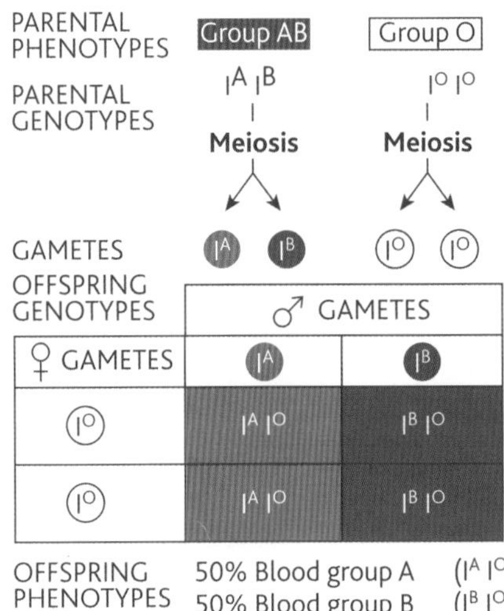

Figure 4 *Cross between an individual of blood group AB and one of blood group O*

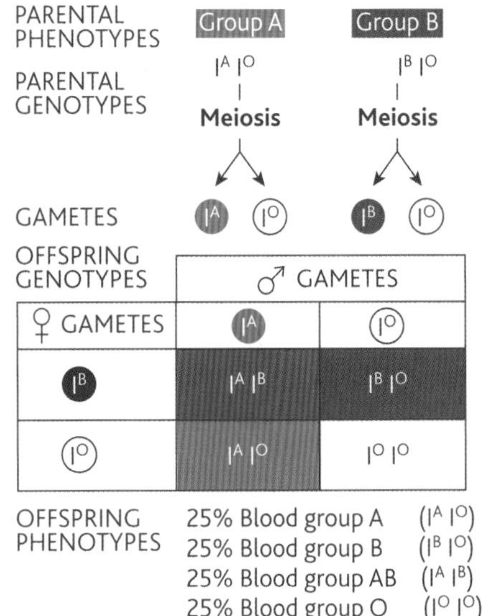

Figure 5 *Cross between an individual of blood group A and one of blood group B*

Multiple alleles and a dominance hierarchy

In blood groups, alleles I^A and I^B are co-dominant and I^O is recessive to both. Sometimes, however, there may be more than three alleles, each of which is arranged in a hierarchy with each allele being dominant to those below it and recessive to those above it. One example is coat colour in rabbits. The gene for coat colour (C) has four alleles. In order of dominance they are (most dominant first):

Agouti coat (C^A) Chinchilla coat (C^{Ch}) Himalayan coat (C^H) Albino coat (C^a)

Table 2 shows the possible genotypes of rabbits with each of these coat colours.

Table 2 *Coat colour in rabbits*

Coat colour	Possible genotypes
Full colour (Agouti)	C^AC^A, C^AC^{Ch}, C^AC^H, C^AC^a
Chinchilla	$C^{Ch}C^{Ch}$, $C^{Ch}C^H$, $C^{Ch}C^a$
Himalayan	C^HC^H, C^HC^a
Albino	C^aC^a

Summary questions

1 In the example of coat colour in rabbits above, list all the possible genotypes of a rabbit with an agouti coat.

2 A man claims not to be the father of a child. The man is blood group O while the mother of the child is blood group A and the child is blood group AB. State, with your reasons, whether you think the man could be the father of the child.

3 In some breeds of domestic fowl, the gene controlling feather shape has two alleles that are co-dominant. When homozygous, the allele A^S produces straight feathers while the allele A^F produces frizzled feathers. The heterozygote for feather shape gives mildly frizzled feathers. Draw a genetic diagram to show the genotypes and phenotypes resulting from a cross between a mildly frizzled cockerel and a frizzled hen. The gene for feather shape is **not** sex-linked.

Application

Coat of many colours

Cats come in a variety of colours, two of the commonest being black and ginger. Coat colour in these cats is controlled by a gene that has its locus on the X chromosome. This gene has two alleles. Allele O, which gives the orange coat of ginger cats and allele B which gives a black coat. The two alleles are co-dominant. Heterozygous individuals have a coat that is mostly a mixture of orange and black. This is known as tortoiseshell (see Figure 6).

Figure 6 *Tortoiseshell coats occur in cats that are heterozygous for the co-dominant alleles of black coat and orange coat*

A male cat with a black coat was mated with a female cat with an orange coat.

1 Draw a genetic diagram to show the possible genotypes and phenotypes of kittens produced from this cross.

2 Explain why it is not possible to produce male cats with tortoiseshell coats.

The phenotype of an organism often results from the influence of the environment on its genotype.

The fur on the ears, face, feet and tail of Siamese cats is darker in colour than the rest of the coat. This is due to the presence of a pigment. The production of this pigment is controlled by the action of an enzyme called tyrosinase. The action of tyrosinase is temperature-dependent.

3 Suggest an explanation why Siamese kittens are born with a completely light-coloured coat and only develop their characteristic markings some days later.

8.5 Allelic frequencies and population genetics

Learning objectives:

- What are meant by the terms 'gene pool' and 'allelic frequency'?
- What is the Hardy–Weinberg principle?
- How can the Hardy–Weinberg principle be used to calculate allele, genotype and phenotype frequencies?

Specification reference: 3.4.8

We have so far looked at how genes and their alleles are passed between individuals in a population. Let us now consider the genes and alleles of an entire population.

All the alleles of all the genes of all the individuals in a population at any one time are known as the **gene pool**. Sometimes the term is used to refer to all the alleles of one particular gene in a population, rather than all the genes. The number of times an allele occurs within the gene pool is referred to as the **allelic frequency**.

Let us look at this more closely by considering just one gene that has two alleles, one of which is dominant and the other recessive. An example is the gene responsible for cystic fibrosis, a disease of humans in which the mucus produced by affected individuals is thicker than normal. The gene has a dominant allele (F) that leads to normal mucus production and a recessive allele (f) that leads to the production of thicker mucus and hence cystic fibrosis. Any individual human has two of these alleles in every one of their cells, one on each of the pair of homologous chromosomes on which the gene is found. As these alleles are the same in every cell, we only count one pair of alleles per gene per individual when considering a gene pool. If there are 10 000 people in a population, there will be twice as many (20 000) alleles in the gene pool *of this gene*.

The pair of alleles of the cystic fibrosis gene has three different possible combinations, namely homozygous dominant (FF), homozygous recessive (ff) and heterozygous (Ff). When we look at allele frequencies, however, it is important to appreciate that the heterozygous combination can exist in two different arrangements, namely Ff and fF. (It is just conventional to put the dominant allele first in all cases.)

In any population the total number of alleles is taken to be 1.0. In our population of 10 000 people, if everyone had the genotype FF, then the frequency of the dominant allele (F) would be 1.0 and the frequency of the recessive allele (f) would be 0.0. If everyone was heterozygous (Ff), the frequency of the dominant allele (F) would be 0.5 and the frequency of the recessive allele (f) would be 0.5. Of course, in practice, the population is not made up of one genotype but of a mixture of all three, the proportions of which vary from population to population. How then can we work out the allele frequency of these mixed populations?

Hint

Whether an allele is recessive or dominant has nothing to do with it being harmful or beneficial. People with type O blood group have two recessive alleles for the gene but as it is the most common blood group it can hardly be harmful. Also, Huntingdon's disease is a fatal condition due to a dominant allele.

The Hardy–Weinberg principle

The Hardy–Weinberg principle provides a mathematical equation that can be used to calculate the frequencies of the alleles of a particular gene in a population. The principle predicts that the proportion of dominant and recessive alleles of any gene in a population remains the same from one generation to the next provided that five conditions are met:

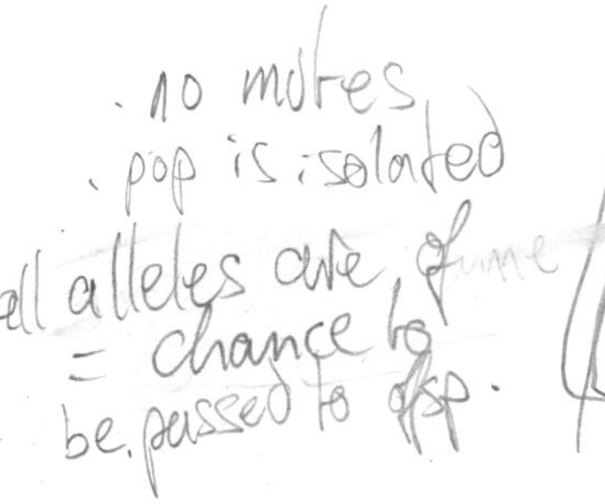

- No mutations arise.
- The population is isolated, that is, there is no flow of alleles into or out of the population.
- There is no selection, that is, all alleles are equally likely to be passed to the next generation.

- The population is large.
- Mating within the population is random.

Although these conditions are probably never totally met in a natural population, the Hardy–Weinberg principle is still useful when studying gene frequencies.

To help us understand the principle let us consider a gene that has two alleles: a dominant allele (A) and a recessive allele (a).

Let the frequency of allele A = p

and the frequency of allele a = q

The first equation we can write is:

$$p + q = 1.0$$

because there are only two alleles and so the frequency of one plus the other must be 1.0 (100%).

As there are only four possible arrangements of the two alleles, it follows that the frequency of all four added together must equal 1.0. Therefore we can state that:

AA + Aa + aA + aa = 1.0 or, expressing this as a frequency:

$$p^2 + 2pq + q^2 = 1.0$$

We can now use these equations to determine the frequency of any allele in a population. For example, suppose that a particular characteristic is the result of the recessive allele a, and we know that one person in 25 000 displays the character.

- The character, being recessive, will only be observed in individuals who have two recessive alleles aa.
- The frequency of aa must be 1/25 000 or 0.00004.
- The frequency of aa is q^2.
- If $q^2 = 0.00004$, then $q = \sqrt{0.00004}$ or 0.00063 approximately.
- We know that the frequency of both alleles A and a is $p + q$ and is equal to 1.0.
- If $p + q = 1.0$, and $q = 0.00063$ then:
- $p = 1.0 - 0.00063 = 0.9937$, that is, the frequency of allele A = 0.9937.
- We can now calculate the frequency of the heterozygous individuals in the population.
- From the Hardy–Weinberg equation we know that the frequency of the heterozygotes is $2pq$.
- In this case, $2pq = (2 \times 0.9937 \times 0.0063) = 0.0125$.
- In other words, 125 individuals in 10 000 carry the allele for the character. This is the equivalent of 313 in our population of 25 000.
- These individuals act as a reservoir of recessive alleles in the population, although they themselves do not express the allele in their phenotype.

Summary questions

1. Define the terms:
   a gene pool
   b allelic frequency.
2. What does the Hardy–Weinberg principle predict?
3. What **five** conditions need to be met for this prediction to hold true?
4. If the frequency p of a dominant allele is 0.942. Calculate the frequency of the heterozygous genotype in the population. Show your working and express your answer as a percentage of the population.

Application

Not as black and white as it seems.

A gene that controls wing colour in the peppered moth has two alleles. The expression of the dominant allele produces moths with light-coloured wings while the expression of the recessive allele produces moths with dark-coloured wings. Scientists sampled a population of moths by catching them in a trap and recording their sex and wing colour. The numbers in Table 1 show how many of the 2215 moths caught were of each type.

Figure 1 *Dark- and light-coloured wing forms of the peppered moth*

Table 1

	Light-coloured wings	Dark-coloured wings
Male	836	269
Female	817	293
Total	**1653**	**562**

1 State, with your reasons, whether you think the gene for wing colour is sex-linked.

2 What proportion of the total sample has two recessive alleles?

3 In the Hardy–Weinberg equation ($p^2 + 2pq + q^2 = 1.0$), p = the frequency of the dominant allele and q = the frequency of the recessive allele. For the population of moths that were caught, use this equation to calculate:

a the frequency (q) of the recessive allele;

b the frequency (p) of the dominant allele;

c the percentage of heterozygotes.

4 Some scientists wanted to estimate the size of the total moth population. Describe how they might do this.

Hint

Unless an allele leads to a phenotype with an advantage or a disadvantage compared with other phenotypes, its allele frequency in a population will stay the same from one generation to the next.

8.6 Selection

Learning objectives:

- How does reproductive success affect allele frequency within a gene pool?
- What is selection?
- What environmental factors exert selection pressure?
- What are stabilising and directional selection?

Specification reference: 3.4.8

We saw in Topic 8.5 that the Hardy–Weinberg principle predicted that the proportion of dominant and recessive alleles of any gene in a population will remain the same from one generation to the next provided certain conditions were met. One such condition is that 'there is no selection', that is, all alleles are equally likely to be passed to the next generation. In practice, not all alleles of a population are equally likely to be passed to the next generation. This is because only certain individuals are reproductively successful and so pass on their alleles.

Reproductive success and allele frequency

Differences between the reproductive success of individuals affects **allele frequency** in populations. The process works like this:

- All organisms produce more offspring than can be supported by the supply of food, light, space, etc.
- Despite overproduction of offspring, most populations remain relatively constant in size.
- This means there is competition between members of a species to be the ones that survive.
- Within any population of a species there will be a **gene pool** containing a wide variety of alleles.
- Some individuals will possess combinations of alleles that make them better able (fitter) to survive in their competition with others.
- These individuals are more likely to obtain the available resources and so grow more rapidly and live longer. As a result, they will have a better chance of successfully breeding and producing more offspring.
- Only those individuals that successfully reproduce will pass on their alleles to the next generation.
- Therefore it is the alleles that gave the parents an advantage in the competition for survival that are most likely to be passed on to the next generation.
- As these new individuals have 'advantageous' alleles, they in turn are more likely to survive, and so reproduce successfully.
- Over many generations, the number of individuals with the 'advantageous' alleles will increase at the expense of the individuals with the 'less advantageous' alleles.
- Over time, the frequency of the 'advantageous' alleles in the population increases while that of the 'non-advantageous' ones decreases.

It must be stressed that what is 'advantageous' depends upon the environmental conditions at any one time. For example, alleles for black body colour may be 'advantageous' as camouflage against a smoke-blackened wall, but 'non-advantageous' against a snowy landscape.

To illustrate the process let us look at the example of the peppered moth. This normally has a light colour that camouflages it against the light background of the lichen-covered trees on which it rests. From time to time black mutant forms of the moth arise. These mutants are

Link

Competition between members of the same species is called intraspecific competition and is covered in Topic 1.4.

highly conspicuous against their light background. As a result, the black mutants are subjected to greater predation from insect-eating birds than the better-camouflaged, normal light forms.

When a dark form of the peppered moth arose in Manchester around 1848, most buildings, walls and trees were blackened by the soot from 50 years of industrial development. Against this black background the dark form was less, not more, conspicuous than the light natural form. As a result, the light form was eaten by birds more frequently than the dark form. More black-coloured moths than light-coloured moths survived and successfully reproduced. Over many generations, the frequency of the 'advantageous' dark-colour allele increased at the expense of the 'less advantageous' light-colour allele. By 1895, 98 per cent of Manchester's population of the moth was of the black type.

Let us now look in more detail at how selection affects a population.

Types of selection

Selection is the process by which organisms that are better adapted to their environment survive and breed, while those that are less well adapted fail to do so. Every organism is subjected to a process of selection, based on its suitability for surviving the conditions that exist at the time. Different environmental conditions favour different characteristics in the population. Depending on which characteristics are favoured, selection will produce a number of different results.

- Selection may favour individuals that vary in one direction from the mean of the population. This is called **directional selection** and changes the characteristics of the population.
- Selection may favour average individuals. This is called **stabilising selection** and preserves the characteristics of a population.

In this chapter, we have been considering characteristics that are influenced by a single gene. In reality, most characteristics are influenced by more than one gene (**polygenes**). We learnt in the AS Biology course that this type of characteristic is influenced by the environment. The effect of the environment on polygenes produces individuals in a population that vary about the mean. When we plot this variation on a graph we get a **normal distribution curve.** Let us look at how these two types of selection affect this curve.

Directional selection

If the environmental conditions change, so will the **phenotypes** needed for survival. Some individuals, which fall to either the left or right of the mean, will possess a phenotype more suited to the new conditions. These individuals will be more likely to survive and breed. They will therefore contribute more offspring (and the alleles these offspring possess) to the next generation than other individuals. Over time, the mean will then move in the direction of these individuals.

To explain, let us imagine a population of a mammal in which there is a range of fur lengths.

- At an average environmental temperature of 10°C the optimum fur length for survival is 15mm. This is the mean fur length of the population, with a number of individuals distributed either side of it (Figure 1a).
- If the average environmental temperature falls to 5°C, individuals with longer fur (say 20mm or more) will be better insulated from

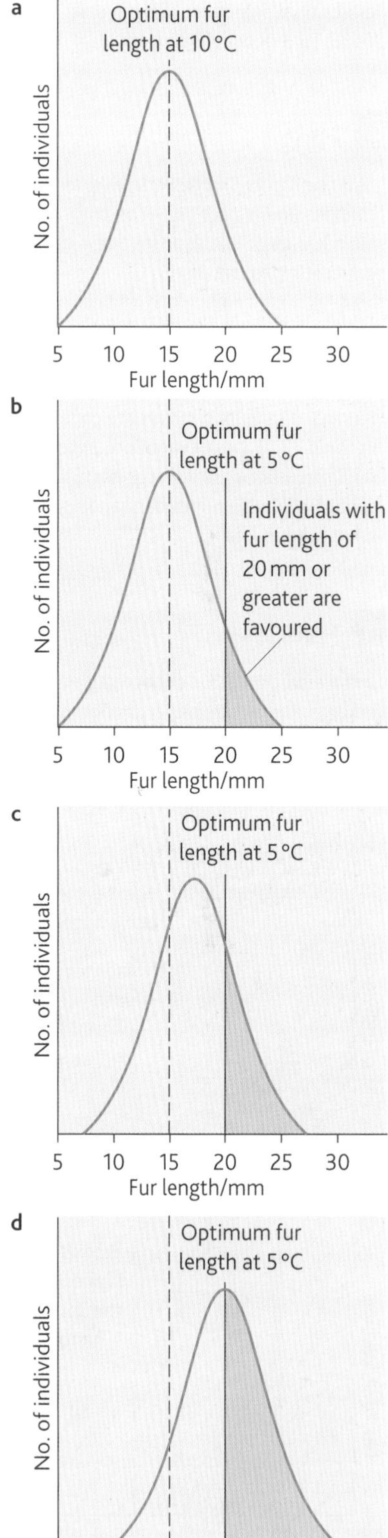

Figure 1 *Directional selection*

Link

Antibiotic resistance is an example of directional selection. It is covered in Topic 16.3 of the *AS Biology* book.

Hint

Selection acts on phenotypes and this has an indirect effect on the inheritance of alleles from one generation to the next.

Link

Normal distribution curves are covered in Topic 7.2, 'Types of variation', in the *AS Biology* book. Knowledge of this spread will help your understanding of what follows here.

Hint

If the environmental change is great enough, there may be no phenotype suited to the new conditions, in which case the population will die out.

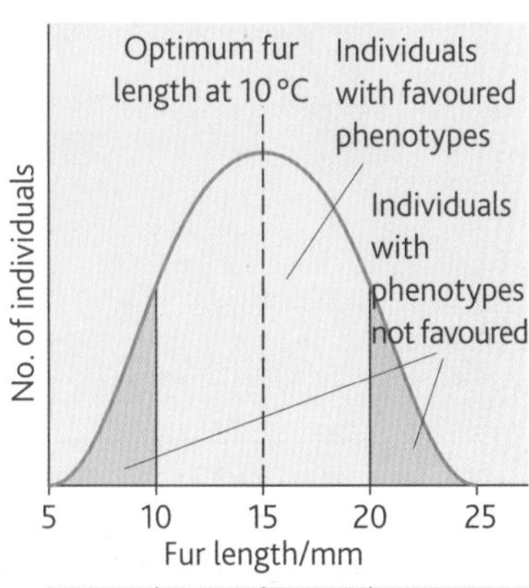

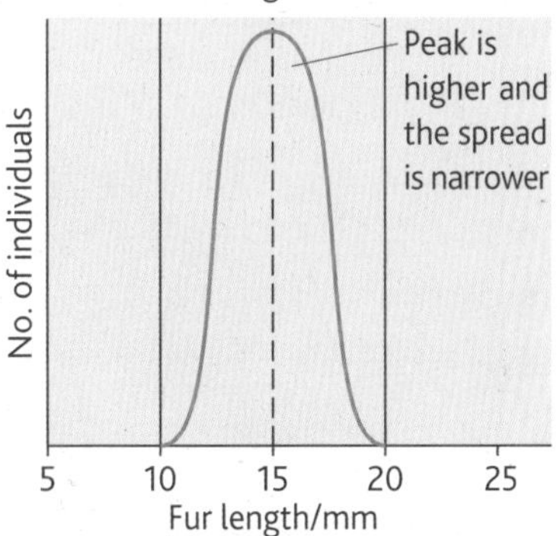

Figure 2 *Stabilising selection*

the cold. These individuals are more likely to survive and so produce more offspring. Those with shorter fur are less likely to survive and so produce fewer offspring (Figure 1b).

- Over many generations, the mean fur length of the population increases as more individuals with longer fur survive, and more individuals with shorter fur die from the cold (Figure 1c). The proportion of alleles in the population for longer fur is increasing at the expense of those for shorter fur.
- Over further generations, the shift in mean fur length continues until it reaches 20 mm – the optimum for the new average environmental temperature of 5 °C (Figure 1d).

Directional selection therefore results in phenotypes at one extreme of the population being selected for and those at the other extreme being selected against.

Stabilising selection

If environmental conditions remain stable, it is the individuals with phenotypes closest to the mean that are favoured. These individuals are more likely to pass their alleles on to the next generation. Those individuals with phenotypes at the extremes are less likely to pass on their alleles. Stabilising selection therefore tends to eliminate the phenotypes at the extremes.

To take our example of mammalian fur length again:

- In years when the average environmental temperature is hotter than usual, individuals with shorter fur will be favoured because they can lose body heat more rapidly.
- In colder years, the opposite is true and individuals with longer fur will be favoured because they are better insulated from the cold.
- Therefore if the temperature fluctuates from year to year (that is, it is unstable) individuals at both extremes will survive. This is because there are some years in which each can thrive at the expense of the other.
- If, however, the environmental temperature is constant, say at 10 °C, individuals at the extremes will never be at an advantage. They will be selected against in favour of those with the mean fur length.
- The mean will remain the same but there will be fewer individuals with fur length at either extreme.

Stabilising selection therefore results in phenotypes around the mean of the population being selected for and those at both extremes being selected against. These events are summarised in Figure 2.

Application

Early selection

The body mass at birth of babies born at a hospital was measured over a 12-year period. In the graph in Figure 3 on the next page the percentage of births in the population (*y*-axis on the left) is plotted against birth mass of the infants as a histogram.

Over the same period, the infant mortality (death) rate was also recorded. The infant mortality rate is measured on a logarithmic scale (*y*-axis on the right) and plotted against infant body mass at birth as a line graph.

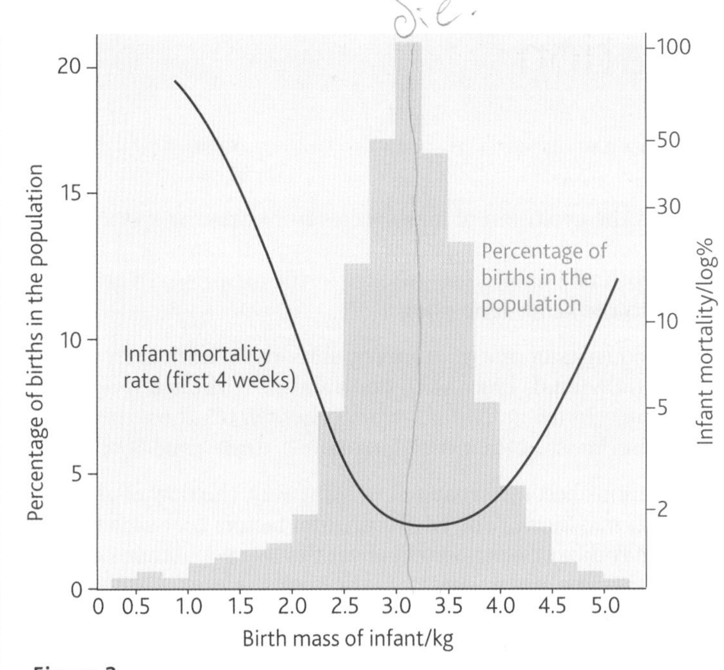

Figure 3

1. Describe the relationship between infant birth mass and
   a the percentage of births in the population
   b infant mortality.
2. Which type of selection is shown by the data? Give reasons for your answer.

Summary questions

1. What is selection?
2. Distinguish between directional selection and stabilising selection.
3. A severe cold spell in 1996 killed over 50% of swallows living on cliffs in Nebraska. Biologists collected nearly 2000 dead swallows from beneath the cliffs and captured around 1000 living ones. By measuring the body mass of the birds, they found that birds with a larger than average body mass survived the cold spell better than ones with a smaller than average body mass. State, giving your reasons, which type of selection was taking place here.

Application

They must be cuckoo!

Cuckoos lay their eggs in the nests of other birds. The host birds will often raise these parasite chicks alongside their own.

In many valleys in southern Spain, great cuckoos and common magpies have lived together for hundreds of years. In some valleys, however, magpies have been around for centuries but cuckoos have only recently arrived.

Scientists placed artificial cuckoo eggs into magpie nests in both types of valley. Where cuckoos and magpies had lived together for a long period, 78 per cent of the magpies removed the cuckoo eggs from their nests. Where cuckoos had only recently colonised the valleys, only 14 per cent of the magpies removed the cuckoo eggs.

It would appear that, in the valleys where cuckoos are well established, selection has favoured those magpies that removed the cuckoo eggs.

1. Suggest **one** advantage to the magpies of removing cuckoo eggs from their nest.
2. Explain how removing cuckoo eggs increases the probability of the alleles for this type of behaviour being passed on to subsequent generations.
3. Suggest why this form of behaviour is not shown by magpies in those valleys where cuckoos have only recently arrived.
4. State, with your reasons, which type of selection is taking place here.

Figure 4 *Cuckoos lay their eggs (bottom row) in the nests of magpies whose eggs are shown on the upper row*

8.7 Speciation

Learning objectives:

- What is speciation?
- What is geographical isolation?
- How can geographical isolation lead to the formation of new species?

Specification reference: 3.4.8

Link

The concept of a species is covered in Topic 14.1 of the *AS Biology* book.

Speciation is the evolution of new species from existing species. A species is a group of individuals that share similar **genes** and are capable of breeding with one another to produce fertile offspring. In other words they belong to the same **gene pool**.

Any species consists of one or more populations. Within each population of a species, individuals breed with one another. Although it is possible for them to breed with individuals from other populations, they breed with each other most of the time. Therefore a single gene pool still exists.

If two populations become separated in some way, the flow of **alleles** between them may cease. The environmental factors that each group encounters may differ. Selection will affect the two populations in different ways and so the type and frequency of the alleles in each will change. Each population will evolve along separate lines. In time, the gene pools of the two populations may become so different that, even if reunited, they would be incapable of successfully breeding with each other. They would have become separate species, each with its own gene pool. Speciation would have taken place. Speciation depends on groups within a population becoming isolated in some way. One such way is geographical isolation.

Geographical isolation

Geographical isolation occurs when a physical barrier prevents two populations from breeding with one another. Such barriers include oceans, rivers, mountain ranges and deserts. What proves a barrier to one species may be no problem for another. While an ocean may isolate populations of hedgehogs it can easily be crossed by species of marine fish. Even the smallest stream may separate two groups of woodlice, whereas the whole Pacific Ocean may fail to isolate certain bird populations. To see how geographical separation of populations can lead to the formation of new species, let us imagine a species X living in an area of forest.

- The individuals of species X form a single gene pool and freely interbreed.
- Climate changes over the centuries lead to drier conditions which reduce the area of forest and separate it into two regions that are many hundreds of kilometres apart.
- Further climate changes cause one forest region (A) to become much colder and wetter and the other forest region (B) to become warmer and drier.
- In region A, **phenotypes** are selected that are better able to survive in colder, wetter conditions.
- In region B, different phenotypes are selected – ones that are better able to survive in warmer, drier conditions.
- The type and frequency of the alleles in the gene pools of each group of species X become increasingly different.
- In time, the differences between the two gene pools become so great that they are, in effect, separate species.
- Further climate change and regrowth of the forest may lead to the two species being reunited. However, they will not be able to interbreed as they are now different species.

These events are summarised in Figure 1, opposite.

1 Species X occupies a forest area. Individuals within the forest form a single gene pool and freely interbreed.

2 Climatic changes to drier conditions reduce the size of the forest to two isolated regions. The distance between the two regions is too great for the two groups of species X to cross to each other.

Arid grassland

3 Further climatic changes result in one region (Forest A) becoming colder and wetter. Group X_1 adapts to these new conditions. The other region (Forest B) becomes warmer and drier. Group X_2 adapts to these conditions.

4 Continued adaptation leads to evolution of new species – Y and Z.

5 A return to the original climatic conditions results in regrowth of forest. Forests A and B merge and the two groups of species are reunited. The two groups are no longer capable of interbreeding. They are now two species, Y and Z, each with its own gene pool.

Figure 1 *Speciation due to geographical isolation*

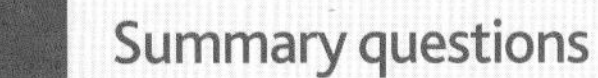

Summary questions

1. What is a species?
2. What is speciation?
3. What is meant by geographical isolation?
4. Explain how geographical isolation of two populations of a species can result in the accumulation of differences in their gene pools.

AQA Examination-style questions

1 (a) Explain what is meant by:

(i) a *recessive* allele;

(ii) *co-dominant* alleles. *(2 marks)*

(b) Chickens homozygous for black feathers (F^BF^B) were crossed with chickens homozygous for white feathers (F^WF^W). These colours are determined by alleles of a single gene. All the F_1 offspring had blue feathers.

When the blue-feathered F_1 chickens were crossed with each other, there were black-feathered, white-feathered and blue-feathered chickens in the F_2 offspring.

(i) Draw a genetic diagram to explain how the F_1 and F_2 phenotypes were produced.

(ii) The number of black-feathered, white-feathered and blue-feathered chickens in the F_2 offspring was counted. The observed ratio of black : white : blue was similar to the ratio expected from theory but not the same.
Explain why observed ratios are not often the same as the expected ratios. *(5 marks)*

AQA, 2003

2 The Old Order Amish of Lancaster County, Pennsylvania, are an isolated human population. Marriages occur almost exclusively within the population. Nearly all can trace their ancestry back to a small group of people who settled in the area in the 18th century.

Microcephaly is a condition which occurs in this population with a frequency of 1 in every 480 births. It is caused by a recessive allele of a single gene. Sufferers usually die within six months of birth.

(a) The incidence of microcephaly in this population is very high compared to non-isolated populations. Suggest **two** reasons for this high incidence. *(2 marks)*

(b) (i) A student used the Hardy–Weinberg equation to estimate the percentage of parents who are heterozygous for microcephaly in this population. What answer should the student have obtained? Show your working.

(ii) The answer to part (b)(i) is likely to be lower than the actual percentage of parents heterozygous for microcephaly in this population. Explain why. *(4 marks)*

AQA, 2007

3 The Hawaiian Islands are 3000 km from the nearest continent. The islands were formed relatively recently by volcanic activity. They have patches of forest separated by wide lava flows. Due to high mountains, the climate varies greatly over short distances. Five hundred species of fruitfly are found in Hawaii.

(a) Explain how the large number of fruitfly species might have evolved in Hawaii. *(6 marks)*

(b) There are 21 833 species of insects in Britain but only 6 500 in Hawaii. Britain however, has 32 species of *Drosophila* (fruitfly) but Hawaii has 500.
Suggest an evolutionary explanation for the difference in the number of species of *Drosophila*. *(4 marks)*

AQA, 2007

4 A species of snake breeds on the shore of a very large lake and on the shore of an isolated island in the lake. There are two forms, a banded form (**B**) and an unbanded form (**U**). These are shown in **Figure 1**.

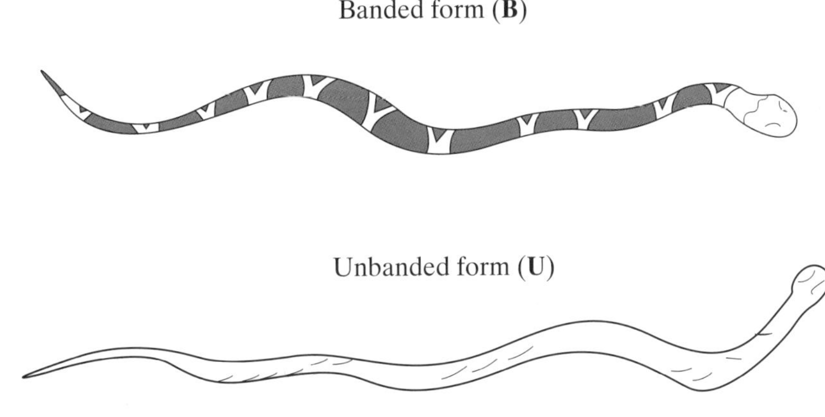

Figure 1

The frequency of the phenotypes was investigated in the two areas. The results are shown in **Figure 2**.

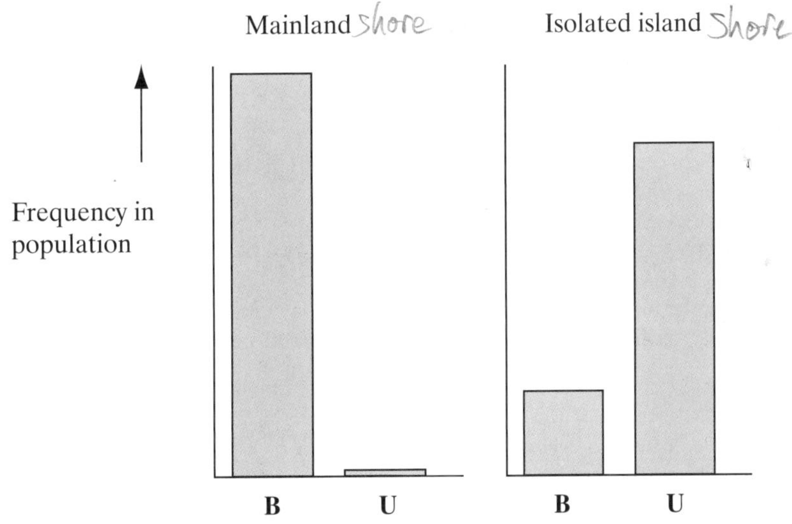

Figure 2

(a) The snakes are preyed upon by birds. Explain how natural selection may have produced the differences in frequency shown in the bar charts. *(4 marks)*

(b) On the island, banded snakes have remained at approximately the same frequency for many generations. Suggest **two** explanations for this. *(2 marks)*

AQA, 2007

Examination-style questions

Unit 4 questions: Populations and the environment

1 (a) What is meant by a community? *(2 marks)*

(b) A farmer stopped using a field for growing crops. Scientists studied succession in the field over the next 30 years. **Figure 1** shows the number of species of Hemiptera (an order of insects) present during that period.

Explain the increase in the number of species of Hemiptera. *(3 marks)*

(c) To calculate a diversity index at a given time, it is necessary to know the number of insects in each population. Name **one** method that could be used to estimate the total number of insects in a population. *(1 mark)*

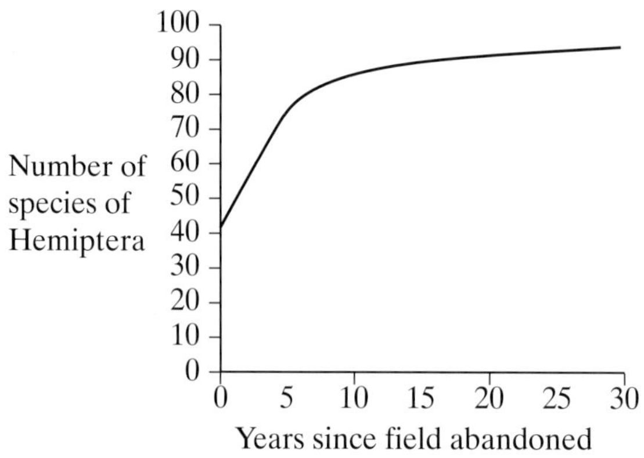

Figure 1

AQA, 2005

2 The table shows some information about the population of Japan and Nigeria.

Country	Population in 1999 / millions	Estimated population in 2015 / millions	Fertility rate / children per woman	Life expectancy at birth / years	Under-5 mortality / deaths per 1000
Japan	126.5	126.1	1.4	80	4
Nigeria	108.9	153.3	5.1	50	187

(a) **Figure 2** shows a population pyramid for Japan in 1999.

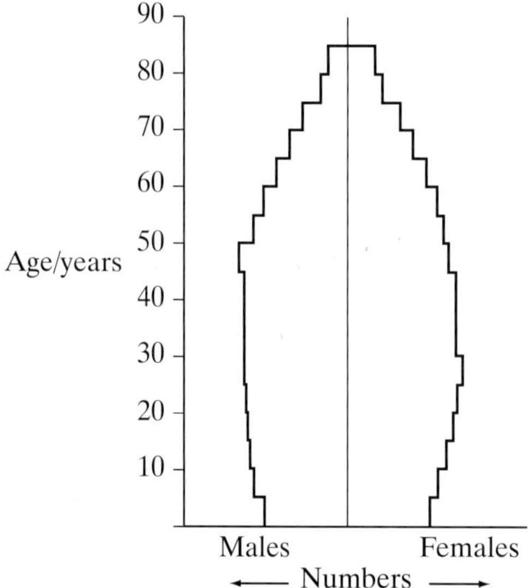

Figure 2

On a copy of **Figure 2**, sketch the population pyramid as you would expect it to appear in 2015. *(2 marks)*

(b) (i) Use information in the table to explain why the population of Nigeria is expected to increase to over 150 million by 2015.

(ii) Suggest and explain **two** factors that could result in the actual increase in the population of Nigeria being less than estimated. *(4 marks)*

AQA, 2003

3 **Figure 3** shows a summary of the light-independent reaction of photosynthesis.

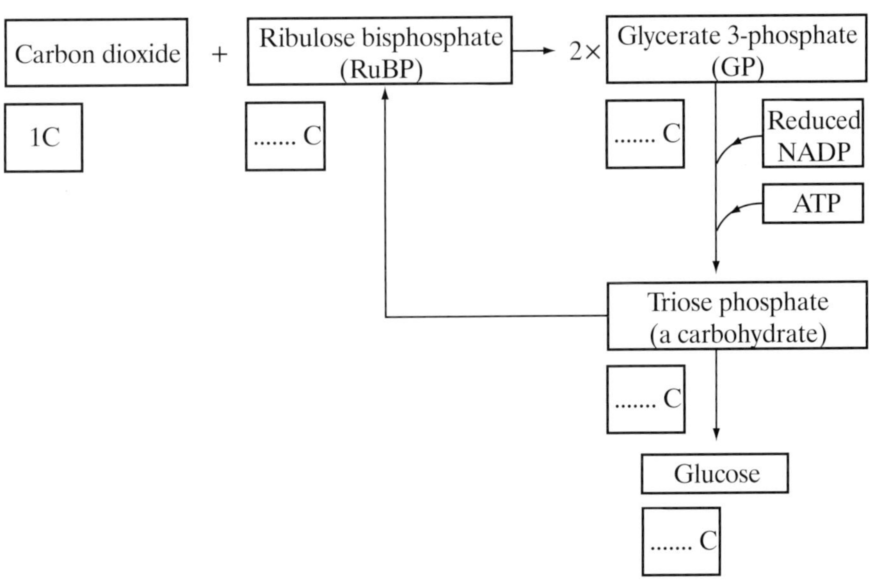

Figure 3

(a) (i) Copy **Figure 3** and complete the boxes to show the number of carbon atoms in the molecules.

(ii) In which part of a chloroplast does the light-independent reaction occur?

(iii) Which process is the source of the ATP used in the conversion of glycerate 3-phosphate (GP) to triose phosphate?

(iv) What proportion of triose phosphate molecules is converted to ribulose bisphosphate (RuBP)? *(5 marks)*

(b) Lowering the temperature has very little effect on the light-dependent reaction, but it slows down the light-independent reaction. Explain why the light-independent reaction slows down at low temperatures. *(2 marks)*

AQA, 2004

4 (a) Mitochondria in muscle cells have more cristae than mitochondria in skin cells.
Explain the advantage of mitochondria in muscle cells having more cristae. *(2 marks)*

(b) Substance **X** enters the mitochondrion from the cytoplasm. Each molecule of substance **X** has three carbon atoms.

(i) Name substance **X**.

(ii) In the link reaction, substance **X** is converted to a substance with molecules effectively containing only two carbon atoms. Describe what happens in this process. *(3 marks)*

(c) The Krebs cycle, which takes place in the matrix, releases hydrogen ions. These hydrogen ions provide a source of energy for the synthesis of ATP, using coenzymes and carrier proteins in the inner membrane of the mitochondrion.
Describe the roles of the coenzymes and carrier proteins in the synthesis of ATP. *(4 marks)*

AQA, 2004

5 (a) The table on the next page shows the effect of different concentrations of inorganic nitrogen fertiliser on the yield of spinach plants.

Nitrogen fertiliser applied/ kg ha^{-1}	40	110	160	210	260	310	360
Yield/ tonnes ha^{-1}	8	13	18	22	24	26	26

(i) Describe and explain the effect of adding inorganic nitrogen fertiliser on the yield of spinach.

(ii) Give **two** advantages of using an inorganic fertiliser, rather than manure. *(4 marks)*

(b) Fertilisers may leach out of farmland into freshwater streams and lakes.

Explain how this can be harmful to the environment. *(6 marks)*

A laboratory investigation was carried out into the relationship between a population of glasshouse whitefly and its wasp parasite. The results are shown in **Figure 4**.

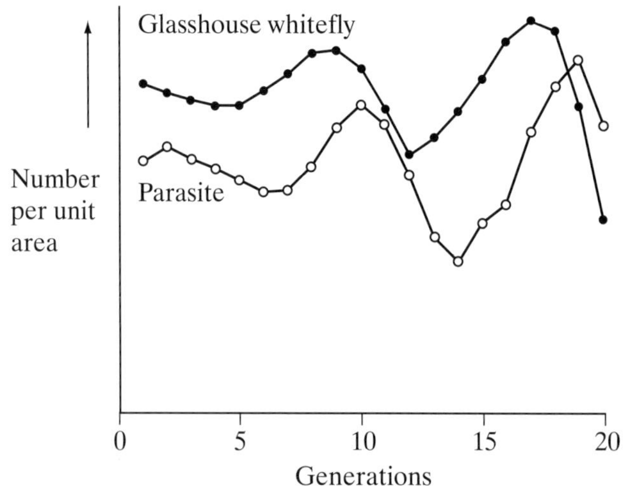

Figure 4

(c) (i) Explain the changes which took place in both populations between generations 6 and 13.

(ii) The wasp parasite could be used as a biological control agent for the glasshouse whitefly. What is *biological control*? *(5 marks)*

AQA, 2005

6 In a sand dune succession the pioneer community (**A**) colonises bare sand. This community is replaced over time by other communities (**B** and **C**) until a climax community of woodland (**D**) is formed.

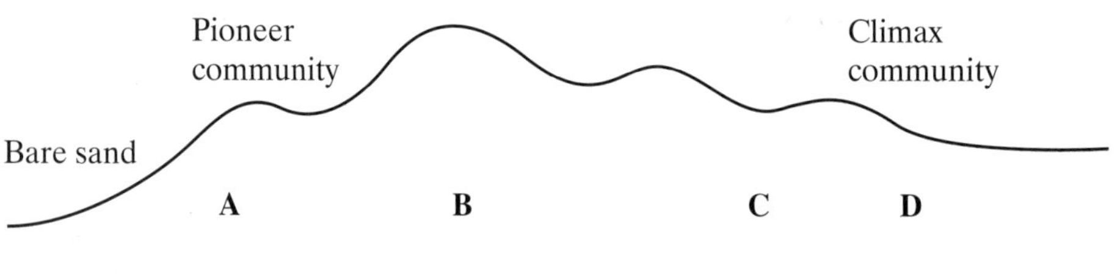

Figure 5

(a) The communities **A** to **D** are composed of different species. Explain how the change in species composition occurs in a succession. *(3 marks)*

(b) Which community, **A** to **D**, is the most stable? Explain what makes this the most stable community. *(2 marks)*

(c) Many species in the pioneer community are xerophytes. Suggest and explain how having sunken stomata is an advantage to these plants. *(3 marks)*

(d) Explain why it would be more appropriate to use a transect rather than random quadrats when investigating this succession. *(1 mark)*

AQA, 2006

7 (a) A protein found on red blood cells, called antigen G, is coded for by a dominant allele of a gene found on the X chromosome. There is no corresponding gene on the Y chromosome.

The members of one family were tested for the presence of antigen G in the blood. The antigen was found in the daughter, her father and her father's mother, as shown in the genetic diagram below. No other members had the antigen.

	Grandmother (has antigen G)	Grandfather	Grandmother	Grandfather
Genotypes	.... or	....	....	....
Gamete genotypes	.... or	....	....	

	Father (has antigen G)	Mother
Genotypes	....	....
Gamete genotypes	....	....

	Daughter (has antigen G)
Genotype	....

(i) One of the grandmothers has two possible genotypes. Copy the genetic diagram and write these on using the symbol $\mathbf{X^G}$ to show the presence of the allele for antigen G on the X chromosome, and $\mathbf{X^g}$ for its absence.

(ii) Complete the rest of the diagram.

(iii) The mother and the father have a son. What is the probability of this son inheriting antigen G? Explain your answer. *(6 marks)*

(b) During meiosis, when the X and Y chromosomes pair up, they do not form a typical bivalent as do other chromosomes. Explain why. *(2 marks)*

AQA, 2004

8 (a) Explain how natural selection produces changes within a species. *(4 marks)*

(b) Malaria is caused by a parasite that attacks red blood cells, producing repeated bouts of serious illness and often causing death. The allele for normal haemoglobin in red cells is $\mathbf{Hb^A}$. In the West African country of Burkina Faso, twenty percent of people are heterozygous for a different allele, $\mathbf{Hb^C}$, which has no effect on their health. People homozygous for $\mathbf{Hb^C}$ suffer a very mild anaemia. **Figure 6** shows how the $\mathbf{Hb^C}$ allele affects the chance of getting malaria.

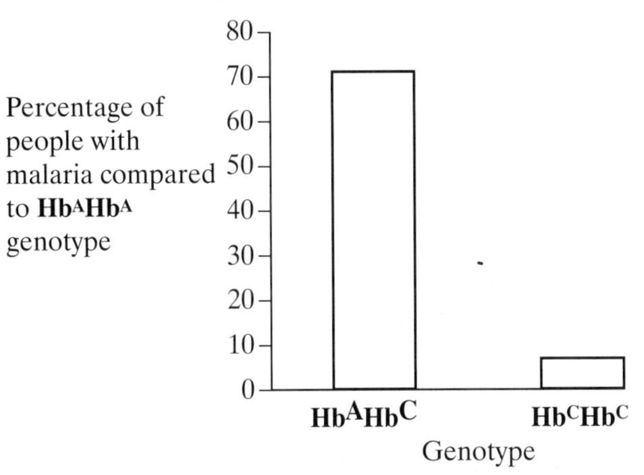

Figure 6

(i) Use the bar chart in **Figure 6** to describe how the $\mathbf{Hb^C}$ allele affects the chance of getting malaria.

(ii) Malaria is very common in Burkina Faso. The $\mathbf{Hb^C}$ allele is increasing in frequency in this part of Africa. Suggest an explanation for this. *(4 marks)*

AQA, 2003

UNIT 5

Control in cells and in organisms

Chapters in this unit

Introduction

Multicellular organisms are able to respond to stimuli that originate both from outside and from within their bodies. By so doing, they can avoid harmful environments while at the same time maintaining an internal environment that provides the optimum conditions for their metabolism. These organisms control their activities through a combination of growth factors, hormones and nervous impulses. At a cellular level, control is achieved by regulating which genes of the genome are transcribed and translated, and when this takes place. The ability of humans to manipulate the transcription and translation of genes has many medical and technological applications.

This unit explores how science works using both information embedded in the text and specific applications designed to cover particular elements of this aspect of the specification.

Response to stimuli describes how organisms detect and respond to both internal and external stimuli. It includes tropisms, taxes and kineses, as well as the reflex arc, the role of internal receptors in controlling heart rate and how organisms respond to external mechanical stimuli.

Coordination considers the differences between nervous and hormonal control. It gives an account of a nerve impulse and how it is transmitted along the axon of a neurone and across a synapse.

Muscle contraction is an account of the structure of skeletal muscle and the mechanism by which it contracts.

Homeostasis explores the principles of how a constant internal environment is maintained and describes how body temperature and blood glucose levels are regulated. Diabetes and its control are also covered.

Feedback mechanisms compares negative and positive feedback within the context of the control of the oestrous cycle in mammals.

Genetic control of protein structure and function gives an account of the structure of RNA and its role in transcription and translation during polypeptide synthesis. This leads on to an account of how genes can mutate and the consequences of such mutations, including cancer.

Control of gene expression explains how DNA translation is controlled. It looks at totipotent cells, such as stem cells, and how cells become specialised to a particular function. The regulation of transcription and translation is also considered.

DNA technology describes how genes can be cloned and the use of such genes in medical techniques such as gene therapy. Other aspects include the use of DNA probes and DNA hybridisation in medical diagnosis, as well as the use of genetic fingerprinting for medical, forensic and breeding purposes.

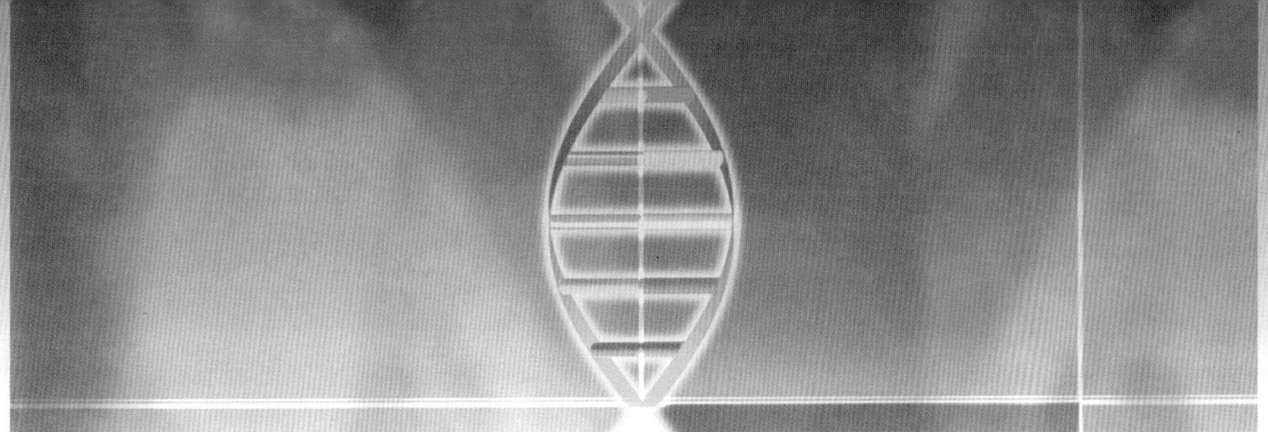

What you already know

While the material in this unit is intended to be self-explanatory, there is certain information from GCSE that will prove very helpful to the understanding of the content of this unit. In addition to all AS material, a knowledge of the following elements of GCSE will be of assistance:

- The nervous system including receptors, the nerve impulse, synapses and reflex actions.
- Hormones and their role in regulating blood sugar levels and the menstrual cycle.
- The control of body temperature.
- Diabetes and its treatment.
- Cell differentiation.
- Gene cloning and genetic engineering.

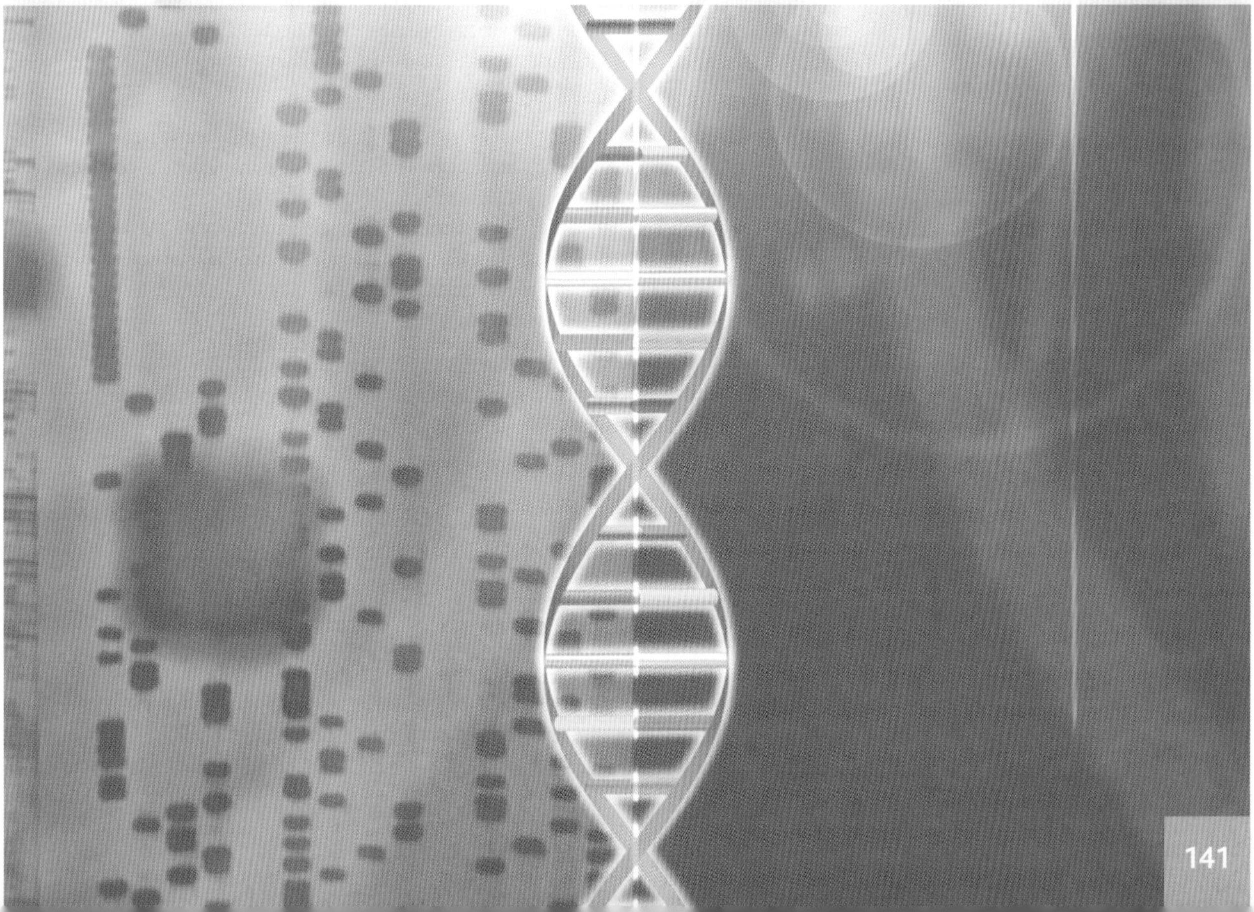

9 Response to stimuli

9.1 Sensory reception

Learning objectives:

- What are a stimulus and a response?
- What is the advantage to organisms of being able to respond to stimuli?
- What are taxes, kineses and tropisms?
- How does each type of response increase an organism's chances of survival?

Specification reference: 3.5.1

In this chapter we shall consider how internal and external stimuli are detected by organisms and how they lead to a response.

Stimulus and response

A **stimulus** is a detectable change in the internal or external environment of an organism that produces a **response** in the organism.

The ability to respond to stimuli increases the chances of survival for an organism. For example, to be able to detect and move away from harmful stimuli, such as predators and extremes of temperature, or to detect and move towards a source of food clearly aid survival. Those organisms that survive have a greater chance of raising offspring and of passing their **alleles** to the next generation. There is always, therefore, a **selection pressure** favouring organisms with more appropriate responses.

Stimuli are detected by cells or organs known as **receptors**. Receptors transform the energy of a stimulus into some form of energy that can be processed by the organism and leads to a response. The response is carried out by one or more of a range of different cells, tissues, organs and systems. These are known as **effectors**. Receptors and effectors are often some distance apart and therefore some form of communication between the two is needed if the organism is to respond effectively. One means of communication occurs via chemicals called hormones (see Topic 10.1), which is a relatively slow process found in both plants and animals.

Animals have another, more rapid, means of communication: the nervous system. Their nervous system usually has many different receptors and effectors. Each receptor and effector is linked to a central **coordinator** of some type. The coordinator acts like a 'switchboard', connecting information from each receptor with the appropriate effector. The sequence of events from stimulus to response can therefore involve either chemical control or nerve cells and may be summarised as:

stimulus ⟶ receptor ⟶ coordinator ⟶ effector ⟶ response

Let us look first at the simplest forms of response to stimuli and how they can increase an organism's chances of survival.

Hint

Plants respond to stimuli, but their receptors produce chemicals and not nerve impulses, and their effectors respond by growing and not by muscle contraction. This means that plants usually respond more slowly than animals.

Taxes

A **taxis** is a simple response whose direction is determined by the direction of the stimulus. As a result, a motile organism (or a separate motile part of it) responds directly to environmental changes by moving its whole body either towards a favourable stimulus or away from an unfavourable one. Taxes are classified according to whether the movement is towards the stimulus (positive taxis) or away from the stimulus (negative taxis) and also by the nature of the stimulus. Some examples are given on the next page:

- Single-celled algae will move towards light (positive phototaxis). This increases their chances of survival since, being photosynthetic, they need light to manufacture their food.
- Earthworms will move away from light (negative phototaxis). This increases their chances of survival because it takes them into the soil, where they are better able to conserve water, find food and avoid predators.
- Some species of bacteria will move towards a region where glucose is more highly concentrated (positive chemotaxis). This increases their chances of survival because they use glucose as a source of food.

Kineses

A **kinesis** is a form of response in which the organism does not move towards or away from a stimulus. Instead, the more unpleasant the stimulus, the more rapidly it moves and the more rapidly it changes direction. A kinesis therefore results in an increase in random movements. This type of response is designed to bring the organism back into favourable conditions. It is important when a stimulus is less directional. Humidity and temperature, for example, do not always produce a clear gradient from one extreme to another.

An example of a kinesis occurs in woodlice. Woodlice lose water from their bodies in dry conditions. When they are in a dry area they move more rapidly and change direction more often. This increases their chances of moving into a different area. If this different area happens to be moist, they slow down and change direction less often. This means that they are likely to stay where they are. In this way they spend more time in favourable moist conditions than in less favourable drier ones. This prevents them drying out and so increases their chances of survival.

Figure 1 *Woodlice exhibit a behaviour called kinesis, which ensures that they spend most of their time in the dark moist conditions that prevent them from drying out and hence aid their survival*

Tropisms

A **tropism** is a growth movement of part of a plant in response to a directional stimulus. In almost all cases the plant part grows towards (positive response) or away from (negative response) the stimulus. Again, the type of response is named after the stimulus. Examples, and the survival value of the response, include the following:

- Plant shoots grow towards light (positive phototropism) so that their leaves are in the most favourable position to capture light for photosynthesis.
- Plant roots grow away from light (negative phototropism) and towards gravity (positive geotropism). In both cases the response increases the probability that roots will grow into the soil, where they are better able to absorb water and mineral ions.
- Plant roots grow towards water (positive hydrotropism) so that, within the soil, root systems will develop where there is most water.

Figure 2 *Cress seedlings exhibiting phototropism. The seedlings in the pot at the top have been grown with equal light from all directions. The seedlings in the pot at the bottom have been grown with light directed at them from the left-hand side*

Summary questions

For each of the following statements, name the type of response described and the survival value of the response.

1. Some species of bacteria move away from the waste products that they produce.
2. The sperm cells of a moss plant are attracted towards a chemical produced by the female reproductive organ of another moss plant.
3. The young stems of seedlings grow away from gravity.

9.2 Nervous control

Learning objectives:

- How does a simple reflex arc work?
- What roles do sensory, intermediate and motor neurones play in a reflex arc?
- How do reflex arcs prevent damage to the body?

Specification reference: 3.5.1

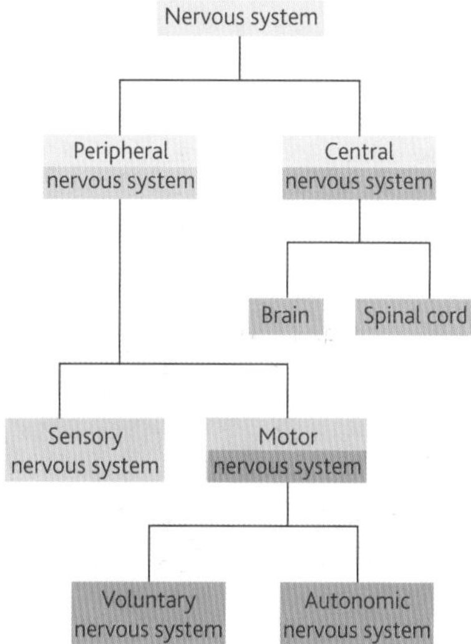

Figure 1 *Nervous organisation*

The simplest type of nervous response to a stimulus is a reflex arc. Before considering how a reflex arc works, it is helpful to understand how the millions of **neurones** in a mammalian body are organised and to be familiar with the structure of the spinal cord.

Nervous organisation

The nervous system has two major divisions:

- the **central nervous system (CNS)**, which is made up of the brain and spinal cord
- the **peripheral nervous system (PNS)**, which is made up of pairs of nerves that originate from either the brain or the spinal cord.

The peripheral nervous system is divided into:

- **sensory neurones**, which carry nerve impulses from receptors towards the central nervous system
- **motor neurones**, which carry nerve impulses away from the central nervous system to **effectors**.

The motor nervous system can be further subdivided as follows:

- the **voluntary nervous system**, which carries nerve impulses to body muscles and is under voluntary (conscious) control
- the **autonomic nervous system**, which carries nerve impulses to glands, smooth muscle and cardiac muscle and is not under voluntary control, that is, it is involuntary (subconscious).

A summary of nervous organisation is given in Figure 1.

The spinal cord

The spinal cord is a column of nervous tissue that runs along the back and lies inside the vertebral column for protection. Emerging at intervals along the spinal cord are pairs of nerves.

The structure of the spinal cord is shown in Figure 2.

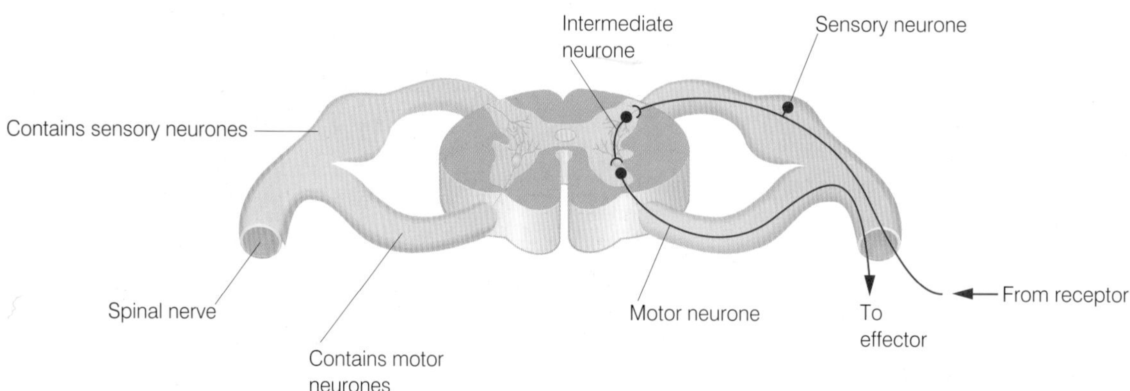

Figure 2 *Section through spinal cord showing the neurones of a reflex arc*

A reflex arc

You will have noticed that you immediately withdraw your hand if you place it on a hot or sharp object. You do not stop to consider any alternative actions. The response is totally involuntary. Indeed, by the time the brain has received nerve impulses from the receptors in the hand, the muscles in the arm have already pulled the hand clear of the danger. This type of involuntary response to a sensory stimulus is called a **reflex**. The pathway of neurones involved in a reflex is known as a **reflex arc**.

Reflex arcs, such as the withdrawal reflex described above, involve just three neurones. One of the neurones is in the spinal cord and so this type of reflex is also called a spinal reflex. The main stages of a spinal reflex arc, such as withdrawing the hand from a hot object, are described below (the numbers relate to the stages shown in Figure 3:

1. the **stimulus** – heat from the hot object
2. a **receptor** – temperature receptors in the skin on the back of the hand, which create a nerve impulse in a sensory neurone
3. a **sensory neurone** – passes the nerve impulse to the spinal cord
4. an **intermediate neurone** – links the sensory neurone to the motor neurone in the spinal cord
5. a **motor neurone** – carries the nerve impulse from the spinal cord to a muscle in the upper arm
6. an **effector** – the muscle in the upper arm, which is stimulated to contract
7. the **response** – pulling the hand away from the hot object.

Hint

Remember the sequence: stimulus, receptor, sensory neurone, intermediate neurone, motor neurone, effector, response.

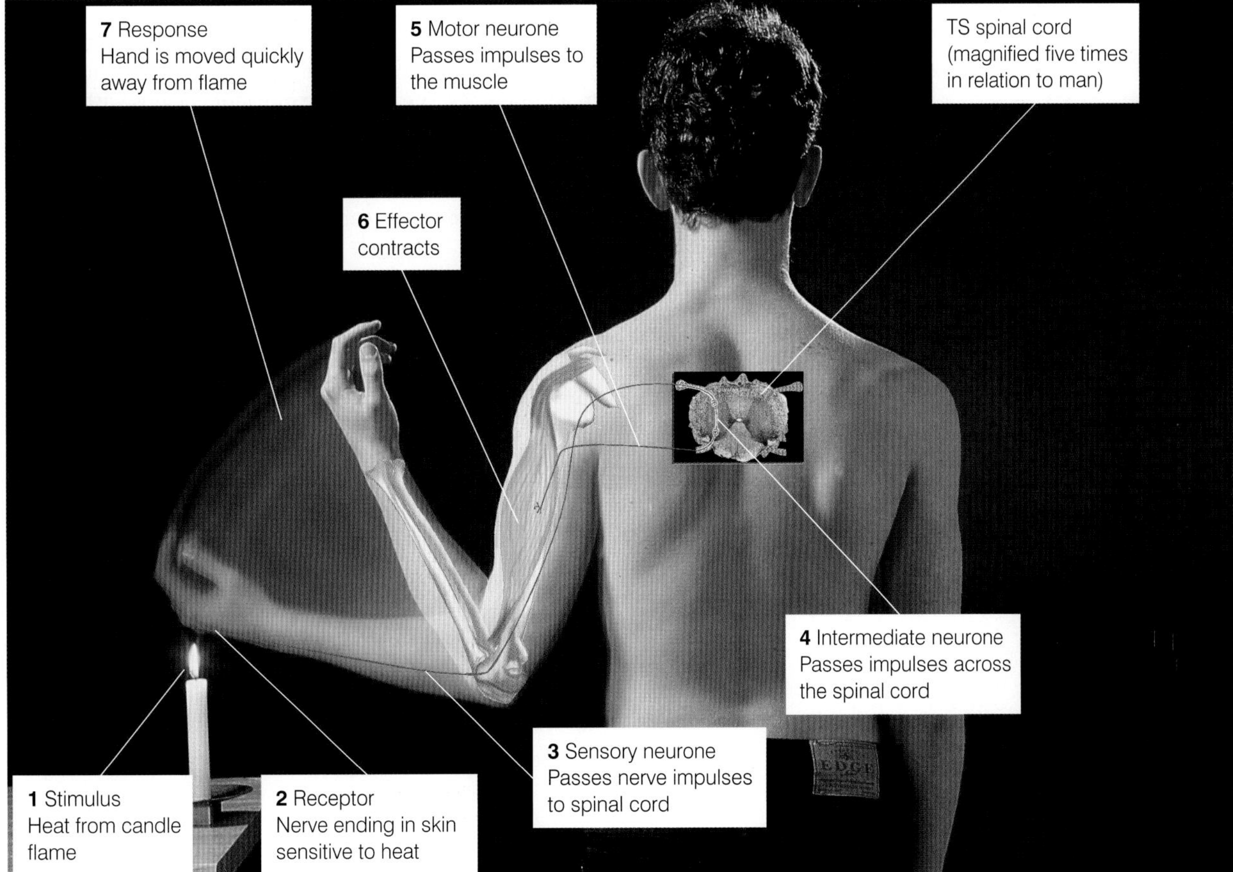

Figure 3 *Reflex arc involved in the withdrawal of the hand from a heat stimulus*

Importance of reflex arcs

Any action that makes survival more likely is clearly of value. Reflexes are involuntary: the actions they control do not need to be 'considered', because there is only one obvious course of action, that is, to remove the hand from the hot object. Reflex actions are important for the following reasons:

- They are involuntary and therefore do not require the decision-making powers of the brain, thus leaving it free to carry out more complex responses. In this way, the brain is not overloaded with situations in which the response is always the same. Some impulses are nevertheless sent to the brain, so that it is informed of what is happening and can sometimes over-ride the reflex if necessary.
- They protect the body from harmful stimuli. They are effective from birth and do not have to be learned.
- They are fast, because the neurone pathway is short with very few, typically one or two, **synapses** (which are the slowest link in a neurone pathway). This is important in withdrawal reflexes.

Summary questions

In the following passage give the word that best replaces the number in brackets.

The nervous system has two main divisions: the central nervous system (CNS), comprising the (1) and (2), and the peripheral nervous system (PNS). The peripheral nervous system is made up of the (3) nerves that carry impulses away from the CNS and (4) nerves that carry impulses towards the CNS. A spinal reflex is an (5) response that involves the spinal cord. An example is the withdrawal reflex, e.g. the withdrawing of the hand from a hot object. The sequence of events begins with the heat from the hot object, which acts as the (6). This is detected by (7) in the skin on the back of the hand, which create a nerve impulse that passes along a (8) neurone into the spinal cord. The impulse then passes to an (9) neurone, in the central region of the spinal cord. The impulse leaves the spinal cord via a (10) neurone. This neurone stimulates a muscle of the upper arm to contract and withdraw the hand from the object. Structures such as these that bring about a response to a stimulus are called (11).

9.3 Control of heart rate

Learning objectives:

- What is the autonomic nervous system?
- How does the autonomic nervous system control heart rate?
- What role do chemical and pressure receptors play in the process?

Specification reference: 3.5.1

Although we are not aware of it, much of the sensory information reaching our central nervous system comes from receptors within our bodies. All the internal systems of our body need to operate efficiently and be ready to adapt to meet the changing demands made upon them. This requires the coordination of a vast amount of information. This information comes from the monitoring of all our internal systems – a process that takes place continuously. Before investigating one example, how heart rate is controlled, let us first look at the part of the nervous system responsible for this type of control: the autonomic nervous system.

The autonomic nervous system

Autonomic means 'self-governing'. The autonomic nervous system controls the involuntary (subconscious) activities of internal muscles and glands. It has two divisions:

- **the sympathetic nervous system**. In general, this stimulates effectors and so speeds up any activity. It acts rather like an emergency controller. It stimulates effectors when we exercise strenuously or experience powerful emotions. In other words, it helps us to cope with stressful situations by heightening our awareness and preparing us for activity (the fight or flight response).
- **the parasympathetic nervous system**. In general, this inhibits effectors and so slows down any activity. It controls activities under normal resting conditions. It is concerned with conserving energy and replenishing the body's reserves.

The actions of the sympathetic and parasympathetic nervous systems normally oppose one another. In other words they are **antagonistic**. If one system contracts a muscle, then the other relaxes it. The activities of internal glands and muscles are therefore regulated by a balance of the two systems. Let us look at one such example: the control of heart rate.

Control of heart rate

The resting heart rate of a typical adult human is around 70 beats per minute. However, it is essential that this rate can be altered to meet varying demands for oxygen. During exercise, for example, the resting heart rate may need to more than double.

Changes to the heart rate are controlled by a region of the brain called the **medulla oblongata**. This has two centres:

- a centre that **increases heart rate**, which is linked to the **sinoatrial node** by the sympathetic nervous system
- a centre that **decreases heart rate**, which is linked to the sinoatrial node by the parasympathetic nervous system.

Which of these centres is stimulated depends upon the information they receive from two types of receptor, which respond to one of the following:

- chemical changes in the blood
- pressure changes in the blood.

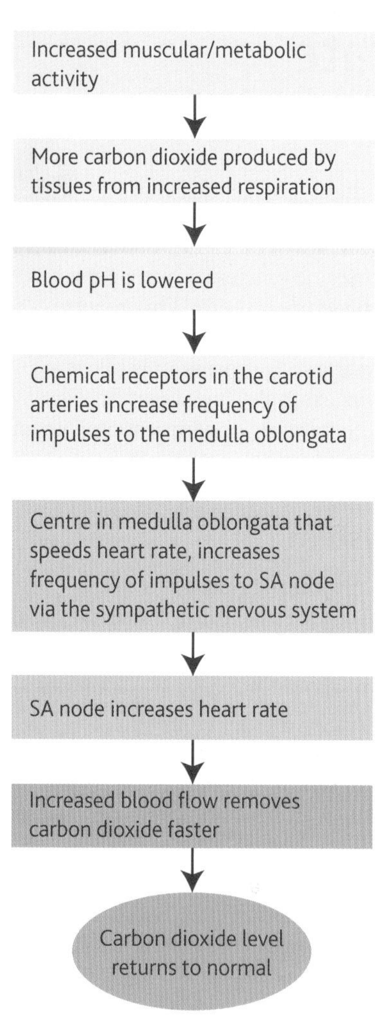

Figure 2 *Effects of exercise on cardiac output (SA node = sinoatrial node)*

Link

The cardiac cycle is covered in Topic 5.2 of the *AS Biology* book. This includes information on how the sinoatrial node controls the heart beat – information that is relevant here.

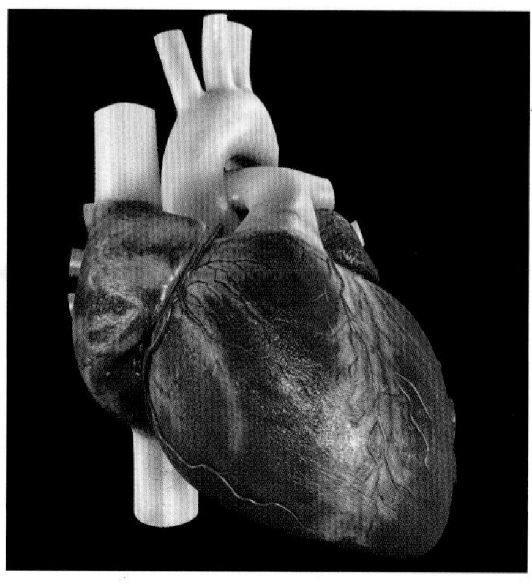

Figure 1 *The rate at which the human heart beats is controlled by receptors that measure the level of carbon dioxide and pressure of the blood*

Control by chemoreceptors

Chemoreceptors are found in the wall of the carotid arteries (the arteries that serve the brain). They are sensitive to changes in the pH of the blood that result from changes in carbon dioxide concentration. In solution, carbon dioxide forms an acid and therefore lowers pH. The process of control works as follows:

- When the blood has a higher than normal concentration of carbon dioxide, its pH is lowered.
- The chemoreceptors in the wall of the carotid arteries and the aorta detect this and increase the frequency of nervous impulses to the centre in the medulla oblongata that increases heart rate.
- This centre increases the frequency of impulses via the sympathetic nervous system to the sinoatrial node which, in turn, increases the heart rate.
- The increased blood flow that this causes leads to more carbon dioxide being removed by the lungs and so the carbon dioxide level of the blood returns to normal.
- As a consequence the pH of the blood rises to normal and the chemoreceptors in the wall of the carotid arteries and aorta reduce the frequency of nerve impulses to the medulla oblongata.
- The medulla oblongata reduces the frequency of impulses to the sinoatrial node, which therefore decreases the heart rate to normal.

This process is summarised in Figure 2, which shows the sequence of events that follows changes in activity levels.

Control by pressure receptors

Pressure receptors occur within the walls of the carotid arteries and the aorta. They operate as follows:

- **When blood pressure is higher than normal,** they transmit a nervous impulse to the centre in the medulla oblongata that decreases heart rate. This centre sends impulses via the parasympathetic nervous system to the sinoatrial node of the heart, which decreases the rate at which the heart beats.
- **When blood pressure is lower than normal**, they transmit a nervous impulse to the centre in the medulla oblongata that increases heart rate. This centre sends impulses via the sympathetic nervous system to the sinoatrial node, which increases the rate at which the heart beats.

Figure 3 summarises the control of heart rate

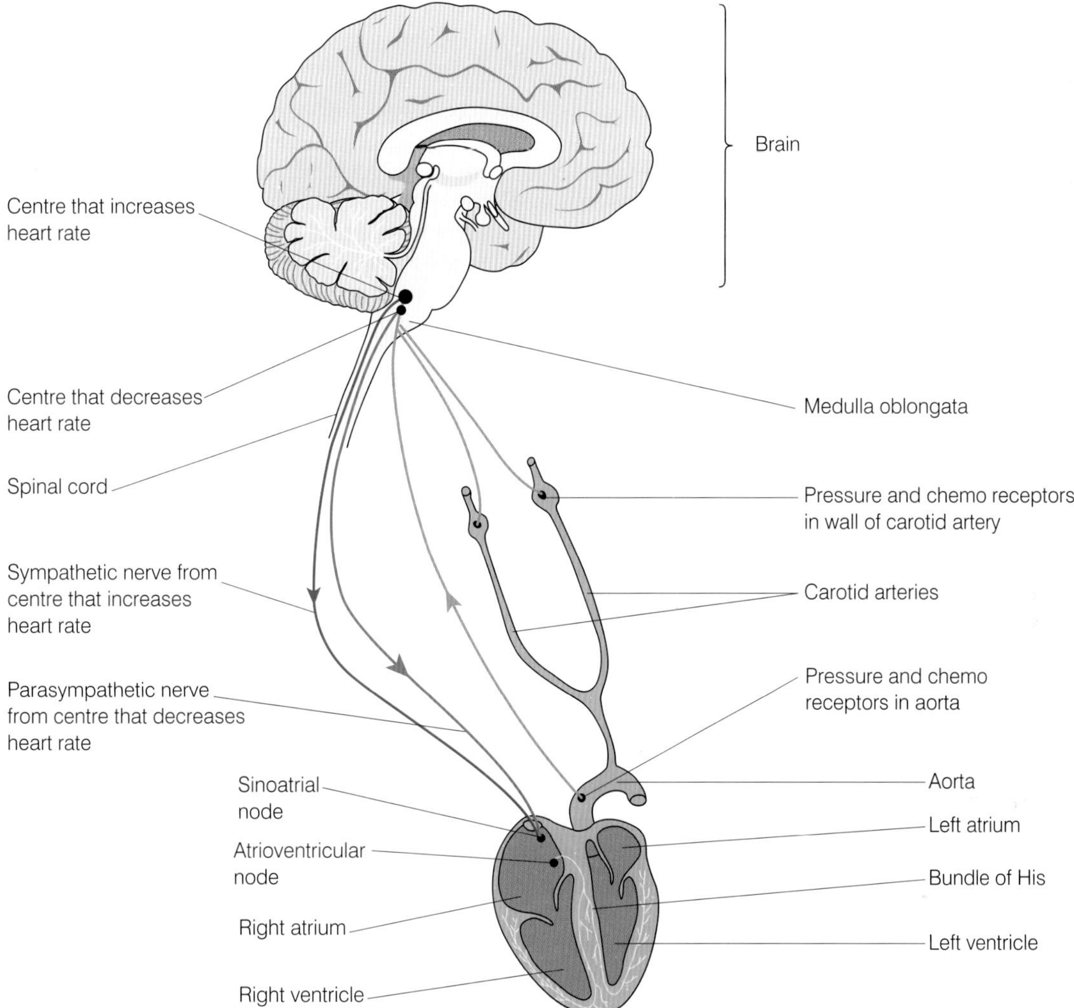

Figure 3 *Control of heart rate*

Summary questions

1. What is the function of the autonomic nervous system?
2. Distinguish between the functions of the sympathetic and parasympathetic nervous systems.
3. Suppose the parasympathetic nerve connections from the medulla oblongata to the sinoatrial node were cut. Suggest what might happen if a person's blood pressure increases above normal.
4. The nerve connecting the carotid artery to the medulla oblongata of a person is cut. This person then undertakes some strenuous exercise. Suggest what might happen to the person's:

   a heart rate;

   b blood carbon dioxide concentration.

   Explain your answers.

9.4 Role of receptors

Learning objectives:

- What are the main features of sensory reception?
- What is a Pacinian corpuscle and how does it work?
- How do receptors work together in the eye?

Specification reference: 3.5.1

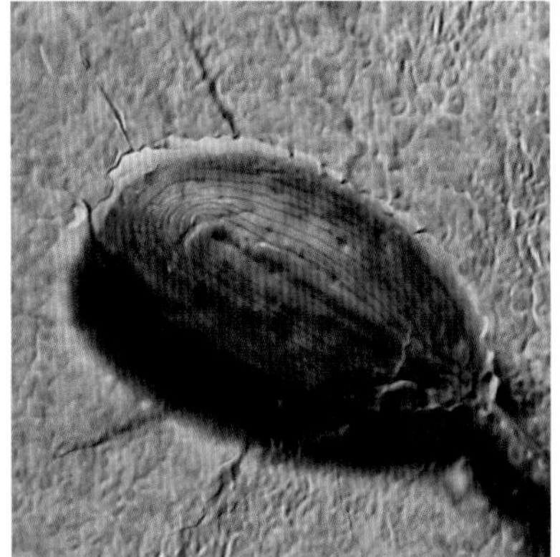

Figure 1 *Pacinian corpuscle*

The central nervous system receives sensory information from its internal and external environment through a variety of sense cells and organs called receptors, each type responding to a different type of stimulus. Sensory reception is the function of these sense organs, whereas sensory perception involves making sense of the information from the receptors. This is largely a function of the brain. The concepts of stimulus and response were covered in Topic 9.1. We shall now look in detail at one receptor: the Pacinian corpuscle.

Features of sensory reception as illustrated by the Pacinian corpuscle

Pacinian corpuscles respond to changes in mechanical pressure. As with all sensory receptors, a Pacinian corpuscle:

- **is specific to a single type of stimulus**. In this case, it responds only to mechanical pressure. It will not respond to other stimuli, such as heat, light or sound.
- **produces a generator potential by acting as a transducer**. All stimuli are forms of information, but unfortunately not forms that the body can understand. It is the role of the transducer to convert the information provided by the stimulus into a form that can be understood by the body, namely nerve impulses. The stimulus is always some form of energy, for example, heat, light, sound or mechanical energy. The nerve impulse is also a form of energy. Receptors therefore convert, or transduce, one form of energy into another. All receptors convert the energy of the stimulus into a nervous impulse known as a **generator potential**. For example, the Pacinian corpuscle, whose action is described below, transduces the mechanical energy of the stimulus into a generator potential.

Structure and function of a Pacinian corpuscle

Pacinian corpuscles respond to mechanical stimuli such as pressure. They occur deep in the skin and are most abundant on the fingers, the soles of the feet and the external genitalia. They also occur in joints, **ligaments** and **tendons**, where they enable the organism to know which joints are changing direction. The single sensory neurone of a Pacinian corpuscle is at the centre of layers of tissue, each separated by a gel. This gives it the appearance of an onion when cut vertically (Figure 2). How does this structure transduce the mechanical energy of the stimulus into a generator potential?

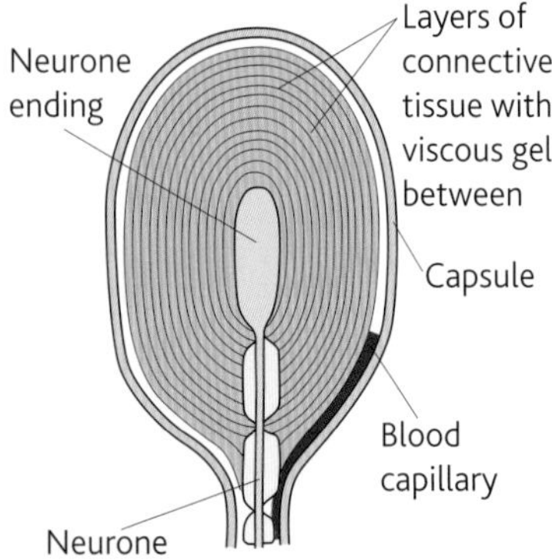

Figure 2 *Structure of a Pacinian corpuscle*

You will have learnt in the AS Biology course that plasma membranes contain proteins that span them. These proteins have channels along which ions can be transported. Some channels carry one specific ion. Sodium channels, for example, carry only sodium ions.

The sensory neurone ending at the centre of the Pacinian corpuscle has a special type of sodium channel in its plasma membrane. This is called a **stretch-mediated sodium channel**. These channels are so-called because their permeability to sodium changes when they change shape, for example, by stretching. The Pacinian corpuscle functions as follows:

- In its normal (resting) state, the stretch-mediated sodium channels of the membrane around the neurone of a Pacinian corpuscle are too narrow to allow sodium ions to pass along them. In this state, the neurone of the Pacinian corpuscle has a resting potential.
- When pressure is applied to the Pacinian corpuscle, it changes shape and the membrane around its neurone becomes stretched (see Figure 3).
- This stretching widens the sodium channels in the membrane and sodium ions diffuse into the neurone.
- The influx of sodium ions changes the potential of the membrane (i.e. it becomes depolarised), thereby producing a generator potential.
- The generator potential in turn creates an action potential (nerve impulse) that passes along the neurone and then, via other neurones, to the central nervous system.

These events are illustrated in Figure 3.

Link

Details of carrier proteins in membranes and how they function to transport ions through them is covered in Topic 3.8, 'Active transport', in the *AS Biology* book. Revision of this material will help you to fully understand what follows here and in the next chapter.

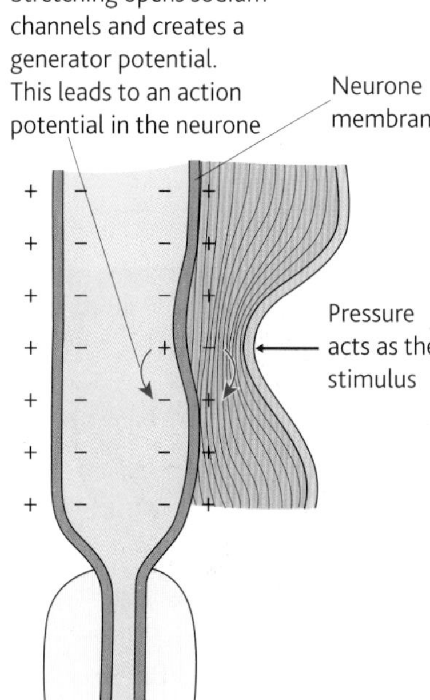

Figure 3 *Creation of a generator potential in a Pacinian corpuscle*

Receptors working together in the eye

We have seen that an individual receptor responds to only one type of stimulus. It also only responds to a certain intensity of stimulus. This means that, if the body is to be able to distinguish between different intensities of a stimulus (e.g. different light intensities), it must have a range of receptors, each responding to a different intensity of stimulus. To illustrate this, let us consider how the eye works.

The light receptor cells of the mammalian eye are found on its innermost layer: the retina. The millions of light receptors found in the retina are of two main types: rod cells and cone cells. Both rod and cone cells act as transducers by converting light energy into the electrical energy of a nerve impulse.

Rod cells

Rod cells cannot distinguish different wavelengths of light and therefore produce images only in black and white. Rod cells are more numerous than cone cells; there are around 120 million in each eye.

Many rod cells share a single sensory neurone (see Figure 4, overleaf). Rod cells can therefore respond to light of very low intensity. This is because a certain threshold value has to be exceeded before a **generator potential** is created in the bipolar cells to which they are attached. As a number of rod cells are attached to a single bipolar cell (= retinal convergence), there is a much greater chance that the threshold value will be exceeded than if only a single rod cell were attached to each bipolar cell. As a result, rod cells allow us to see in low light intensity (i.e. at night), although only in black and white.

In order to create a generator potential, the pigment in the rod cells (rhodopsin) must be broken down. Low-intensity light is sufficient to cause this breakdown. This also helps to explain why rod cells respond to low-intensity light.

A consequence of many rod cells linking to a single bipolar cell is that light received by rod cells sharing the same neurone will only generate a single impulse regardless of how many of the neurones are stimulated. This means that they cannot distinguish between the separate sources of light that stimulated them. Two dots close together will appear as a single blob. Rod cells therefore have low **visual acuity**.

Table 1 *Differences between rod and cone cells*

Rod cells	Cone cells
Rod-shaped	Cone-shaped
Greater numbers than cone cells	Fewer numbers than rod cells
Distribution – more at the periphery of the retina, absent at the fovea	Fewer at the periphery of the retina, concentrated at the fovea
Give poor visual acuity	Give good visual acuity
Sensitive to low-intensity light	Not sensitive to low-intensity light

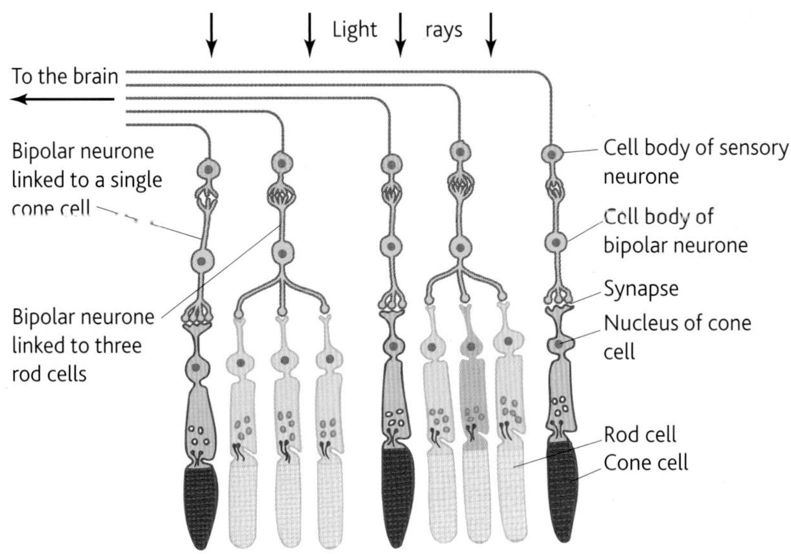

Figure 4 *Microscopic structure of the retina*

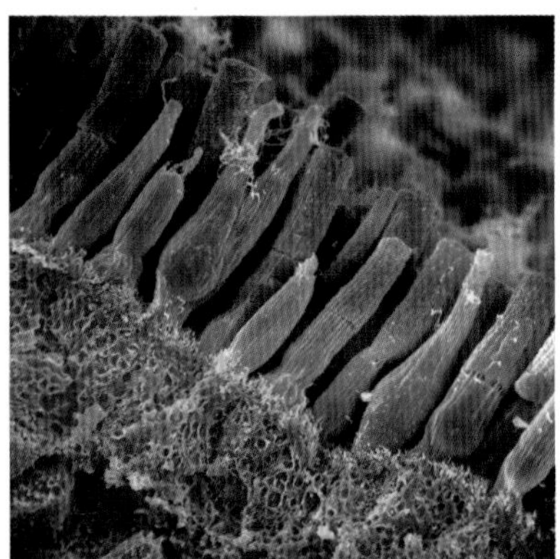

Figure 5 *False colour SEM of rod and cone cells in the retina of the eye. Rod cells (brown) are long nerve cells responding to dim light while cone cells (green) detect colour*

Cone cells

Cone cells are of three different types, each responding to a different wavelength of light. Depending upon the proportion of each type that is stimulated, we can perceive images in full colour.

In each human eye, there are around 6 million cone cells, often with their own separate bipolar cell connected to a sensory neurone (see Figure 4). This means that the stimulation of a number of cone cells cannot be combined to help exceed the threshold value and so create a generator potential. As a result, cone cells only respond to high light intensity and not to low light intensity.

In addition, cone cells contain a different pigment from that found in rod cells. The pigment in cone cells (iodopsin) requires a higher light intensity for its breakdown. Only light of high intensity will therefore break it down and create a generator potential. As only cone cells respond to different wavelengths (colours) of light, this also explains why we cannot see colours in low light intensity (i.e. at night).

Each cone cell has its own connection to a single bipolar cell, which means that, if two adjacent cone cells are stimulated, the brain receives two separate impulses. The brain can therefore distinguish between the two separate sources of light that stimulated the two cone cells. This means that two dots close together will appear as two dots. Therefore cone cells give very accurate vision, that is, they have good visual acuity.

Table 1 summarises the differences between rod and cone cells.

The distribution of rod and cone cells on the retina is uneven. Light is focused by the lens on the part of the retina opposite the pupil. This point is known as the fovea. The fovea therefore receives the highest intensity of light. Therefore cone cells, but not rod cells, are found at the fovea. The concentration of cone cells diminishes further away from the fovea. At the peripheries of the retina, where light intensity is at its lowest, only rod cells are found.

All this shows how the distribution of rod and cone cells, and the connections they make in the optic nerve, can explain the differences in sensitivity and visual acuity in mammals. By having different types of light receptor, each responding to different stimuli, mammals can benefit from good all-round vision both day and night.

Summary questions

1. What is a stretch-mediated sodium channel?
2. Describe the sequence of events by which pressure on a Pacinian corpuscle results in the creation of a generator potential.
3. Explain why brightly coloured objects often appear grey in dim light.
4. At night, it is often easier to see a star in the sky by looking slightly to the side of it rather than directly at it. Suggest why this is so.

AQA Examination-style questions

1 The human body-louse is an insect which lives and feeds on the surface of the skin. A louse was placed in a chamber, half of which was kept at 35 °C and half at 30 °C.

Figure 1 shows the pattern of movement of the louse.

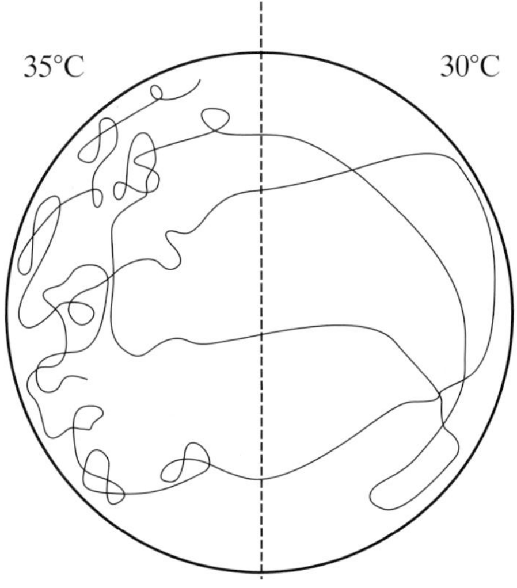

Figure 1

(a) Name the type of behavioural response shown by the body-louse in this investigation. Give evidence for your answer. *(2 marks)*

(b) Suggest and explain **one** advantage of this behaviour to the human body-louse. *(2 marks)*

AQA, 2006

2 A gardener accidentally pricks her finger on a thorn. She quickly pulls the finger away. This reaction results from a simple reflex arc involving three neurones.

(a) **Figure 2** shows part of the pathway involved in this reaction.

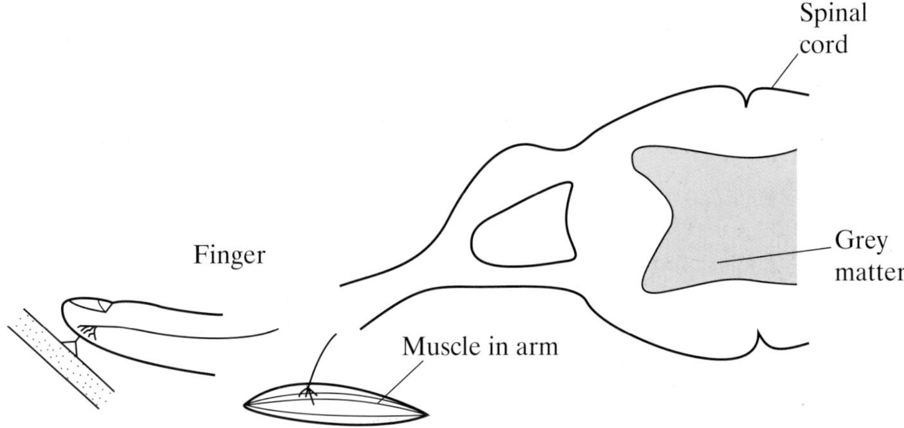

Figure 2

(i) Copy and complete the diagram to show the rest of the simple reflex arc.

On your diagram:

(ii) name and label the three neurones;

(iii) label the effector. *(3 marks)*

(b) (i) What is a reflex?

(ii) Explain the importance of reflex actions. *(4 marks)*

AQA, 2004

3 **Figure 3** shows the results of a sequence of treatments to investigate the control of heart rate by the autonomic nervous system.

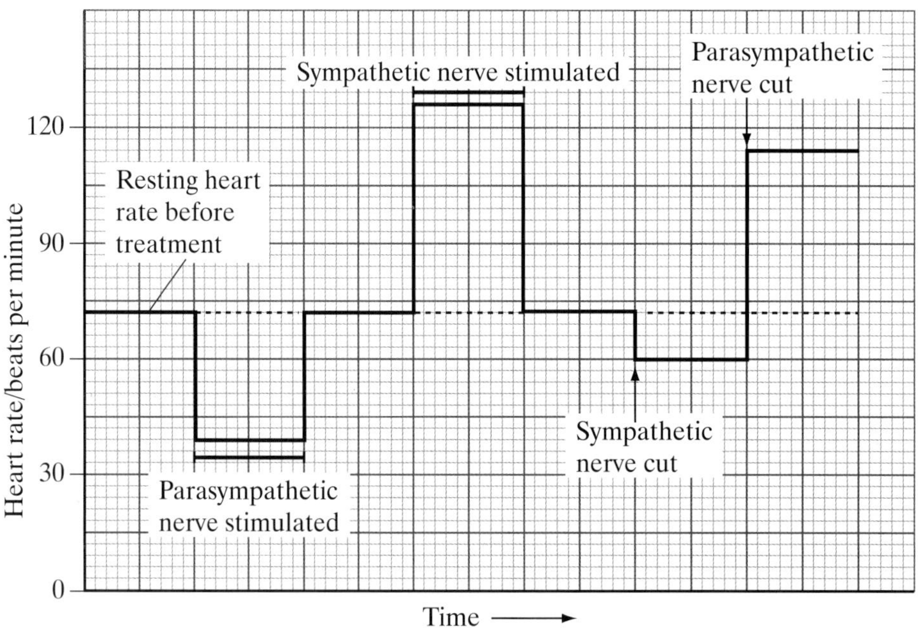

Figure 3

(a) Explain what the results of cutting the sympathetic and parasympathetic nerves demonstrate about the control of resting heart rate. *(3 marks)*

(b) What does **Figure 3** suggest about how a change in heart rate occurs when a person exercises? *(1 mark)*

AQA, 2003

4 (a) Explain how a Pacinian corpuscle produces a generator potential in response to a specific stimulus. *(3 marks)*

(b) Explain how stimulation of chemoreceptors during exercise results in a change in heart rate. *(5 marks)*

5 **Figure 4** shows a section through a human eye. **Figure 5** shows the distribution of rods and cones in the retina of the human eye.

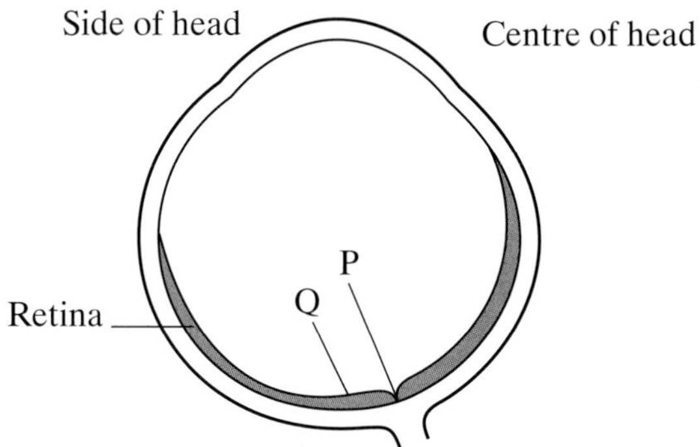

Figure 4

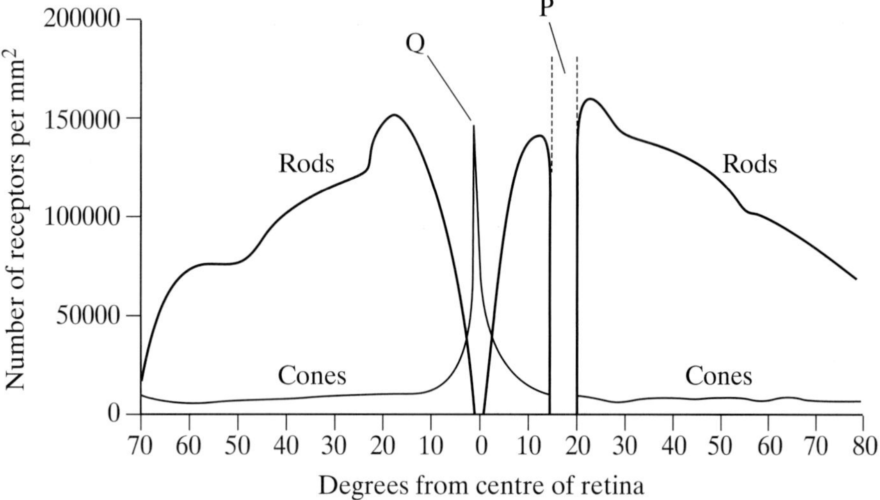

Figure 5

(a) Use **Figures 4** and **5** to explain why:

(i) no image is perceived when rays of light strike the retina at the point marked **P**;

(ii) most detail is perceived when rays of light strike the part of the retina labelled **Q**. *(3 marks)*

(b) Rod cells allow us to see objects in dim light. Explain how the connections of rod cells to neurones in the retina make this possible. *(3 marks)*

AQA, 2003

10 Coordination

10.1 Coordination

Learning objectives:

- How do nervous and hormonal coordination differ?
- What are chemical mediators and how do they work?
- What changes do plants respond too?
- How do plants respond to change?
- What are plant growth factors?

Specification reference: 3.5.2

As species have evolved, their cells have become adapted to perform specialist functions. By specialising in one function, cells have lost the ability to perform other functions. Different groups of cells each carry out their own function. This makes cells dependent upon others to carry out the functions they have lost. Cells specialising in reproduction, for example, depend on other cells to obtain oxygen for their respiration, to provide glucose or to remove their waste products. These different functional systems must be coordinated if they are to perform efficiently. No body system can work in isolation, but all must be integrated in a coordinated fashion. In this chapter we shall look at how this coordination is achieved.

Principles of coordination

There are two main forms of coordination in mammals: the nervous system and the hormonal system:

- **The nervous system** uses nerve cells to pass electrical impulses along their length. They stimulate their target cells by secreting chemicals, known as **neurotransmitters**, directly on to them. This results in rapid communication between specific parts of an organism. The responses produced are often short-lived and restricted to a localised region of the body. An example of nervous coordination is a reflex action, such as the withdrawal of the hand from an unpleasant stimulus. For obvious reasons this type of action, which is covered in Topic 9.2, is rapid, short-lived and restricted to one region of the body.
- **The hormonal system** produces chemicals (hormones) that are transported in the blood plasma to their target cells, which they then stimulate. This results in a slower, less specific form of communication between parts of an organism. The responses are often long-lasting and widespread. An example of hormonal coordination is the control of blood glucose (see Topic 12.3), which produces a slower response but has a more long term and more widespread effect.

Although different, both systems work together and interact with one another. A comparison of the nervous and hormonal systems is given in Table 1.

Hint

The nervous system operates like a telephone system, allowing rapid communication between two specific individuals. The hormonal system can be likened to a nationwide mail shot, sending a slower, more general message to everyone, everywhere, but only those individuals who are sensitive to it respond.

Chemical mediators

The nervous and hormonal systems work well in coordinating the activities of a whole organism. At the cellular level, however, they are complemented by a further form of coordination. This involves substances known as **chemical mediators**. These are chemicals that are released from certain mammalian cells and have an effect on cells in their immediate vicinity. They are typically released by infected or injured cells and cause small arteries and arterioles to dilate. This leads to a rise in temperature and swelling of the affected area – the so-called 'inflammatory response'. Two examples of chemical mediators are:

- **histamine**, which is stored in certain white blood cells and released following injury or in response to an **allergen**, such as pollen. It causes dilation of small arteries and arterioles and increased permeability of capillaries, leading to localised swelling, redness and itching.
- **prostaglandins**, which are found in cell membranes and also cause dilation of small arteries and arterioles. Their release following injury increases the permeability of capillaries. They also affect blood pressure and **neurotransmitters** (substances involved in the transmission of nerve impulses). In so doing, they affect pain sensation.

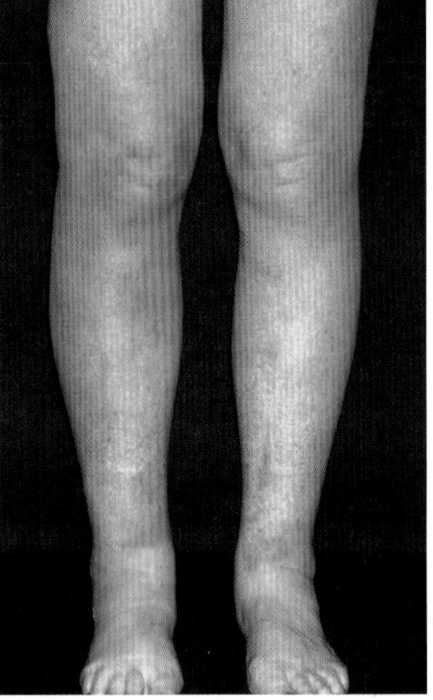

Figure 1 *Inflammation and rash caused by an allergic response*

Table 1 *Comparison of hormonal and nervous systems*

Hormonal system	Nervous system
Communication is by chemicals called hormones	Communication is by nerve impulses
Transmission is by the blood system	Transmission is by neurones
Transmission is usually relatively slow	Transmission is very rapid
Hormones travel to all parts of the body, but only target organs respond	Nerve impulses travel to specific parts of the body
Response is widespread	Response is localised
Response is slow	Response is rapid
Response is often long-lasting	Response is short-lived
Effect may be permanent and irreversible	Effect is temporary and reversible

Plant growth factors

Unlike animals, plants have no nervous system. Nevertheless, in order to survive, plants must respond to changes in both their external and internal environments. For example, plants must respond to:

- **light**. Stems grow towards light (i.e. are positively phototropic) because light is needed for photosynthesis.
- **gravity**. Plants need to be firmly anchored in the soil. Roots are sensitive to gravity and grow in the direction of its pull (i.e. they are positively geotropic).
- **water**. Almost all plant roots grow towards water (i.e. are positively hydrotropic) in order to absorb it for use in photosynthesis and other metabolic processes, as well as for support.

Plants respond to external stimuli by means of plant hormones or, more correctly, **plant growth factors**. The latter term is more descriptive because:

- They exert their influence by affecting growth.
- Unlike animal hormones, they are made by cells located throughout the plant rather than in particular organs.
- Unlike animal hormones, some plant growth factors affect the tissues that release them rather than acting on a distant target organ.

Figure 2 *Geotropism is a plant response to the Earth's gravitational field. This bean plant shows a turn in its stem which occurred after the pot was tipped over. The response also occurs in the dark, showing that it is not phototropism*

Plant growth factors are produced in small quantities. They have their effects close to the tissue that produces them. An example of a plant growth factor is **indoleacetic acid (IAA)**, which, amongst other things, causes plant cells to elongate.

Summary questions

1 State **three** ways in which a response to a hormone differs from a response to a nerve impulse.

2 Name **two** chemical mediators and state the effects that they each have on blood vessels.

3 Suggest **two** advantages to a plant of having roots that respond to gravity by growing in the direction of its pull.

4 State **two** differences between animal hormones and plant growth factors.

Control of tropisms by IAA

We learnt in Topic 9.1 that a tropism is a growth movement of a plant in response to a directional stimulus. In the case of light, we can observe that a young shoot will bend towards light that is directed at it from one side. This response is due to the following sequence of events:

1 Cells in the tip of the shoot produce IAA, which is then transported down the shoot.
2 The IAA is initially transported to all sides as it begins to move down the shoot.
3 Light causes the movement of IAA from the light side to the shaded side of the shoot.
4 A greater concentration of IAA builds up on the shaded side of the shoot than on the light side.
5 As IAA causes elongation of cells and there is a greater concentration of IAA on the shaded side of the shoot, the cells on this side elongate more.
6 The shaded side of the shoot grows faster, causing the shoot to bend towards the light.

IAA also controls the bending of roots in the direction of gravity. However, whereas a high concentration of IAA increases growth in stem cells, it decreases growth in root cells. For example, an IAA concentration of 10 parts per million increases stem growth by 200 per cent but decreases root growth by 100 per cent.

Application and How science works

Discovering the role of IAA in tropisms

The mechanism by which IAA causes shoots to bend towards the light was unravelled in a series of experiments stretching over almost a century. They are a classic example of How science works as follows:

- Scientists use their knowledge and understanding when observing events, in defining a problem, and when questioning the explanations of other scientists (HSW: B).
- Scientists use their observations to form a hypothesis that they test experimentally (HSW: C).
- Evidence from experiments is constantly questioned by scientists and is a stimulus for further investigation involving experimental refinements and new hypotheses (HSW: F).

No less a person than Charles Darwin was one of the earliest scientists to investigate the response of plant shoots to light. He kept birds and so grew grasses that he later fed to them. He observed that young grass shoots bent towards the window. (i.e. they were positively phototropic). Being curious, he proposed the hypothesis that the stimulus of the light was detected by the tip of the shoot, which was therefore the source of the response. He tested his hypothesis in a series of experiments in which he removed the tips of shoots or covered the tips with lightproof covers.

These experiments and the results are summarised in Figure 3 (experiments 1–3).

1 Which of Darwin's three experiments acted as a control?

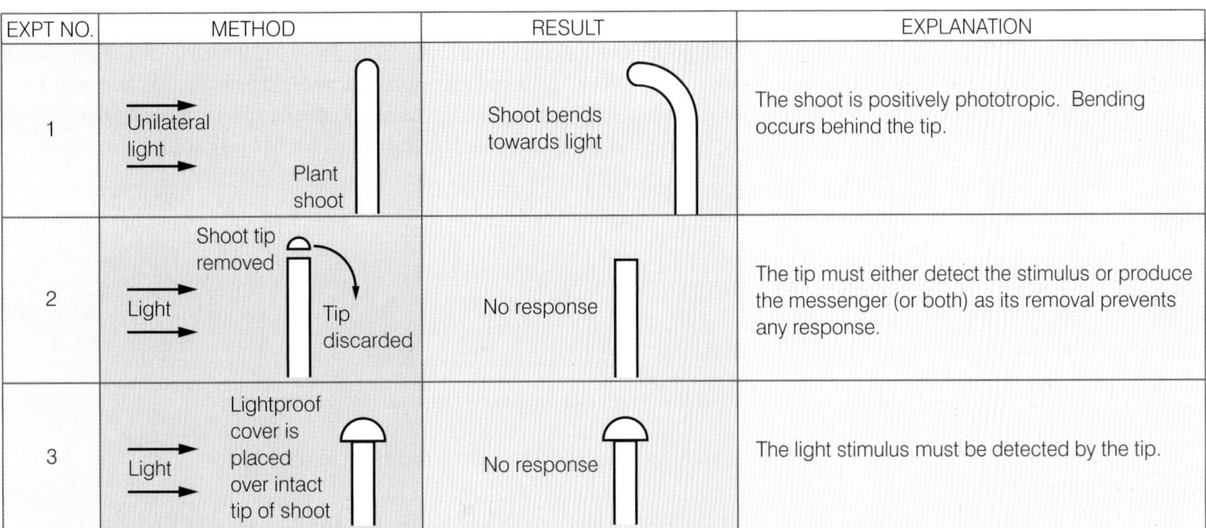

EXPT NO.	METHOD	RESULT	EXPLANATION
1	Unilateral light; Plant shoot	Shoot bends towards light	The shoot is positively phototropic. Bending occurs behind the tip.
2	Light; Shoot tip removed; Tip discarded	No response	The tip must either detect the stimulus or produce the messenger (or both) as its removal prevents any response.
3	Light; Lightproof cover is placed over intact tip of shoot	No response	The light stimulus must be detected by the tip.

Figure 3 *Darwin's experiments to show that it is the tips of shoots that are the source of the phototropic response*

Once it had been shown that the tip is the light-sensitive region of the shoot but that the response (bending) occurs lower down the shoot, some scientists proposed another hypothesis: that a chemical substance was being produced in the tip and transported down the stem, where it caused a response. Others disagreed and put forward an alternative theory: that it was an electrical signal that was passing from the tip and causing the response. One scientist, Peter Boysen-Jensen, carried out a further set of experiments designed to prove the hypothesis that the 'messenger' was a chemical. In these experiments, he used mica, which conducts electricity but not chemicals, and gelatin, which conducts chemicals but not electricity. His experiments and the results are summarised in Figure 4 (experiments 4–6).

EXPT NO.	METHOD	RESULT	EXPLANATION
4	Light; Thin, impermeable barrier of mica	Movement of chemical down shaded side; Bends towards the light	Mica on the illuminated side of the shoot allows the hormone to pass only down the shaded side where it increases growth and causes bending.
5	Light; Mica inserted on shaded side	Movement of chemical down shaded side is prevented by mica; No response	
6	Light; Tip removed, gelatin block inserted and tip replaced; Gelatin block	Movement of chemical down shaded side; Bends towards the light	As gelatin allows chemicals to pass through it, but not electrical messages, the bending which occurs must be due to a chemical passing from the tip.

Figure 4 *Boysen-Jensen's experiments to show the nature of the 'messenger' in the phototropic response*

2 Suggest an explanation for the results in experiment 5.

Boysen-Jenson's experiments stimulated another scientist, Arpad Paal, to investigate how the chemical messenger worked. He removed the tips of shoots and placed them on one side of the cut surface. He kept the shoots in total darkness throughout the experiment. His experiment and its results are shown in Figure 5 (experiment 7).

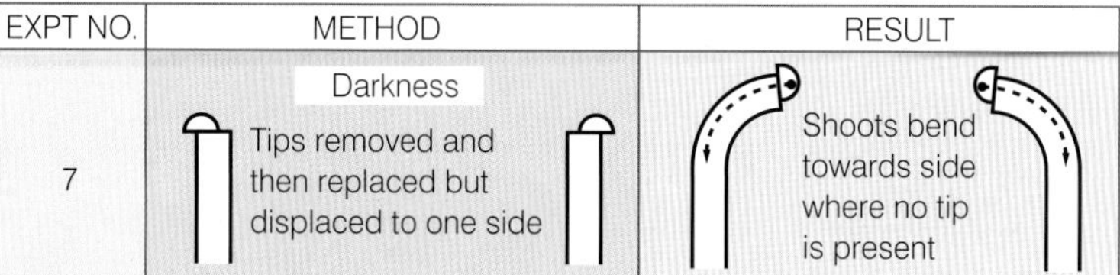

Figure 5 *Paal's experiment on the action of the 'messenger'*

3 **Suggest an explanation for the results in experiment 7.**

So far, it had been established that bending was due to a chemical which was produced in the tip and caused growth on the shaded side of the shoot. This chemical was later shown to be indoleacetic acid (IAA). The next question was how did light cause the uneven distribution of IAA? Different theories were put forward, including:

- Light inhibits IAA production in the tip and so it is only produced on the shaded side.
- Light destroys the IAA as it passes down the light side of the shoot.
- IAA is transported from the light side to the shaded side of the shoot.

This prompted Winslow Briggs and his associates to test these hypotheses. They set up experiments as shown in Figure 6 (experiments 8–10).

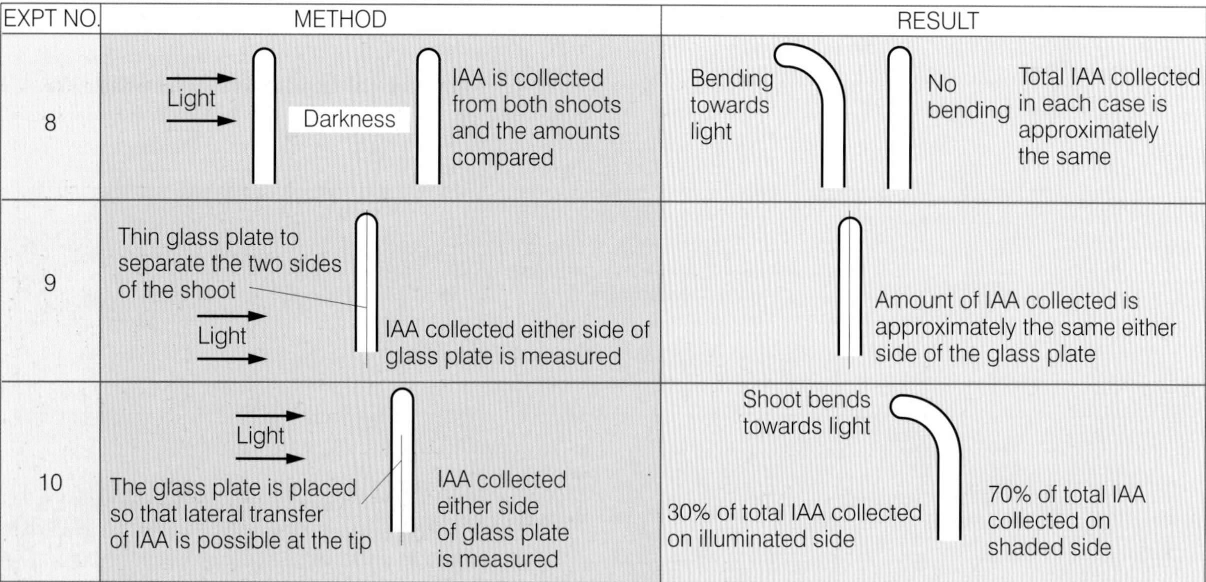

Figure 6 *Briggs's experiments to determine how IAA becomes unevenly distributed*

Study experiments 8, 9 and 10.

4 **Suggest reasons for using a glass plate in experiments 9 and 10.**

5 **State which of the three theories the results tend to support. Give reasons for your answer.**

10.2 Neurones

Learning objectives:

- What is a neurone?
- What is the structure of a myelinated motor neurone?
- What are the different types of neurone?

Specification reference: 3.5.2

Neurones (nerve cells) are specialised cells adapted to rapidly carrying electrochemical changes called **nerve impulses** from one part of the body to another.

The structure of neurones

A mammalian neurone is made up of:

- a **cell body**, which contains a nucleus and large amounts of rough endoplasmic reticulum. This is associated with the production of proteins and **neurotransmitters**
- **dendrons**, small extensions of the cell body which subdivide into smaller branched fibres, called **dendrites**, that carry nerve impulses towards the cell body
- an **axon**, a single long fibre that carries nerve impulses away from the cell body
- **Schwann cells**, which surround the axon, protecting it and providing electrical insulation. They also carry out **phagocytosis** (the removal of cell debris) and play a part in nerve regeneration. Schwann cells wrap themselves around the axon many times, so that layers of their membranes build up around it
- a **myelin sheath**, which forms a covering to the axon and is made up of the membranes of the Schwann cells. These membranes are rich in a lipid known as **myelin**. Neurones with a myelin sheath are called myelinated neurones. Some neurones lack a myelin sheath and are called unmyelinated neurones. Myelinated neurones transmit nerve impulses faster than unmyelinated neurones
- **nodes of Ranvier**, gaps between adjacent Schwann cells where there is no myelin sheath. The gaps are 2–3 µm long and occur every 1–3 mm in humans (see Figure 1).

The structure of a myelinated motor neurone is illustrated in Figure 2 overleaf.

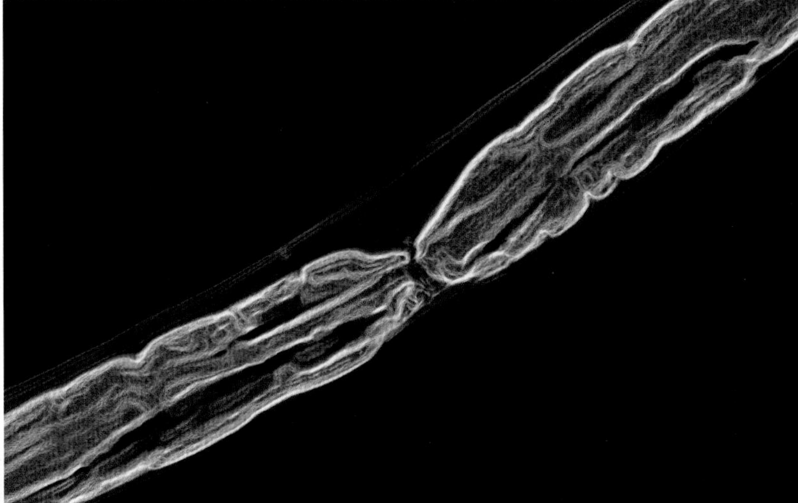

Figure 1 *LM of a node of Ranvier in a neurone. The node is the gap in the centre. The gap is a small area without myelin in an otherwise myelinated nerve fibre*

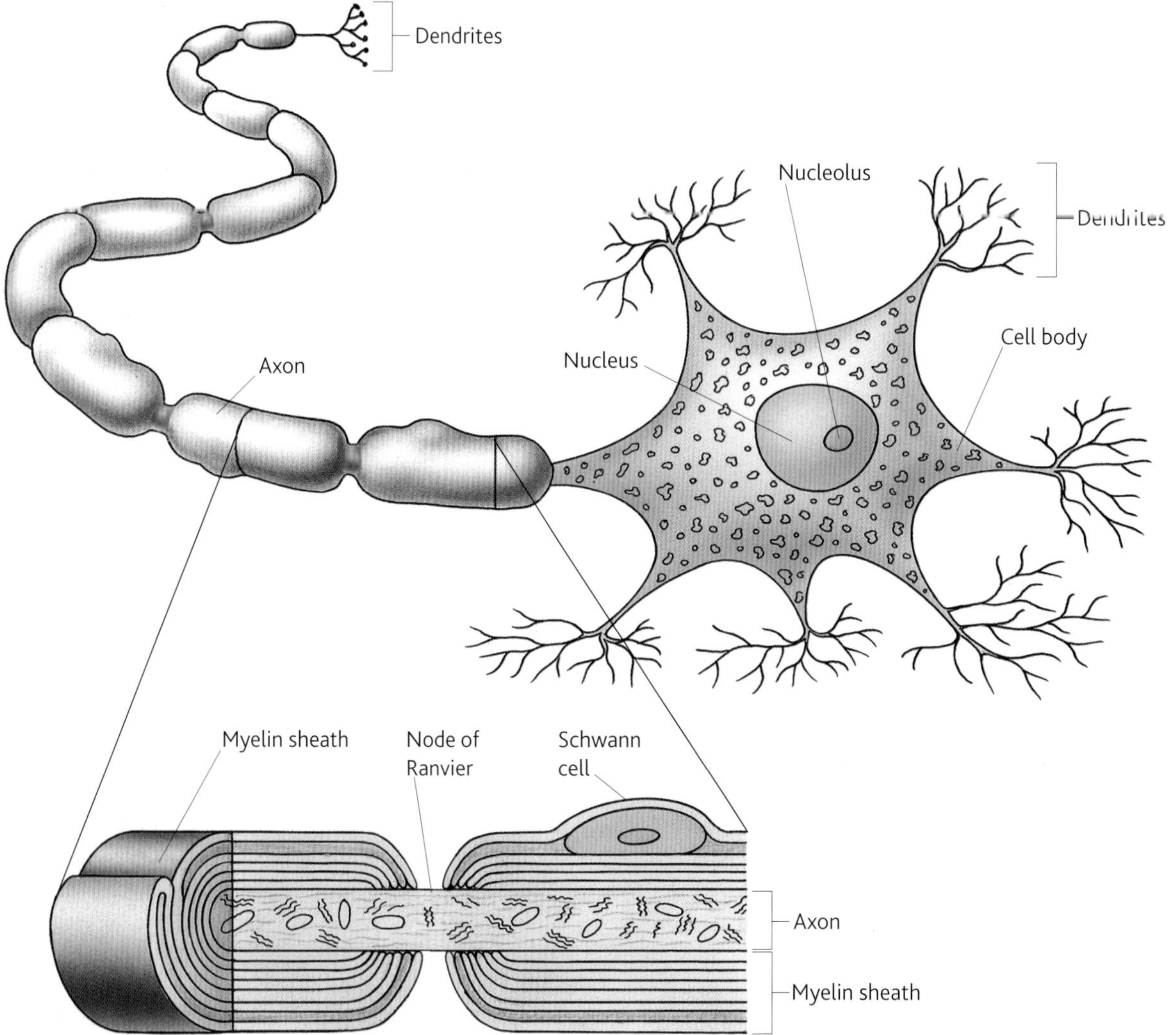

Figure 2 *Myelinated motor neurone*

Neurones can be classified according to their function:

- **Sensory neurones** transmit nerve impulses from a **receptor** to an intermediate or motor neurone. They have one dendron that carries the impulse towards the cell body and one axon that carries it away from the cell body.
- **Motor neurones** transmit nerve impulses from an intermediate or sensory neurone to an **effector**, such as a gland or a muscle. They have a long axon and many short dendrites.
- **Intermediate neurones** transmit impulses between neurones, for example, from sensory to motor neurones. They have numerous short processes.

Figure 3 overleaf, shows the structure of all three types of neurone.

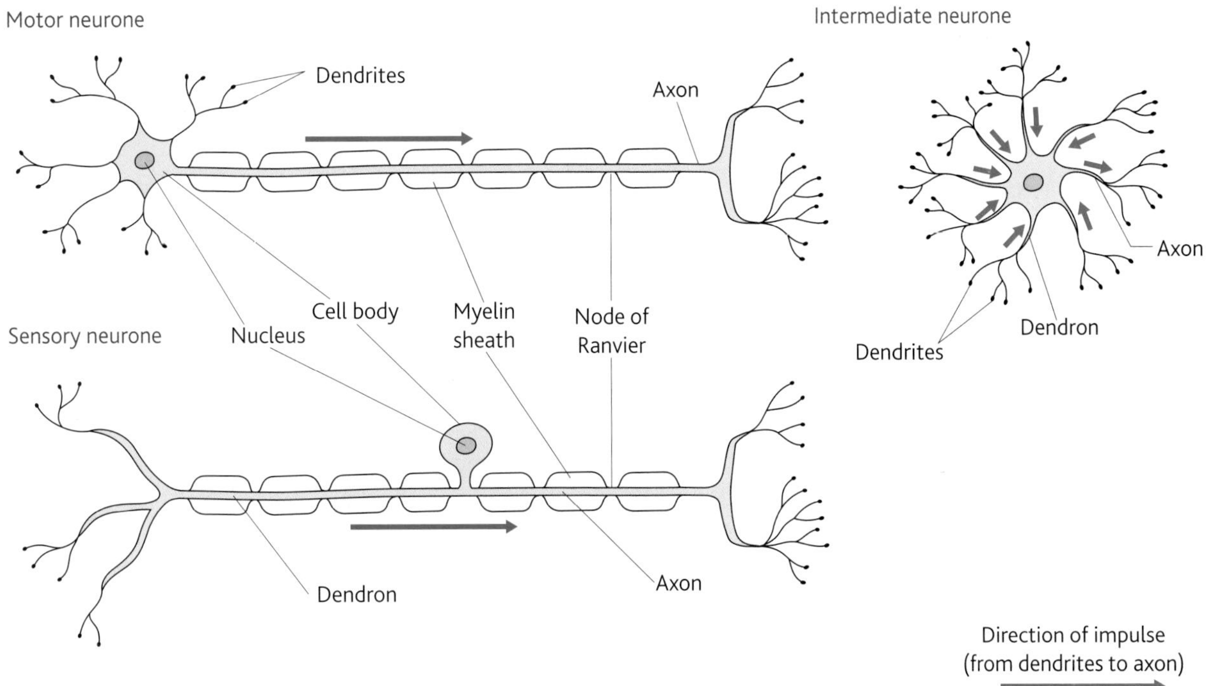

Figure 3 *Types of neurone*

Summary questions

In the following passage give the word that best replaces the numbers in brackets.

Neurones are adapted to carry electrochemical charges called (1). Each neurone comprises a cell body that contains a (2) and large amounts of (3) , which is used in the production of proteins and neurotransmitters. Extending from the cell body is a single long fibre called an axon and smaller branched fibres called (4). Axons are surrounded by (5) cells, which protect and provide (6) because their membranes are rich in a lipid known as (7). There are three main types of neurone. Those that carry nerve impulses to an effector are called (8) neurones. Those that carry impulses from a receptor are called (9) neurones and those that link the other two types are called (10) neurones.

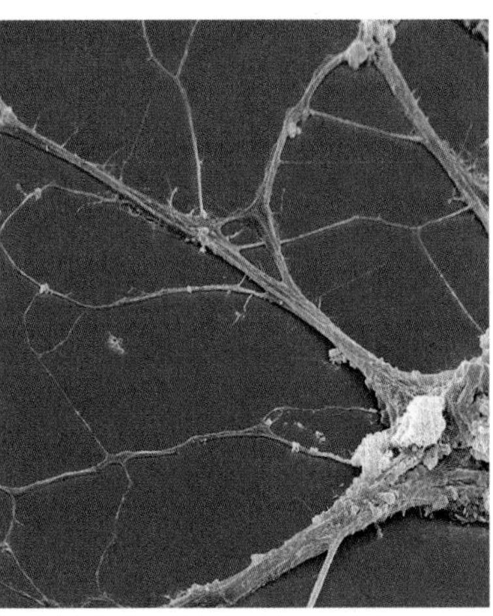

Figure 4 *SEM of a neurone with the cell body at its centre and dendrites radiating from it*

10.3 The nerve impulse

Learning objectives:

- What is resting potential?
- How is resting potential established in a neurone?
- What is an action potential?

Specification reference: 3.5.2

A nerve impulse may be defined as a self-propagating wave of electrical disturbance that travels along the surface of the axon membrane. It is not, however, an electrical current, but a temporary reversal of the electrical potential difference across the axon membrane. This reversal is between two states, called the **resting potential** and the **action potential**.

Resting potential

The movement of ions, such as sodium ions (Na^+) and potassium ions (K^+), across the axon membrane is controlled in a number of ways:

- The **phospholipid** bilayer of the axon plasma membrane prevents sodium and potassium ions diffusing across it.
- Molecules of proteins, known as **intrinsic proteins**, span this phospholipid bilayer. These proteins contain channels, called ion channels, which pass through them. Some of these channels have 'gates', which can be opened or closed in order to allow sodium or potassium ions to move through them at any one time, but prevent their movement on other occasions. There are different 'gated' channels for sodium and potassium ions. Some channels, however, remain open all the time, allowing the sodium and potassium ions to diffuse through them unhindered.
- Some intrinsic proteins actively transport potassium ions into the axon and sodium ions out of the axon. This process is called a **sodium–potassium pump**.

As a result of these various controls, the inside of an axon is negatively charged relative to the outside. This is known as the **resting potential** and ranges from 50 to 90 millivolts (mV), but is usually 65 mV. In this condition the axon is said to be **polarised**. The establishment of this potential difference (the difference in charge between the inside and outside of the axon) is due to the following events:

- Sodium ions are actively transported **out** of the axon by the sodium–potassium pumps.
- Potassium ions are actively transported **into** the axon by the sodium–potassium pumps.
- The active transport of sodium ions is greater than that of potassium ions, so three sodium ions move out for every two potassium ions that move in.
- Although both sodium and potassium ions are positive, the outward movement of sodium ions is greater than the inward movement of potassium ions. As a result, there are more sodium ions in the tissue fluid surrounding the axon than in the cytoplasm, and more potassium ions in the cytoplasm than in the tissue fluid, thus creating a chemical gradient.
- The sodium ions begin to diffuse back naturally into the axon while the potassium ions begin to diffuse back out of the axon.
- However, most of the 'gates' in the channels that allow the potassium ions to move through are open, while most of the 'gates' in the channels that allow the sodium ions to move through are closed.

Link

To understand the nerve impulse requires a thorough knowledge and understanding of plasma membranes, particularly the structure of plasma membranes and the role of their intrinsic proteins in the sodium–potassium pump. It would be useful to revise Topics 3.5 and 3.8 of the *AS Biology* book as a starting point for this section.

Hint

As the phospholipid bilayer does not allow diffusion of sodium and potassium ions, they diffuse back through those sodium and potassium 'gates' that are permanently open.

- As a result the axon membrane is 100 times more permeable to potassium ions, which therefore diffuse back out of the axon faster than the sodium ions diffuse back in. This further increases the potential difference (difference in charge) between the negative inside and the positive outside of the axon.
- Apart from the chemical gradient that causes the movement of the potassium and sodium ions, there is also an electrical gradient. As more and more potassium ions diffuse out of the axon, so the outside of the axon becomes more and more positive. Further outward movement of potassium ions therefore becomes difficult because, being positively charged, they are attracted by the overall negative state inside the axon, which compels them to move into the axon, and repelled by the overall positive state of the surrounding tissue fluid, which prevents them from moving out of the axon.
- An equilibrium is established in which the chemical and electrical gradients are balanced and there is no net movement of ions.

These events are summarised in Figure 1.

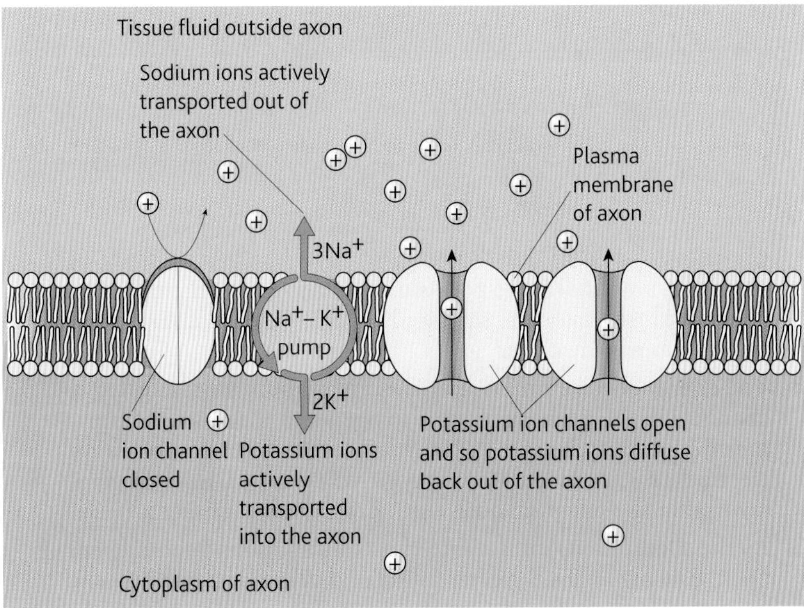

Figure 1 *Distribution of ions at resting potential*

The action potential

When a stimulus is received by a receptor or nerve ending, its energy causes a temporary reversal of the charges on the axon membrane. As a result, the negative charge of -65 mV inside the membrane becomes a positive charge of around $+40$ mV. This is known as the **action potential**, and in this condition the membrane is said to be **depolarised**. This depolarisation occurs because the channels in the axon membrane change shape, and hence open or close, depending on the voltage across the membrane. They are therefore called voltage-gated channels. The sequence of events is described below (the numbers relate to the stages illustrated in Figure 2):

Hint

Make sure that you learn the sequence of events during an action potential in terms of the intrinsic (trans-membrane) proteins that actively transport sodium and potassium ions, i.e. in terms of sodium ion channel proteins, potassium ion channel proteins and sodium–potassium pumps.

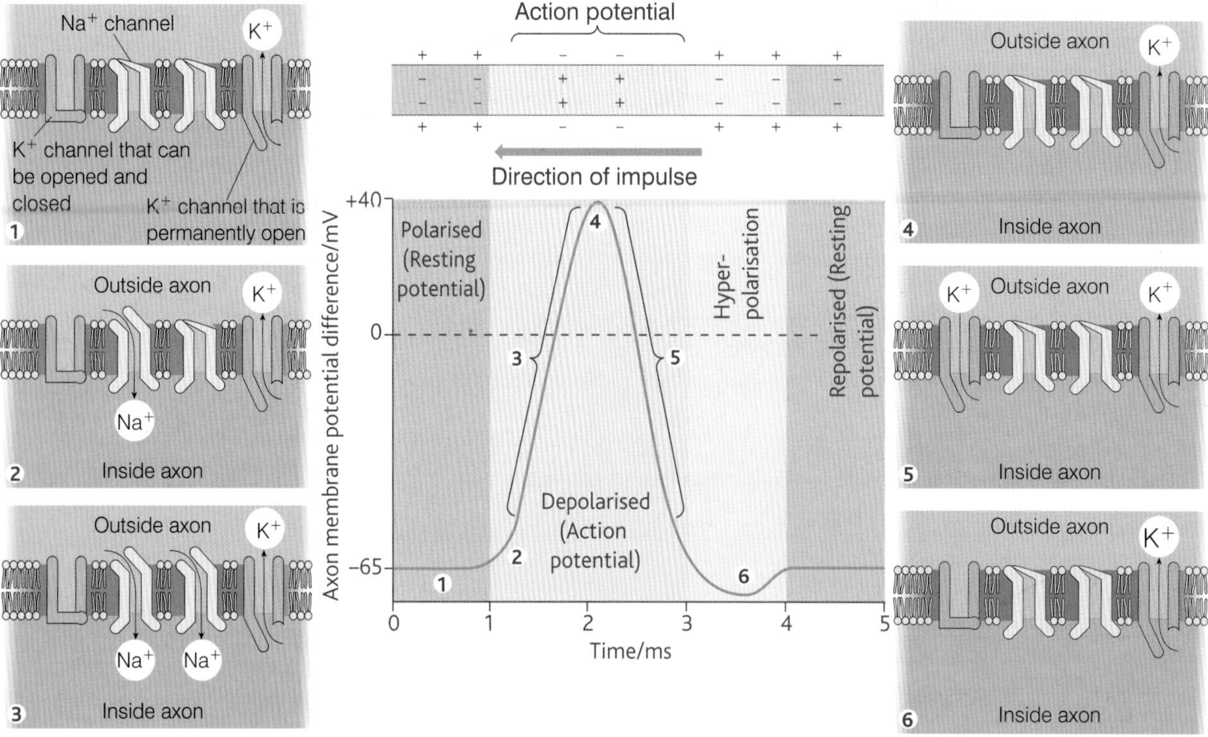

Figure 2 *The action potential*

Hint

The unit of time given on the *y*-axis of the graph (see Figure 2) is the millisecond (ms). A millisecond is 0.001 of a second. There are therefore 1000 milliseconds in a second. At 2 ms each, action potentials are very short-lived!

1 At resting potential some potassium voltage-gated channels are open (namely those that are permanently open) but the sodium voltage-gated channels are closed.

2 The energy of the stimulus causes some sodium voltage-gated channels in the axon membrane to open and therefore sodium ions diffuse into the axon through these channels along their electrochemical gradient. Being positively charged, they trigger a reversal in the potential difference across the membrane.

3 As the sodium ions diffuse into the axon, so more sodium channels open, causing an even greater influx of sodium ions by diffusion.

4 Once the action potential of around +40 mV has been established, the voltage gates on the sodium ion channels close (thus preventing further influx of sodium ions) and the voltage gates on the potassium ion channels begin to open.

5 With some potassium voltage-gated channels now open, the electrical gradient that was preventing further outward movement of potassium ions is now reversed, causing more potassium ion channels to open. This means that yet more potassium ions diffuse out, causing repolarisation of the axon.

6 The outward diffusion of these potassium ions causes a temporary overshoot of the electrical gradient, with the inside of the axon being more negative (relative to the outside) than usual (= hyperpolarisation) The gates on the potassium ion channels now close and the activities of the sodium–potassium pumps once again cause sodium ions to be pumped out and potassium ions in. The resting potential of −65 mV is re-established and the axon is said to be **repolarised**.

The terms 'action potential' and 'resting potential' can be misleading because the movement of sodium ions inwards during the action potential is purely due to diffusion – which is a passive process – while the resting potential is maintained by active transport – which is an active process. The term 'action potential' simply means that the axon membrane is transmitting a nerve impulse, whereas 'resting potential' means that it is not.

Table 1

	Resting	Beginning to depolarise	Repolarising
Membrane potential/mV	–70	–50	–20
Na^+ channels in axon membrane	A	B	C
K^+ channels in axon membrane	D	E	F

Application and How science works

Measuring action potentials

The plasma membrane of an axon will transmit an action potential when stimulated to do so. The action potential involves changes in the electrical potential across the membrane due to the movement of positive ions.

1 Which **two** positive ions are responsible for this change in electrical potential?

Figure 3 shows two action potentials that were recorded using an instrument called an oscilloscope.

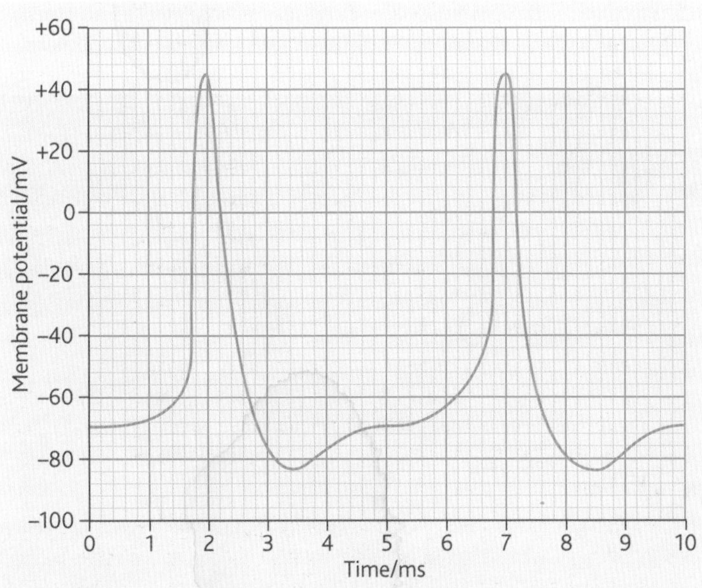

Figure 3

2 Between 0.5 and 2.0 ms there is a considerable change in membrane potential. Explain how this change is brought about.

3 How many action potentials will occur in 1 second if the frequency shown on the graph is maintained for this period? Show your working.

10 ms = 0.1 second.
10 × 10 = 100 200

Summary questions

1 Describe how the movement of ions establishes the resting potential in an axon.

2 Table 1 shows the membrane potential of an axon at different stages of an action potential. The table refers to those channels that can be open and closed, not those that remain permanently open. For each of the letters A–F, indicate the state of the relevant channels, i.e. 'open' or closed'.

10.4 Passage of an action potential

Learning objectives:

- How does an action potential pass along an unmyelinated axon?
- How does an action potential pass along a myelinated axon?

Specification reference: 3.5.2

Once it has been created, an **action potential** 'moves' rapidly along an **axon**. The size of the action potential remains the same from one end of the axon to the other. Strictly speaking, nothing physically 'moves' from place to place along the axon of the neurone but rather the reversal of electrical charge is reproduced at different points along the axon membrane. As one region of the axon produces an action potential and becomes depolarised, it acts as a stimulus for the **depolarisation** of the next region of the axon. In this manner, action potentials are regenerated along each small region of the axon membrane. In the meantime, the previous region of the membrane returns to its resting potential, that is, it undergoes **repolarisation**.

The process can be likened to the 'Mexican wave' that often takes place in a crowded stadium during a sporting event. Although the wave of people standing up and raising their hands (the action potential) moves around the stadium, the people themselves do not move from seat to seat with the wave (i.e. they do not physically pass around the stadium until they reach their original seat again). Instead, their individual actions of standing and raising their hands are stimulated by the action of the person on one side of them and are reproduced by the person on the other side.

Passage of an action potential along an unmyelinated axon

It is easier to understand how a nerve impulse is propagated in a myelinated axon if we first look at how it is propagated in an unmyelinated one. This process is described and illustrated in Figure 2, opposite.

AQA Examiner's tip

Do not assume that that every question refers to human neurones – other animals have them too. Always use the information given in a question and answer accordingly.

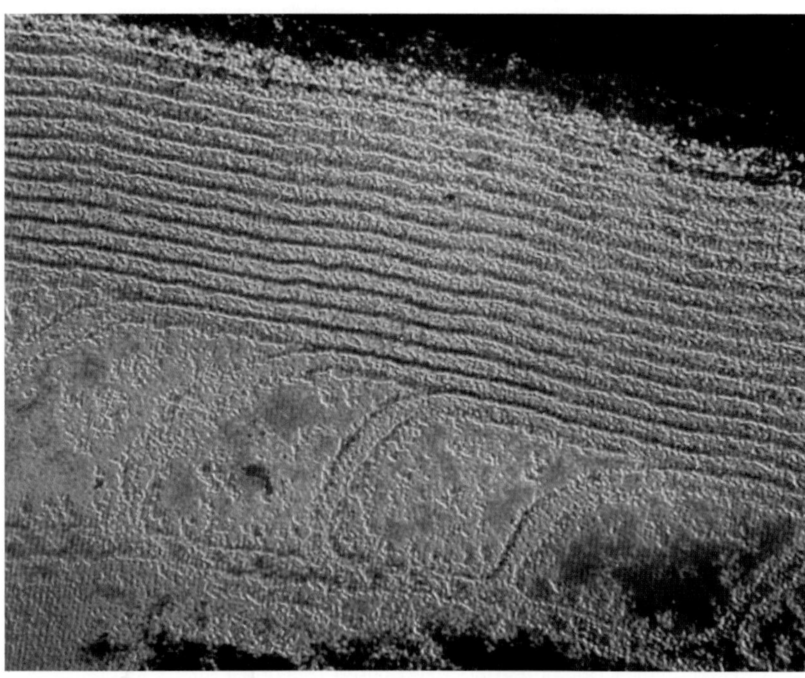

Figure 1 *False-colour TEM of the myelin sheath (orange bands at top) around the axon (bottom)*

Polarised

Depolarised

Repolarised

1

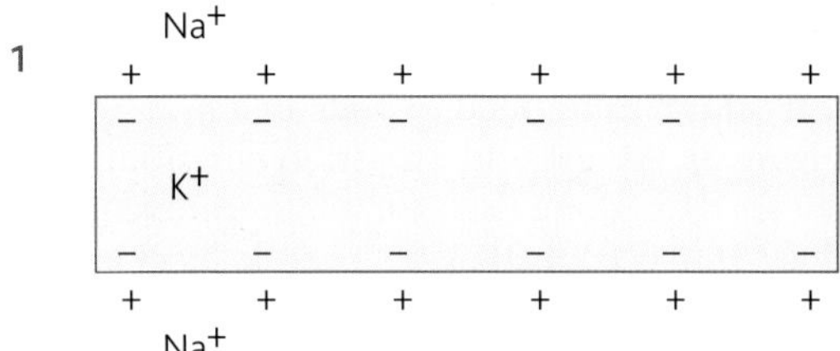

1 At resting potential the concentration of sodium ions outside the axon membrane is high relative to the inside, whereas that of the potassium ions is high inside the membrane relative to the outside. The overall concentration of positive ions is, however, greater on the outside, making this positive compared with the inside. The axon membrane is polarised. In our Mexican wave analogy, this is equivalent to the whole stadium being seated, i.e. at rest.

2

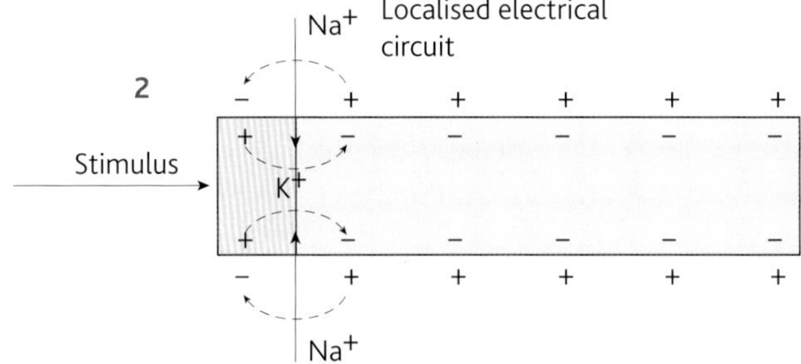

2 A stimulus causes a sudden influx of sodium ions and hence a reversal of charge on the axon membrane. This is the action potential and the membrane is depolarised. In our analogy, a prompt leads a vertical line of people to stand and wave their arms, i.e. they are stimulated into action.

3

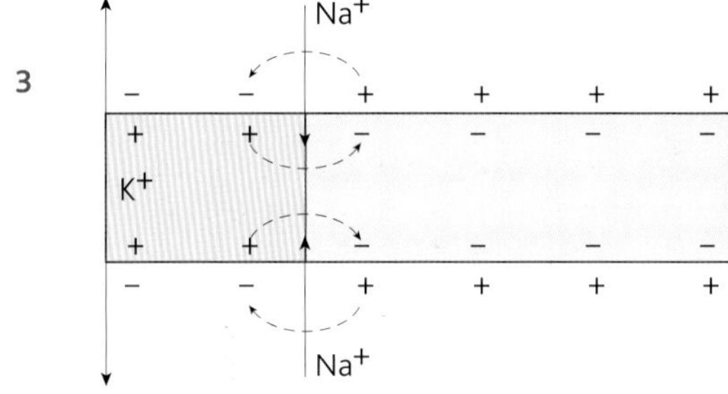

3 The localised electrical circuits established by the influx of sodium ions cause the opening of sodium voltage-gated channels a little further along the axon. The resulting influx of sodium ions in this region causes depolarisation. Behind this new region of depolarisation, the sodium voltage-gated channels close and the potassium ones open. Potassium ions begin to leave the axon along their electrochemical gradient. The sight of the person next to them standing and waving prompts the person in the adjacent seat to stand and wave. A new vertical line of people stands and waves, while the original line of people begin to sit down again.

4

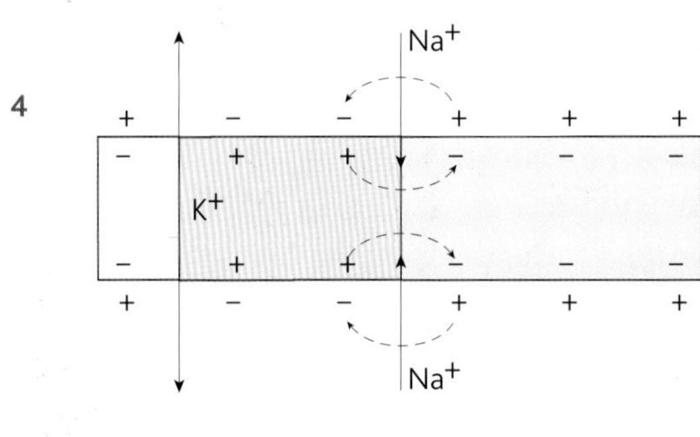

4 The action potential (depolarisation) is propagated in the same way further along the neurone. The outward movement of the potassium ions has continued to the extent that the axon membrane behind the action potential has returned to its original charged state (positive outside, negative inside), i.e. it has been repolarised. The second line of people standing and waving prompts the third line of people to do the same. Meanwhile, the first line have now resumed their original positions, i.e. they are re-seated.

5

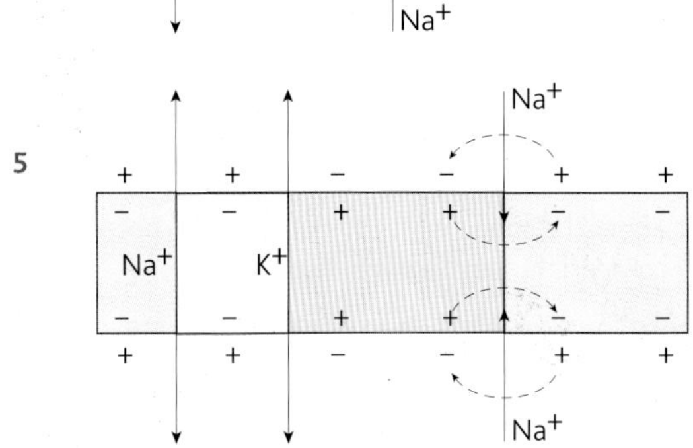

5 Repolarisation of the neurone allows sodium ions to be actively transported out, once again returning the neurone to its resting potential in readiness for a new stimulus if it comes. The people who have just sat down settle back in their seats and readjust themselves in readiness to repeat the process should they be prompted to do so again.

Figure 2 *Passage of an impulse along the axon of an unmyelinated neurone*

Passage of an action potential along a myelinated axon

In myelinated axons, the fatty sheath of myelin around the axon acts as an electrical insulator, preventing action potentials from forming. At intervals of 1–3 mm there are breaks in this myelin insulation, called nodes of Ranvier (see Topic 10.2). Action potentials can occur at these points. The localised circuits therefore arise between adjacent nodes of Ranvier and the action potentials in effect 'jump' from node to node in a process known as saltatory conduction (Figure 3). As a result, an action potential passes along a myelinated neurone faster than along the axon of an unmyelinated one. In our Mexican wave analogy, this is equivalent to a whole block of spectators leaping up simultaneously, followed by the next block and so on. Instead of the wave passing around the stadium in hundreds of small stages, it passes around in 20 or so large ones and is consequently more rapid.

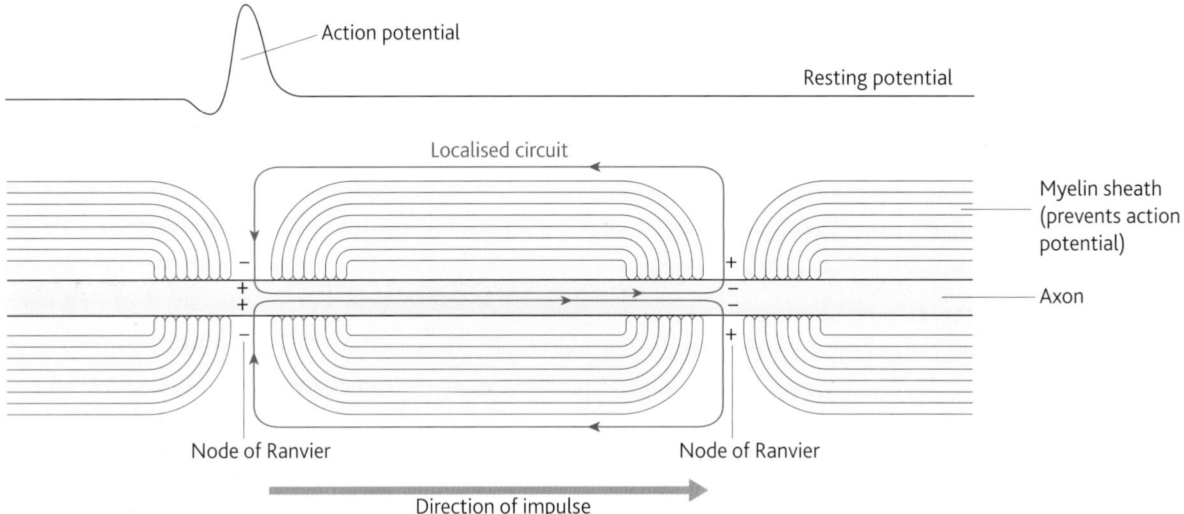

Figure 3 *Passage of an action potential along a myelinated axon. Action potentials are produced only at nodes of Ranvier. Depolarisation therefore skips from node to node (= saltatory conduction)*

Hint

The term 'saltatory' in saltatory conduction comes from the Latin word *saltare*, meaning 'to jump'.

Summary questions

1 In a myelinated axon, sodium and potassium ions can only be exchanged at certain points along it.

a What is the name given to these points?

b Explain why ions can only be exchanged at these points.

c What effect does this have on the way an action potential is conducted along the axon?

d What name is given to this type of conduction?

e How does it affect the speed with which the action potential is transmitted compared to an unmyelinated axon?

2 What happens to the size of an action potential as it moves along an axon?

10.5 Speed of the nerve impulse

Learning objectives:

- What factors affect the speed of conductance of an action potential?
- What is the refractory period?
- What is its role in separating one impulse from the next?
- What is meant by the all-or-nothing principle?

Specification reference: 3.5.2

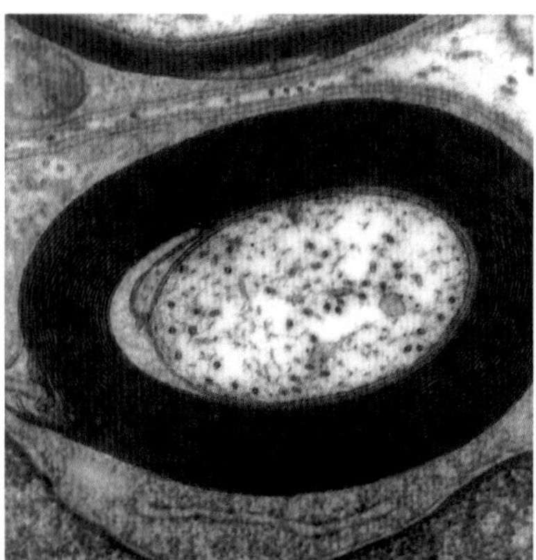

Figure 1 *False-coloured TEM of a section through a myelinated neurone and Schwann cell. Myelin (black) surrounds the axon (purple), increasing the speed at which nerve impulses travel. It is formed when Schwann cells (green) wrap around the axon, depositing layers of myelin between each coil*

Once an **action potential** has been set up, it moves rapidly from one end of the axon to the other without any decrease in size. In other words, the final action potential at the end of the axon is the same size as the first action potential. This transmission of the action potential along the axon of a neurone is the **nerve impulse**.

Factors affecting the speed at which an action potential travels

A number of factors affect the speed at which the action potential passes along the axon. Depending upon these factors, an action potential may travel at a speed of as little as $0.5\,\mathrm{m\,s^{-1}}$ or as much as $120\,\mathrm{m\,s^{-1}}$. These factors include:

- **the myelin sheath**. We saw in Topic 10.4 that the myelin sheath acts as an electrical insulator, preventing an action potential forming in the part of the axon covered in myelin. It does, however, jump from one node of Ranvier to another (**saltatory conduction**). This increases the speed of conductance from $30\,\mathrm{m\,s^{-1}}$ in an unmyelinated neurone to $90\,\mathrm{m\,s^{-1}}$ in a similar myelinated one.
- **the diameter of the axon**. The greater the diameter of an axon, the faster the speed of conductance. This is due to less leakage of ions from a large axon (leakage makes membrane potentials harder to maintain).
- **temperature**. This affects the rate of diffusion of ions and therefore the higher the temperature the faster the nerve impulse. The energy for active transport comes from respiration. Respiration, like the **sodium–potassium pump**, is controlled by enzymes. Enzymes function more rapidly at higher temperatures up to a point. Above a certain temperature, enzymes and the plasma membrane proteins are denatured and impulses fail to be conducted at all. Temperature is clearly an important factor in response times in cold-blooded (ectothermic) animals, whose body temperature varies in accordance with the environment.

The refractory period

Once an action potential has been created in any region of an axon, there is a period afterwards when inward movement of sodium ions is prevented because the sodium **voltage-gated channels** are closed. During this time it is impossible for a further action potential to be generated. This is known as the **refractory period** (Figure 2, overleaf).

The refractory period serves three purposes:

- **It ensures that an action potential is propagated in one direction only**. An action potential can only pass from an active region to a resting region. This is because an action potential cannot be propagated in a region that is refractory, which means that it can only move in a forward direction. This prevents the action potential from spreading out in both directions, which it would otherwise do.
- **It produces discrete impulses**. Due to the refractory period, a new action potential cannot be formed immediately behind the first one. This ensures that action potentials are separated from one another.

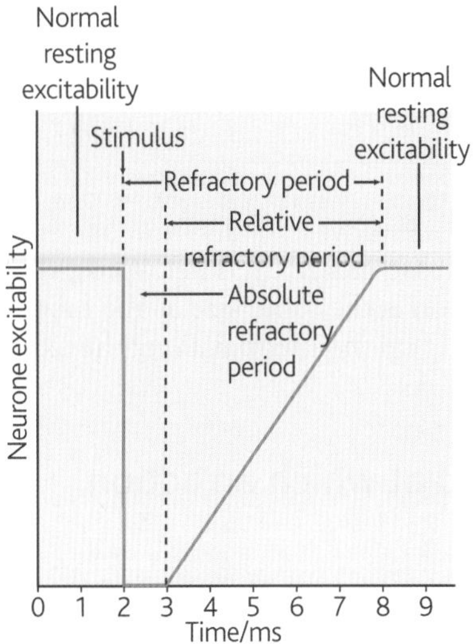

Figure 2 *Graph illustrating neurone excitability before and after a nerve impulse*

Hint

The brain would be overloaded with information if it became aware of every little stimulus. The all-or-nothing nature of the action potential acts as a filter, preventing minor stimuli from setting up nerve impulses and thus preventing the brain becoming overloaded.

- **It limits the number of action potentials**. As action potentials are separated from one another this limits the number of action potentials that can pass along an axon in a given time.

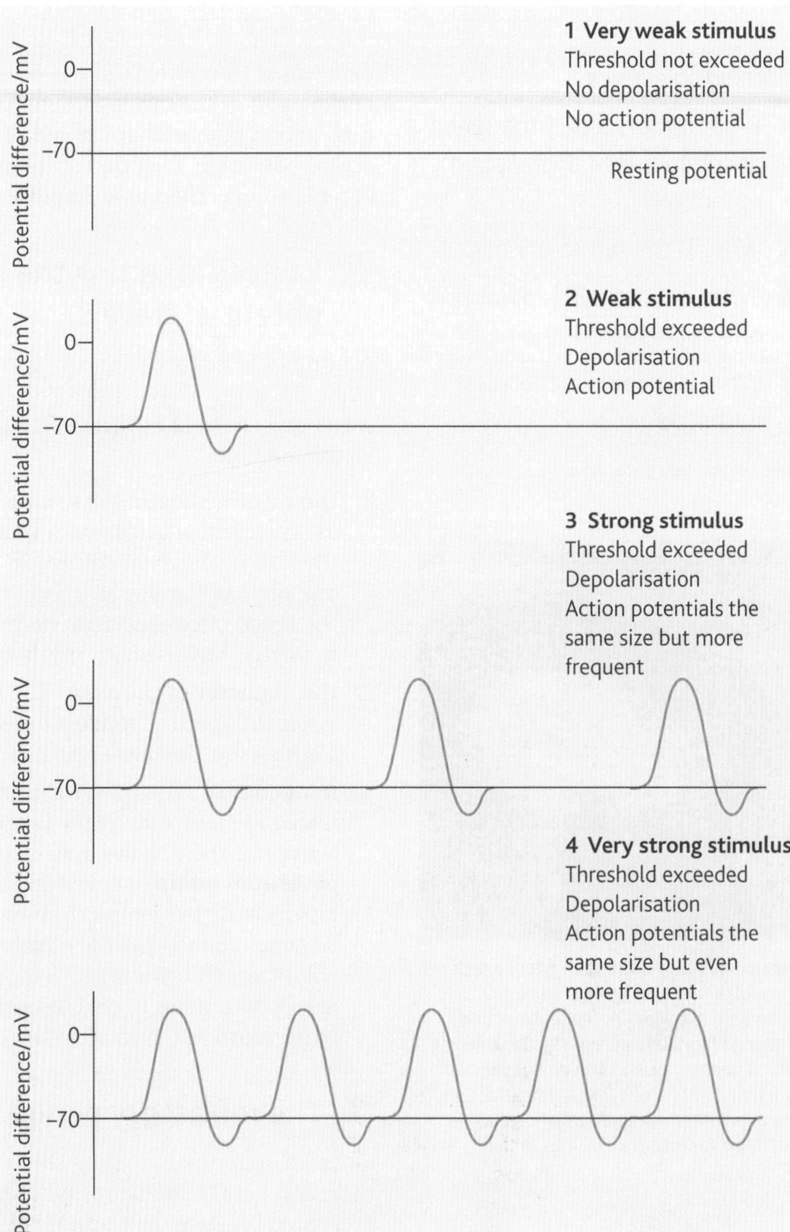

Figure 3 *Effect of stimulus intensity on impulse frequency*

All-or-nothing principle

Nerve impulses are described as **all-or-nothing** responses. There is a certain level of stimulus, called the **threshold value**, which triggers an action potential. Below the threshold value, no action potential, and therefore no impulse, is generated. Any stimulus, of whatever strength, that is below the threshold value will fail to generate an action potential – this is the 'nothing' part. Any stimulus above the threshold value will succeed in generating an action potential. It does not matter how much above the threshold a stimulus is, it will still only generate one action

Summary questions

1. Explain how the refractory period ensures that nerve impulses are kept separate from one another.
2. What is the all-or-nothing principle?

potential – this is the 'all' part. How then can an organism perceive the size of a stimulus? This is achieved in two ways:

- by the number of impulses passing in a given time. The larger the stimulus, the more impulses that are generated in a given time (Figure 3)
- by having different neurones with different threshold values. The brain interprets the number and type of neurones that pass impulses as a result of a given stimulus and thereby determines its size.

Application

Different axons, different speeds

Table 1 below shows the speeds at which different axons conduct action potentials.

Table 1

Axon	Myelin	Axon diameter /mm	Transmission speed/m s^{-1}
Human motor axon to leg muscle	Yes	20	120
Human sensory axon from skin pressure receptor	Yes	10	50
Squid giant axon	No	500	25
Human motor axon to internal organ	No	1	2

1 Using data from the table, describe the effect of axon diameter on the speed of conductance of an action potential.

2 The data show that a myelinated axon conducts an action potential faster than an unmyelinated axon. Explain why this is so.

3 What is the name of the cells whose membranes make up the myelin sheath around some types of axon?

4 State which has the greater effect on the speed of conductance of an action potential: the presence of myelin or the diameter of the axon. Use information from Table 1 to explain your answer.

5 The squid is an ectothermic animal. This means that its body temperature fluctuates with the temperature of the water in which it lives. Suggest how this might affect the speed at which action potentials are conducted along a squid axon.

10.6 Structure and function of synapses

Learning objectives:

- What is a synapse?
- What functions do synapses perform?

Specification reference: 3.5.2

A synapse is the point where the **axon** of one **neurone** connects with the **dendrite** of another or with an **effector**. They are important in linking different neurones together and therefore coordinating activities.

Structure of a synapse

Synapses transmit impulses from one neurone to another by means of chemicals known as **neurotransmitters**. Neurones are separated by a small gap, called the **synaptic cleft**, which is 20–30 nm wide. The neurone that releases the neurotransmitter is called the **presynaptic neurone**. The axon of this neurone ends in a swollen portion known as the **synaptic knob**. This possesses many mitochondria and large amounts of endoplasmic reticulum. These are required in the manufacture of the neurotransmitter. Once made, the neurotransmitter is stored in the **synaptic vesicles**. Once the neurotransmitter is released from the vesicles it diffuses across to the postsynaptic neurone, which possesses receptor molecules on its membrane to receive it. The structure of a chemical synapse is illustrated in Figure 1.

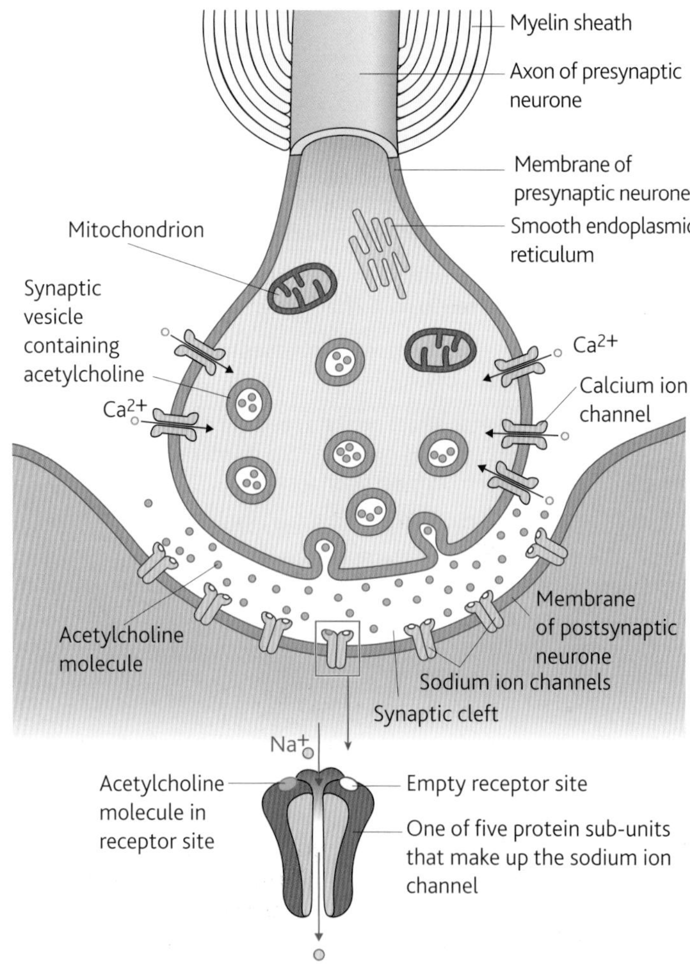

Figure 1 *Structure of a synapse*

Functions of synapses

Synapses transmit impulses from one neurone to another. In so doing, they act as junctions, allowing:

- a single impulse along one neurone to be transmitted to a number of different neurones at a synapse. This allows a single stimulus to create a number of simultaneous responses
- a number of impulses to be combined at a synapse. This allows stimuli from different receptors to interact in order to produce a single response.

We shall look in more detail at how synapses transmit nerve impulses in Topic 10.7. However, to understand the basic functioning of synapses as described here, it is sufficient to appreciate the following:

- A chemical (the neurotransmitter) is made **only** in the presynaptic neurone and not in the postsynaptic neurone.
- The neurotransmitter is stored in synaptic vesicles and released into the synapse when an **action potential** reaches the synaptic knob.
- When released, the neurotransmitter diffuses across the synapse to receptor molecules on the postsynaptic neurone.
- The neurotransmitter binds with the receptor molecules and sets up a new action potential in the postsynaptic neurone.

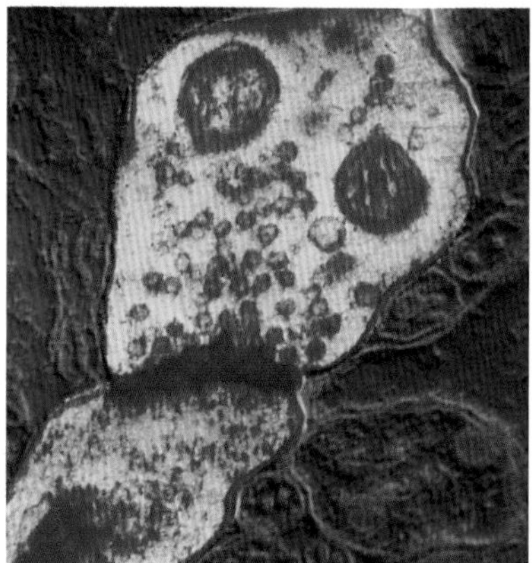

Figure 2 *TEM of synapse. The synaptic cleft between the two neurones (centre) appears deep red. The cell above the cleft has many small vesicles (red–yellow spheres) containing neurotransmitter, whereas the two larger spheres above the vesicles are mitochondria*

Features of synapses

The basic way in which synapses function means they have a number of different features.

Unidirectionality

Synapses can only pass impulses in one direction: from the presynaptic neurone to the postsynaptic neurone. In this way, synapses act like valves.

Summation

Low-frequency action potentials often produce insufficient amounts of neurotransmitter to trigger a new action potential in the postsynaptic neurone. They can, however, be made to do so by a process called summation. This entails a build-up of neurotransmitter in the synapse by one of two methods:

- **spatial summation**, in which a number of different presynaptic neurones together release enough neurotransmitter to exceed the threshold value of the postsynaptic neurone. Together they therefore trigger a new action potential.
- **temporal summation**, in which a single presynaptic neurone releases neurotransmitter many times over a short period. If the total amount of neurotransmitter exceeds the threshold value of the postsynaptic neurone, then a new action potential is triggered.

Spatial and temporal summation are illustrated in Figure 3, overleaf.

Inhibition

On the postsynaptic membrane of some synapses, the protein channels carrying chloride ions (Cl^-) can be made to open. This leads to an inward diffusion of Cl^- ions, making the inside of the postsynaptic membrane even more negative than when it is at resting potential. This is called hyperpolarisation and makes it less likely that a new action potential will be created. For this reason these synapses are called inhibitory synapses.

Summary questions

1. How is a presynaptic neurone adapted for the manufacture of neurotransmitter?
2. How is the postsynaptic neurone adapted to receive the neurotransmitter?
3. Describe the basic events in the transmission of a nerve impulse from one neurone to another.
4. If a neurone is stimulated in the middle of its axon, an action potential will pass both ways along it to the synapses at each end of the neurone. However, the action potential will only pass across the synapse at one end. Explain why.
5. When walking along a street we barely notice the background noise of traffic. However, we often respond to louder traffic noises, such as the sound of a horn.
   a. From your knowledge of summation, explain this difference.
   b. Suggest an advantage in responding to high-level stimuli but not to low-level ones.
6. Explain why hyperpolarisation reduces the likelihood of a new action potential being created.

Spatial summation

Presynaptic neurone A

No action potential

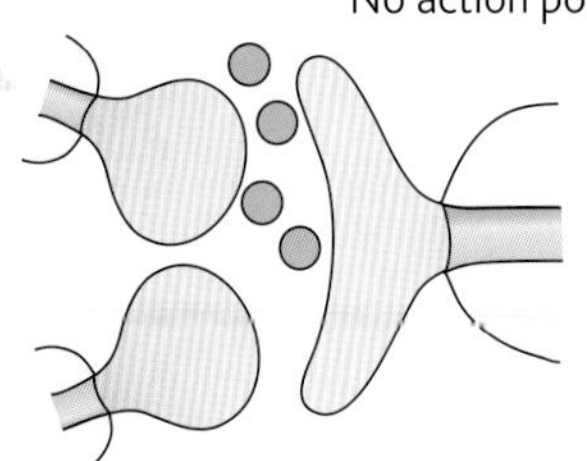

Presynaptic neurone B

Neurone A releases neurotransmitter but quantity is below threshold to trigger action potential in postsynaptic neurone.

No action potential

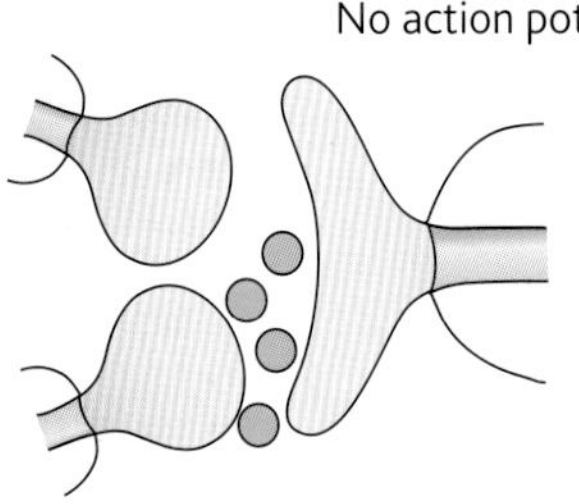

Neurone B releases neurotransmitter but quantity is below threshold to trigger action potential in postsynaptic neurone.

Action potential

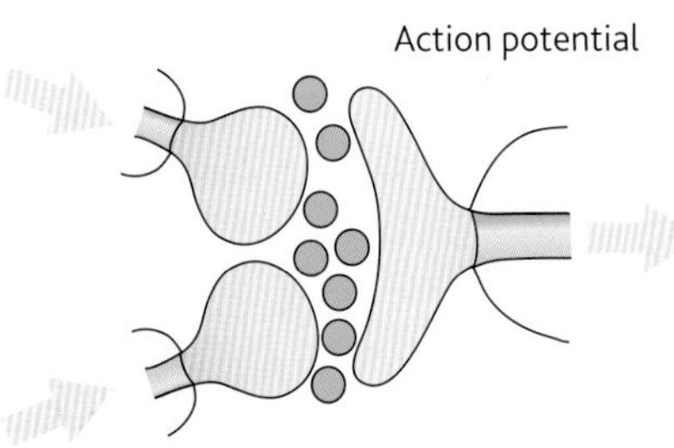

Neurone A and B release neurotransmitter. Quantity is above threshold and so an action potential is triggered in the postsynaptic neurone.

Temporal summation

Low-frequency action potentials

No action potential

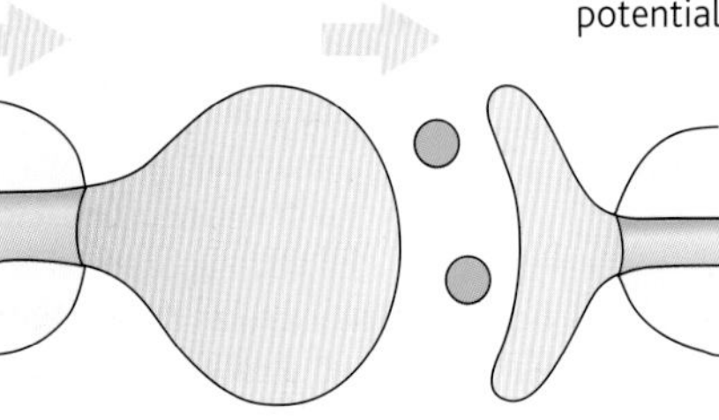

Low-frequency action potentials lead to release of small amount of neurotransmitter. Quantity is below the threshold to trigger an action potential in the postsynaptic neurone.

High-frequency action potentials

Action potential

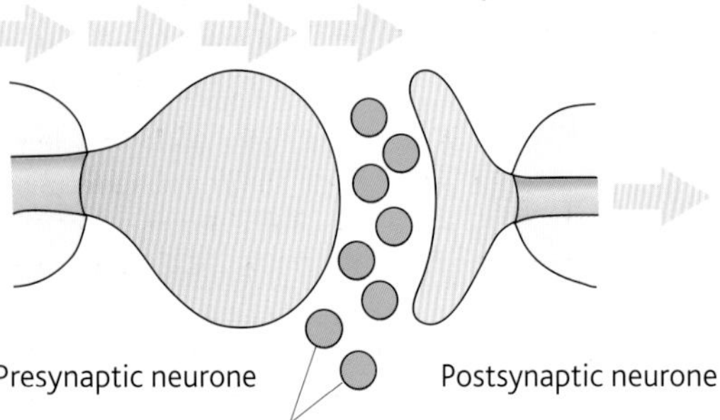

High-frequency action potentials lead to release of large amount of neurotransmitter. Quantity is above the threshold to trigger an action potential in the postsynaptic neurone.

Figure 3 *Spatial and temporal summation*

10.7 Transmission across a synapse

Learning objectives:

- How is information transmitted across a synapse?

Specification reference: 3.5.2

Hint

You need to think in terms of separate 'bursts' of neurotransmitter release from the presynaptic knob. Each one relates to the arrival of an action potential along the neurone.

AQA Examiner's tip

In this topic, there are many possibilities for synoptic questions that bring together other topics. These include membrane structure, enzyme action, mitochondria and ATP production, diffusion (across the synaptic cleft and down channels in membrane proteins) and how molecular shapes fit one another (e.g. a neurotransmitter fitting into receptors on the postsynaptic neurone).

In Topic 10.6 we outlined how **neurotransmitters** transmit an action potential from one neurone to another. Let us now consider this in more detail by looking at a cholinergic synapse.

A **cholinergic** synapse is one in which the neurotransmitter is a chemical called **acetylcholine**. Acetylcholine is made up of two parts: acetyl (more precisely ethanoic acid) and choline. Cholinergic synapses are common in vertebrates, where they occur in the central nervous system and at neuromuscular junctions (junctions between neurones and muscles). Details of the neuromuscular junction are given in Topic 11.1.

The process of transmission across a cholinergic synapse is described in the series of diagrams in Figure 2 on page 179. To simplify matters, only the relevant structures are shown on each diagram.

Application

Effects of drugs on synapses

There are many different neurotransmitters responsible for the exchange of information across a synapse. There are also many different types of receptor on the postsynaptic neurone. Some of these neurotransmitters and receptors are excitatory, i.e. they create a new action potential in the postsynaptic neurone. Others are inhibitory, i.e. they make it less likely that a new action potential will be created in the postsynaptic neurone. Overall, the action of a neurotransmitter depends on the receptor to which it binds.

Given that our perception of the world is through information received by receptors and transferred to the brain by neurones that connect via synapses, it is not surprising that the effects of many medicinal and recreational drugs are due to their actions on synapses. Drugs act on synapses in two main ways:

- **They stimulate the nervous system by creating more action potentials in postsynaptic neurones**. A drug may do this by mimicking a neurotransmitter, stimulating the release of more neurotransmitter or inhibiting the enzyme that breaks down the neurotransmitter. The outcome is to enhance the body's responses to impulses passed along the postsynaptic neurone. For example, if the neurone transmits impulses from sound receptors, a person will perceive the sound as being louder.
- **They inhibit the nervous system by creating fewer action potentials in postsynaptic neurones**. A drug may do this by inhibiting the release of neurotransmitter or blocking the receptors on sodium/potassium ion channels on the postsynaptic neurone. The outcome is to reduce the body's responses to impulses passed along the postsynaptic neurone. In this case, if the neurone transmits impulses from sound receptors, a person will perceive the sound as being quieter.

The effects of a drug on a neurotransmitter depend on the type of transmitter. For example, a drug that inhibits an excitatory neurotransmitter will reduce a particular effect, but a drug that inhibits an inhibitory neurotransmitter will enhance a particular effect. Let us look at some examples of the effects of drugs on synapses.

Endorphins are neurotransmitters used by certain sensory nerve pathways, especially pain pathways. Endorphins block the sensation of pain by binding to pain receptor sites. Drugs such as **morphine**, **codeine** and **heroin** bind to the specific receptors used by endorphins.

1 Suggest the likely effect of drugs like morphine and codeine on the body.

2 Explain how the effect you suggest might be brought about.

Serotonin is a neurotransmitter involved in the regulation of sleep and certain emotional states. Reduced activity of the neurones that release serotonins is thought to be one cause of clinical depression. **Prozac** is an antidepressant drug that affects serotonin within synaptic clefts.

3 Suggest a way that the drug Prozac might affect serotonin within synaptic clefts.

4 Explain how the effect you suggest makes Prozac an effective antidepressant.

GABA is a neurotransmitter that inhibits the formation of action potentials when it binds to postsynaptic neurones. **Valium** is a drug that enhances the binding of GABA to its receptors.

5 Suggest the likely effect of Valium on the nerve pathways that cause muscle contractions.

6 Explain the reasoning for your answer.

7 Epilepsy can be the result of an increase in the activity of neurones in the brain due to insufficient GABA. An enzyme breaks down GABA on the postsynaptic membrane. A drug called Vigabatrin has a molecular structure similar to GABA and is used to treat epilepsy. Suggest a way in which Vigabatrin might be effective in treating epilepsy.

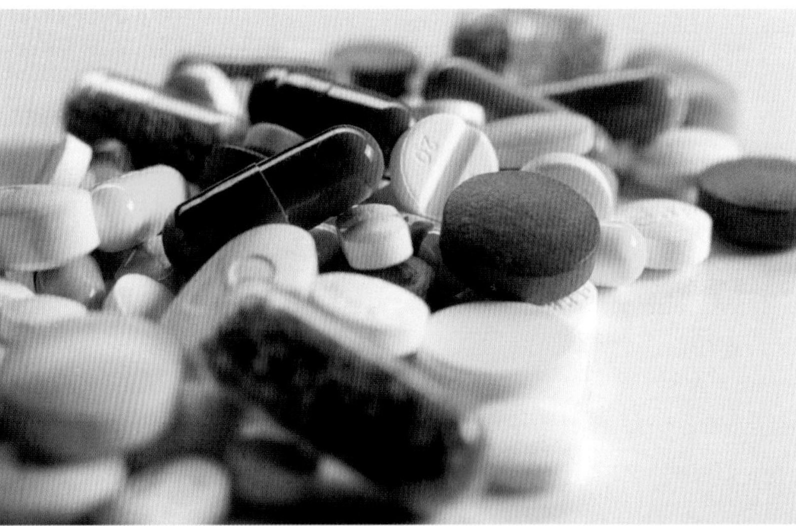

Figure 1 *Many drugs function by acting on synapses*

Summary questions

1 For each of the following, state as accurately as possible the name of the substance described.

a They diffuse into the postsynaptic neurone where they generate an action potential.

b A neurotransmitter found in a cholinergic synapse.

c It is released by mitochondria to enable the neurotransmitter to be re-formed.

d Their influx into the presynaptic neurone causes synaptic vesicles to release their neurotransmitter.

2 Why is it necessary for acetylcholine to be hydrolysed by acetylcholinesterase?

AQA Examiner's tip

The specification doesn't name any particular drugs so you don't have to learn any specific names. You will be given any information you may need in an examination question. All you have to do is apply your knowledge of how synapses work to the information provided.

1 The arrival of an action potential at the end of the presynaptic neurone causes calcium ion channels to open and calcium ions (Ca^{2+}) enter the synaptic knob.

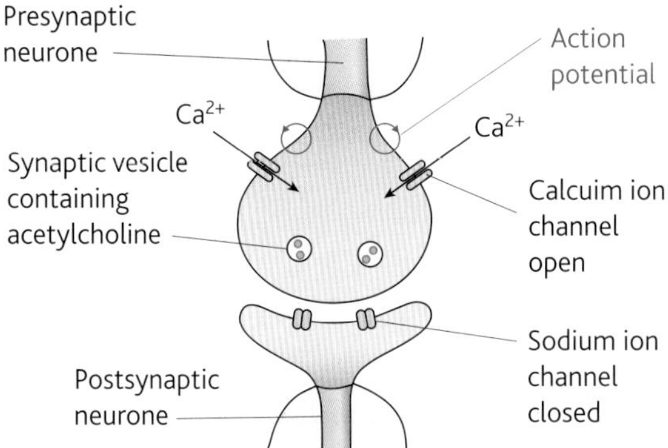

4 The influx of sodium ions generates a new action potential in the postsynaptic neurone.

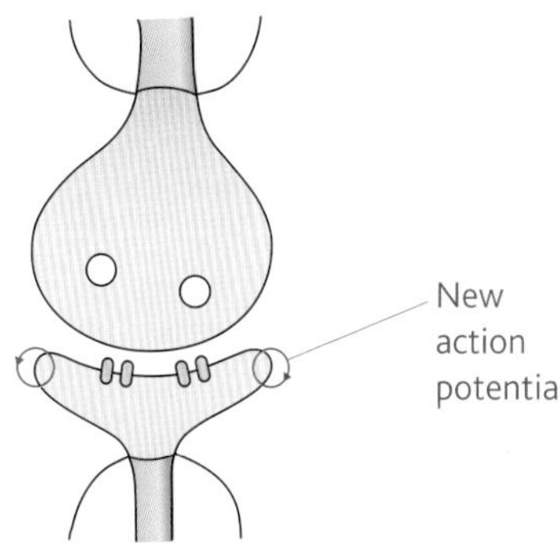

2 The influx of calcium ions into the presynaptic neurone causes synaptic vesicles to fuse with the presynaptic membrane, so releasing acetylcholine into the synaptic cleft.

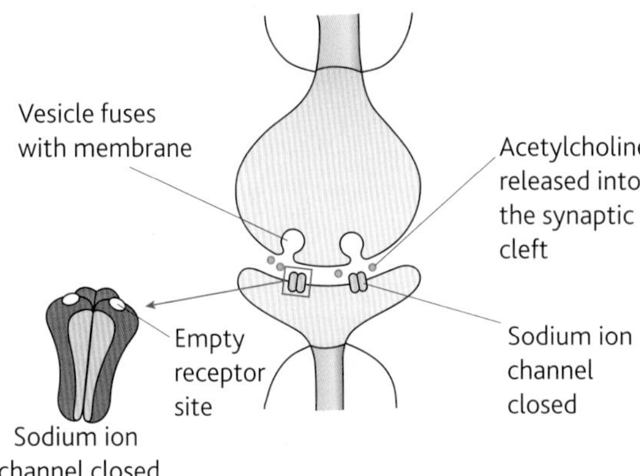

5 Acetylcholinesterase hydrolyses acetylcholine into choline and ethanoic acid (acetyl), which diffuse back across the synaptic cleft into the presynaptic neurone (= recycling). In addition to recycling the choline and ethanoic acid, the breakdown of acetylcholine also prevents it from continuously generating a new action potential in the postsynaptic neurone.

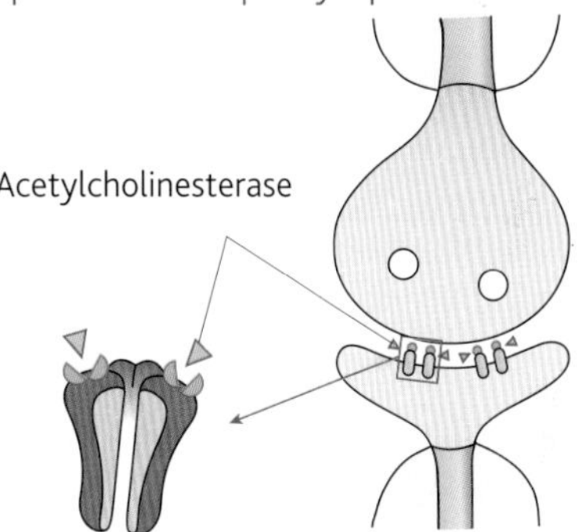

3 Acetylcholine molecules fuse with receptor sites on the sodium ion channel in the membrane of the postsynaptic neurone. This causes the sodium ion channels to open, allowing sodium ions (Na^+) to diffuse in rapidly along a concentration gradient.

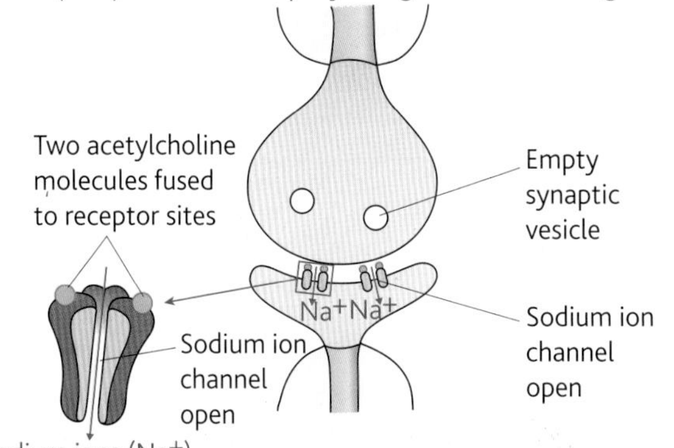

6 ATP released by mitochondria is used to recombine choline and ethanoic acid into acetycholine. This is stored in synaptic vesicles for future use. Sodium ion channels close in the absence of acetylcholine in the receptor sites.

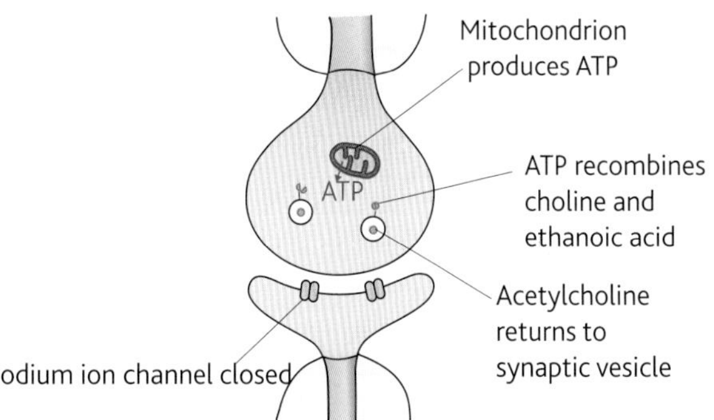

Figure 2 *Mechanism of transmission across a cholinergic synapse*

AQA Examination-style questions

1 IAA is a substance that promotes the growth of cells in plants. It is produced in the tips of shoots. The shoots of seedlings grow towards a unidirectional light source as shown in **Figure 1**.

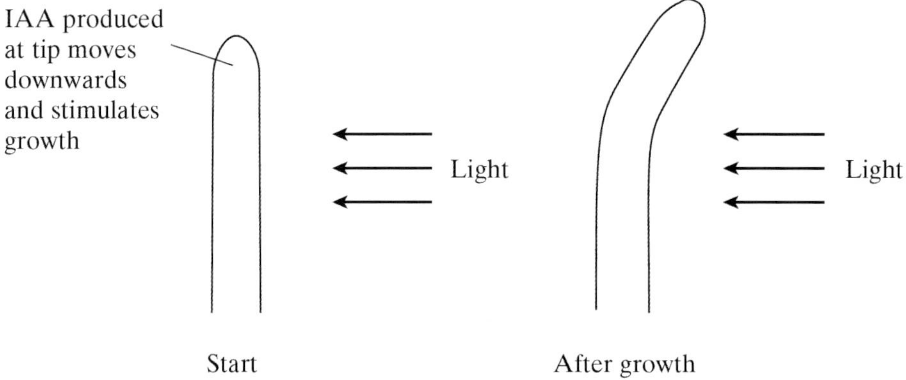

Figure 1

(a) Explain the benefit to the seedling of the response shown in **Figure 1**. *(1 mark)*

Two hypotheses were put forward to explain these responses:

Hypothesis 1 - IAA is inactivated by light;

Hypothesis 2 - IAA moves laterally away from light.

Figure 2 shows the results from four experiments carried out to test these two hypotheses.

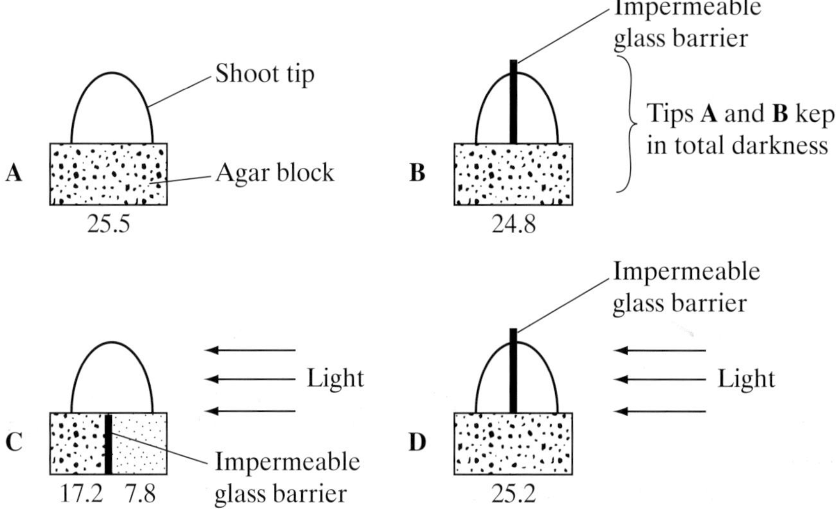

Figure 2

The numbers under each agar block represent the amount of IAA (arbitrary units) that has diffused into the block from the shoot tip.

(b) Suggest **one** variable which needs to be kept constant in the four experiments. *(1 mark)*

(c) Give **one** way in which the results of these experiments provide evidence:

(i) against hypothesis 1;

(ii) in support of hypothesis 2. *(2 marks)*

(d) The shoot tip was removed from a seedling and then an agar block containing IAA was placed on the cut end, as shown in **Figure 3**. Sketch a diagram to show how the seedling would grow following this treatment. *(1 mark)*

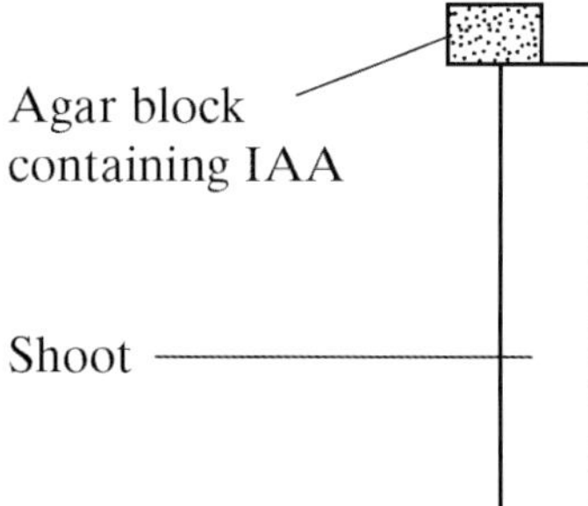

Figure 3

AQA, 2003

2 Acetylcholine is a neurotransmitter which binds to postsynaptic membranes and stimulates the production of nerve impulses. GABA is another neurotransmitter. It is produced by certain neurones in the brain and spinal cord. GABA binds to postsynaptic membranes and inhibits the production of nerve impulses. **Figure 4** shows a synapse involving three neurones.

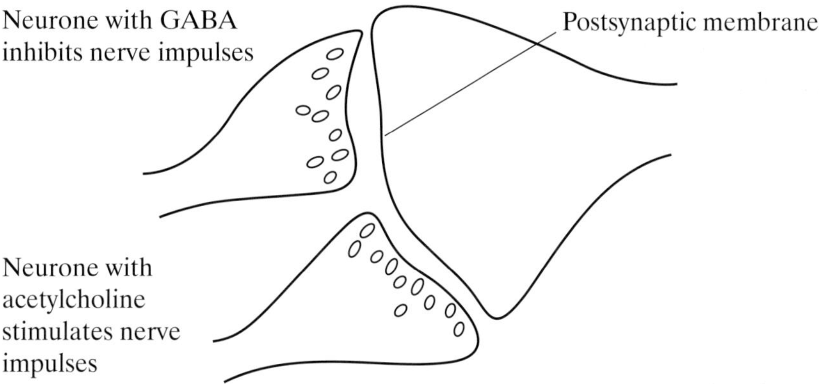

Figure 4

(a) Describe the sequence of events leading to the release of acetylcholine and its binding to the postsynaptic membrane. *(4 marks)*

(b) The binding of GABA to receptors on postsynaptic membranes causes negatively charged chloride ions to enter postsynaptic neurones. Explain how this will inhibit transmission of nerve impulses by postsynaptic neurones. *(3 marks)*

(c) Epilepsy may result when there is increased neuronal activity in the brain.

(i) One form of epilepsy is due to insufficient GABA. GABA is broken down on the postsynaptic membrane by the enzyme GABA transaminase. Vigabatrin is a new drug being used to treat this form of epilepsy. The drug has a similar molecular structure to GABA. Suggest how Vigabatrin may be effective in treating this form of epilepsy.

(ii) A different form of epilepsy has been linked to an abnormality in GABA receptors. Suggest and explain how an abnormality in GABA receptors may result in epilepsy. *(5 marks)*

AQA, 2006

3 **Figure 5** shows the change in the charge across the surface membrane of a non-myelinated axon when an action potential is produced.

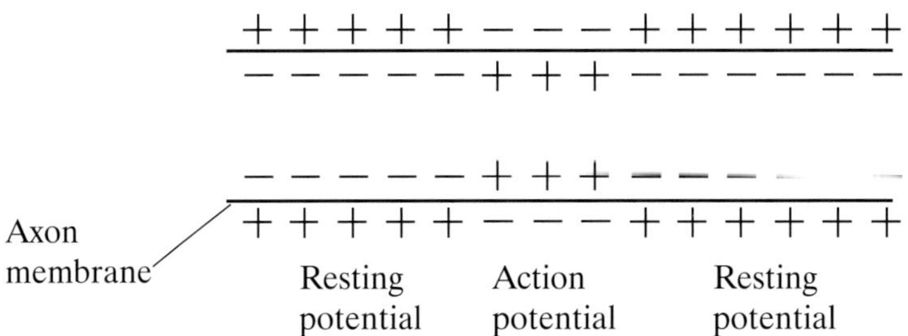

Figure 5

(a) Describe how the change shown in **Figure 5** occurs when an action potential is produced. *(2 marks)*

(b) Explain what causes the conduction of impulses along a non-myelinated axon to be slower than along a myelinated axon. *(3 marks)*

AQA, 2005

4 (a) Describe **two** ways in which hormonal control differs from nervous control. (*2 marks*)

(b) The resting potential of a neurone is maintained at –70mV. A metabolic poison was applied to a neurone and the change in the resting potential was measured over several hours. The results are shown in **Figure 6**.

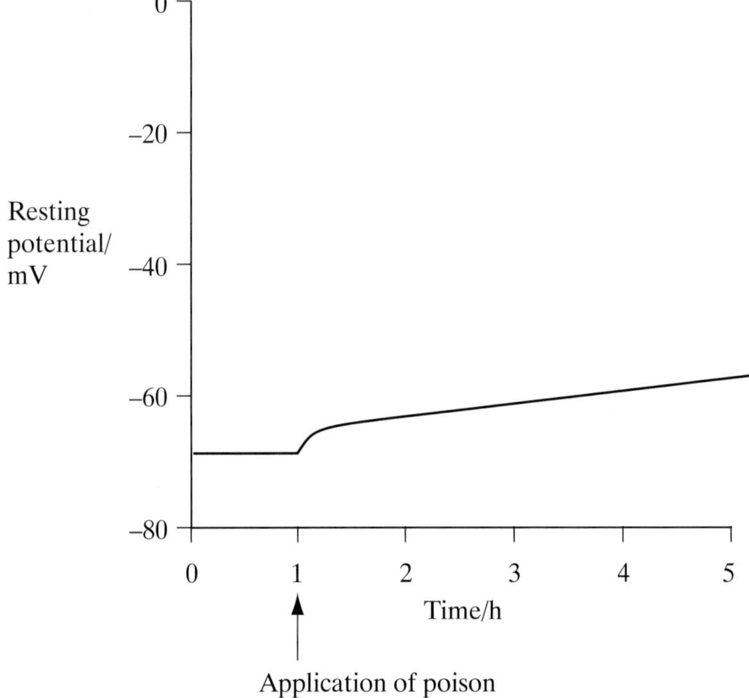

Figure 6

Explain the change in resting potential that takes place after the application of the metabolic poison. (*4 marks*)

AQA, 2007

5 (a) What is a reflex? *(2 marks)*

Figure 7 shows the neurones in a reflex arc.

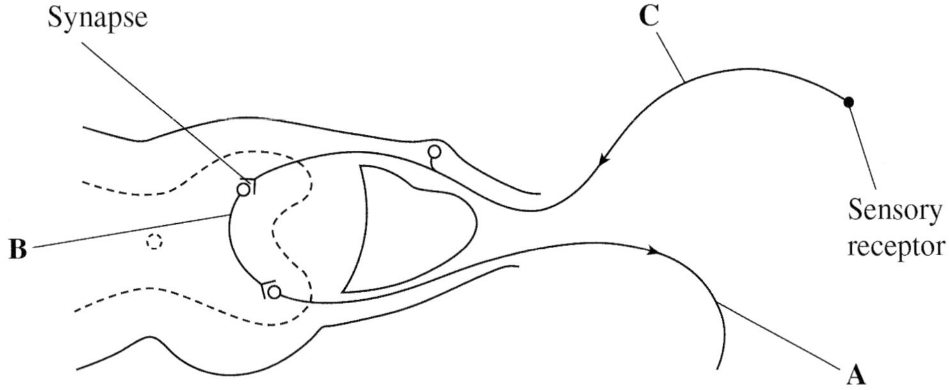

Figure 7

(b) Name the types of neurone labelled **A**, **B** and **C**. *(1 mark)*

(c) Nervous transmission is delayed at synapses. Explain why. *(2 marks)*

(d) The axon of neurone **A** is myelinated. The axon of neurone **B** is non-myelinated. Explain why impulses travel faster along the axon of neurone **A**. *(2 marks)*

AQA, 2008

11 Muscle contraction

11.1 Structure of skeletal muscle

Learning objectives:

- What are the gross and microscopic structure of a skeletal muscle?
- What is the ultrastructure of a myofibril?
- How are actin and myosin arranged within a myofibril?

Specification reference: 3.5.3

Muscles are effector organs that respond to nervous stimulation by contracting and so bring about movement. There are three types of muscle in the body. **Cardiac muscle** is found exclusively in the heart while **smooth muscle** is found in the walls of blood vessels and the gut. Neither of these types of muscle is under conscious control and we remain largely unaware of their contractions. The third type, **skeletal muscle**, makes up the bulk of body muscle in vertebrates. It is attached to bone and acts under voluntary, conscious control.

A rope is made up of millions of separate threads. Each thread has very little individual strength and can easily be snapped. Yet grouped together in a rope, these threads can support a mass running into hundreds of tonnes. In the same way, individual muscles are made up of millions of tiny muscle fibres called **myofibrils**. In themselves, they produce almost no force while collectively they can be extremely powerful. Just as the threads in a rope are lined up parallel to each other in order to maximise its strength, so the myofibrils are arranged in order to give maximum force. And just as the threads of a rope are grouped into strings, the strings are grouped into small ropes and small ropes are grouped into bigger ropes, so muscle is composed of smaller units bundled into progressively larger ones (see Figure 1).

If muscle were made up of individual cells joined end to end it would not be able to perform the function of contraction very efficiently. This is partly because the junction between adjacent cells would be a point of weakness that would reduce the overall strength of the muscle. To overcome this, muscles have a different structure. The separate cells have become fused together into muscle fibres. These muscle fibres share nuclei and also cytoplasm, called **sarcoplasm**, which is mostly found around the circumference of the fibre (Figure 1). Within the sarcoplasm is a large concentration of mitochondria and endoplasmic reticulum.

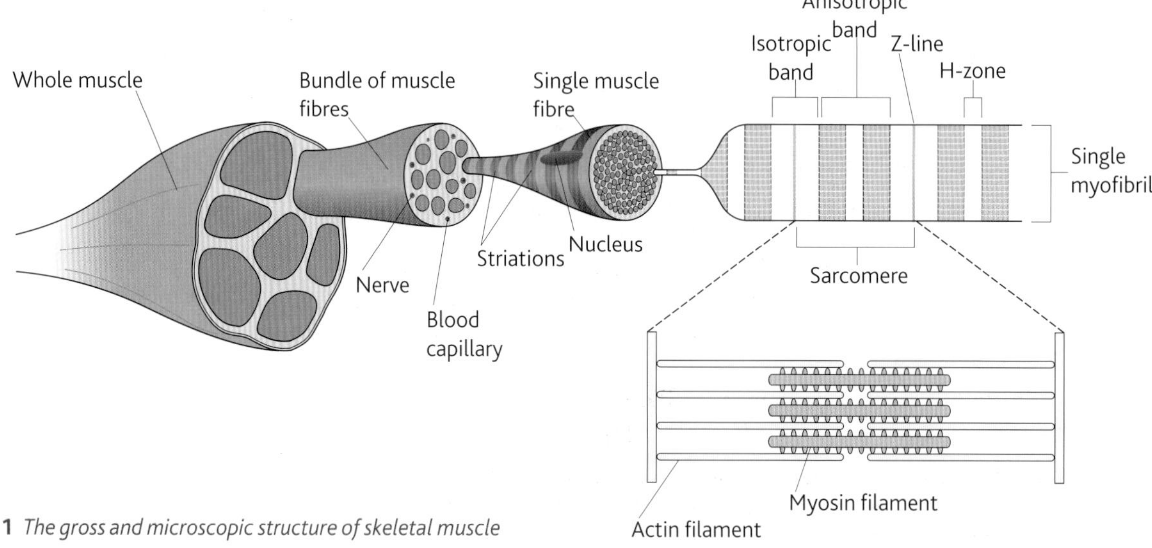

Figure 1 *The gross and microscopic structure of skeletal muscle*

Microscopic structure of skeletal muscle

We can see from Figure 1 that each muscle fibre is made up of myofibrils. Myofibrils are made up of two types of protein filament:

- **actin**, which is thinner and consists of two strands twisted around one another
- **myosin**, which is thicker and consists of long rod-shaped fibres with bulbous heads that project to the side.

The arrangement of these filaments is shown in Figure 2.

Myofibrils appear striped due to their alternating light-coloured and dark-coloured bands. The light bands are called **isotropic bands** (or **I-bands**). They appear lighter because the actin and myosin filaments do not overlap in this region. The dark bands are called **anisotropic bands** (or **A-bands**). They appear darker because the actin and myosin filaments overlap in this region.

At the centre of each anisotropic band is a lighter-coloured region called the **H-zone**. At the centre of each isotropic band is a line called the **Z-line**. The distance between adjacent Z-lines is called a **sarcomere** (Figure 1). When a muscle contracts, these sarcomeres shorten and the pattern of light and dark bands changes (see Topic 11.2).

Two other important proteins are found in muscle:

- **tropomyosin**, which forms a fibrous strand around the actin filament
- a globular protein (**troponin**) involved in muscle contraction.

We will learn more about these in Topic 11.2.

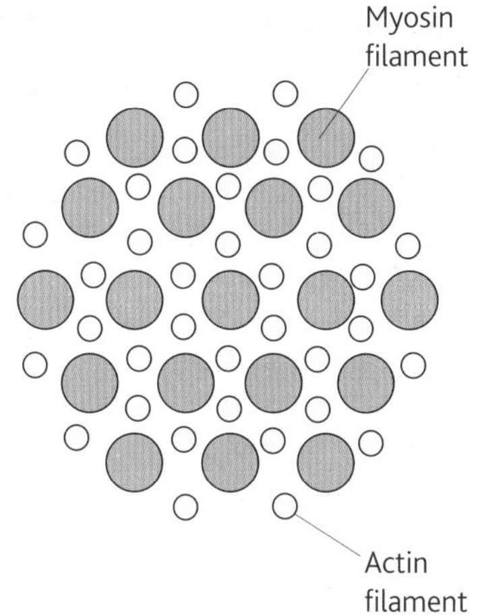

Figure 2 *Transverse section through part of a muscle fibre showing the arrangement of actin and myosin filaments*

Hint

To help you remember which band is the dark band and which is the light band, look at the vowels in 'light' and 'dark'. This vowel is the first letter of the relevant band. Therefore the dark band is the A-band and the light band is the I-band.

Hint

The arrangement of sarcomeres into a long line means that, when one sarcomere contracts a little, the line as a whole contracts a lot! In addition, having the lines of sarcomeres running parallel to each other means that all the force is generated in one direction.

Types of muscle fibre

There are two types of muscle fibre, the proportions of which vary from muscle to muscle and person to person. The two types are:

- **slow-twitch fibres**. These contract more slowly and provide less powerful contractions over a longer period. They are therefore adapted to endurance work, such as running a marathon. In humans they are more common in muscles like the calf muscle, which need to contract constantly to maintain the body in an upright position. They are suited to this role by being adapted for **aerobic** respiration in order to avoid a build-up of lactic acid, which would cause them to function less effectively. These adaptations include having:
  - a large store of myoglobin (a bright red molecule that stores oxygen, which accounts for the red colour of slow-twitch fibres)
  - a supply of glycogen to provide a source of metabolic energy
  - a rich supply of blood vessels to deliver oxygen and glucose
  - numerous mitochondria to produce ATP.
- **fast-twitch fibres**. These contract more rapidly and produce powerful contractions but only for a short period. They are therefore adapted to intense exercise, such as weight-lifting. As a result they are more common in muscles which need to do short bursts of intense activity, like the biceps muscle of the upper arm. Fast-twitch fibres are adapted to their role by having:
  - thicker and more numerous myosin filaments
  - a high concentration of enzymes involved in **anaerobic** respiration
  - a store of phosphocreatine, a molecule that can rapidly generate ATP from ADP in anaerobic conditions and so provide energy for muscle contraction.

Figure 3 *Light micrograph of a neuromuscular junction*

Neuromuscular junctions

A neuromuscular junction is the point where a motor neurone meets a skeletal muscle fibre. There are many such junctions along the muscle. If there were only one junction of this type it would take time for a wave of contraction to travel across the muscle, in which case not all the fibres would contract simultaneously and the movement would be slow. As rapid muscle contraction is frequently essential for survival there are many neuromuscular junctions spread throughout the muscle. This ensures that contraction of a muscle is rapid and powerful when it is simultaneously stimulated by action potentials. All muscle fibres supplied by a single motor neurone act together as a single functional unit and are known as a motor unit. This arrangement gives control over the force that the muscle exerts. If only slight force is needed, only a few units are stimulated. If a greater force is required, a larger number of units are stimulated.

When a nerve impulse is received at the neuromuscular junction, the synaptic vesicles fuse with the presynaptic membrane and release their acetylcholine. The acetylcholine diffuses to the postsynaptic membrane, altering its permeability to sodium ions (Na^+), which enter rapidly, depolarising the membrane. A description of how this leads to the contraction of the muscle is given in Topic 11.2.

The acetylcholine is broken down by acetylcholinesterase to ensure that the muscle is not over-stimulated. The resulting choline and ethanoic acid (acetyl) diffuse back into the neurone, where they are recombined to form acetylcholine using energy provided by the mitochondria found there.

The structure of a neuromuscular junction is shown in Figure 4 and an account of how it functions is provided in Topic 11.2.

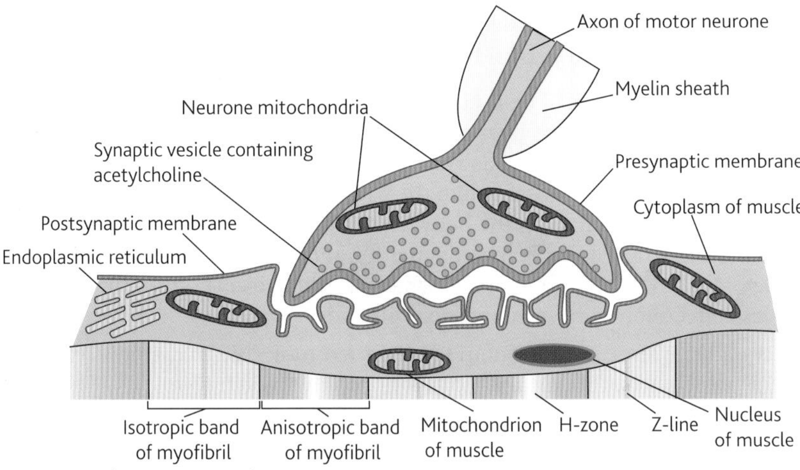

Figure 4 *The neuromuscular junction*

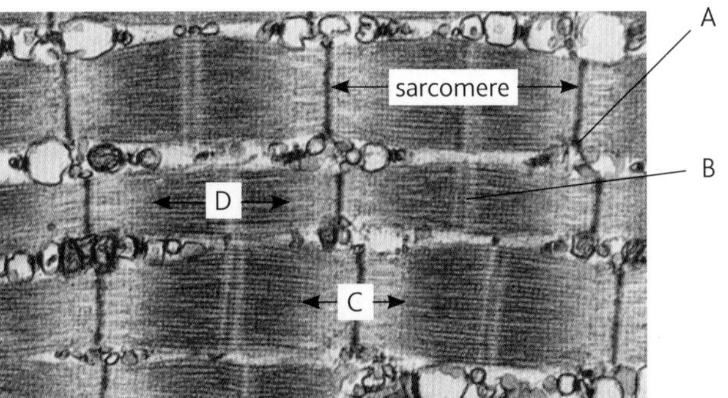

Figure 5 *TEM of skeletal muscle*

Summary questions

1. Suggest a reason why there are numerous mitochondria in the sarcoplasm.
2. Study Figure 5 and name the structures labelled A–D.
3. If we cut across a myofibril at certain points, we see only thick myosin filaments. Cut at a different point we see only thin actin filaments. At yet other points we see both types of filament. Explain why.
4. How do slow-twitch fibres differ from fast-twitch fibres in the way they function?
5. Describe how each type of fibre is adapted to its functions.

11.2 Contraction of skeletal muscle

Learning objectives:

- What evidence supports the sliding filament mechanism of muscle contraction?
- How does the sliding filament mechanism cause a muscle to contract and relax?
- Where does the energy for muscle contraction come from?

Specification reference: 3.5.3

Now we have looked at the structure of skeletal muscle in Topic 11.1, let us turn our attention to how exactly the arrangement of the various proteins brings about contraction of the muscle fibre. The process involves the actin and myosin filaments sliding past one another and is therefore called the **sliding filament mechanism**.

Evidence for the sliding filament mechanism

In Topic 11.1 we saw that myofibrils appear darker in colour where the actin and myosin filaments overlap and lighter where they do not. If the sliding filament mechanism is correct, then there will be more overlap of actin and myosin in a contracted muscle than in a relaxed one. If you look at Figure 1, you will see that, when a muscle contracts, the following changes occur to a **sarcomere**:

- The I-band becomes narrower.
- The Z-lines move closer together or, in other words, the sarcomere shortens.
- The H-zone becomes narrower.

The A-band remains the same width. As the width of this band is determined by the length of the myosin filaments, it follows that the myosin filaments have not become shorter. This discounts the theory that muscle contraction is due to the filaments themselves shortening.

Link

The functioning of muscle depends on the molecular shapes of the four main proteins involved.
The importance of shape on the functioning of proteins is covered in Topic 2.5 of the *AS Biology* book.

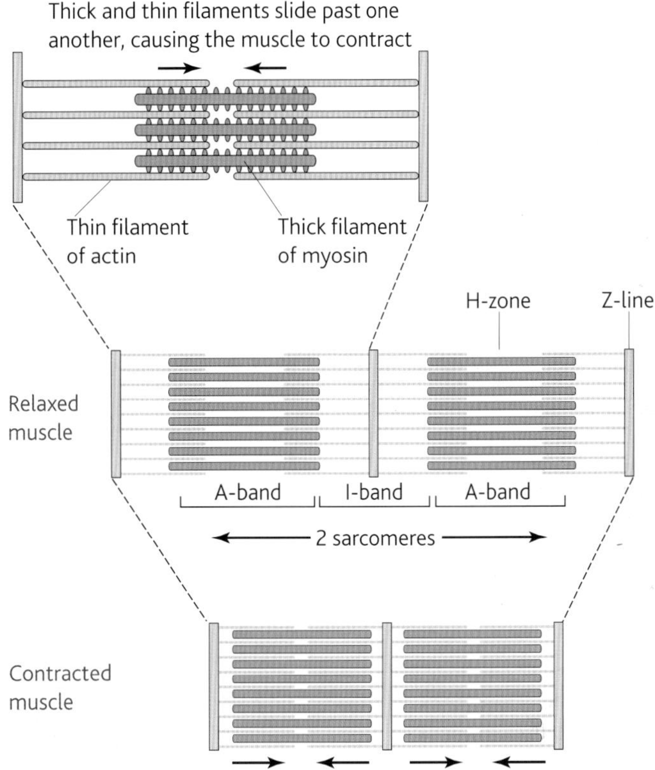

Figure 1 *Comparison of two sarcomeres in a relaxed and a contracted muscle*

Before we look at how the sliding filament mechanism works, let us take a closer look at the three main proteins involved in the process:

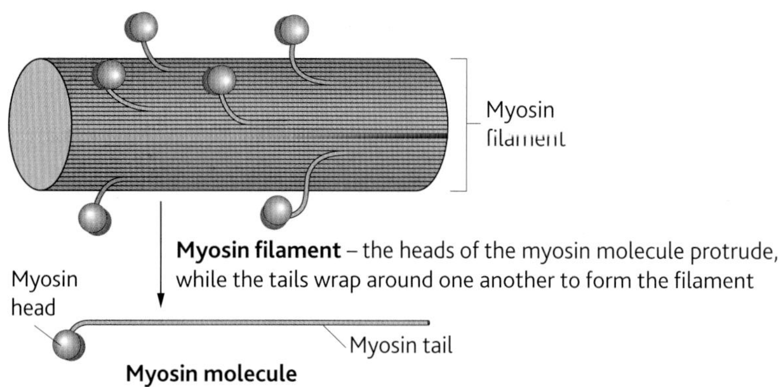

Figure 2 *Structure of myosin*

- Myosin (see Figure 2) is made up of two types of protein:
  - a fibrous protein arranged into a filament made up of several hundred molecules (the tail)
  - a globular protein formed into two bulbous structures at one end (the head).
- Actin is a globular protein whose molecules are arranged into long chains that are twisted around one another to form a helical strand.
- Tropomyosin forms long thin threads that are wound around actin filaments.

The arrangement of the molecules of actin and tropomyosin are shown in Figure 3.

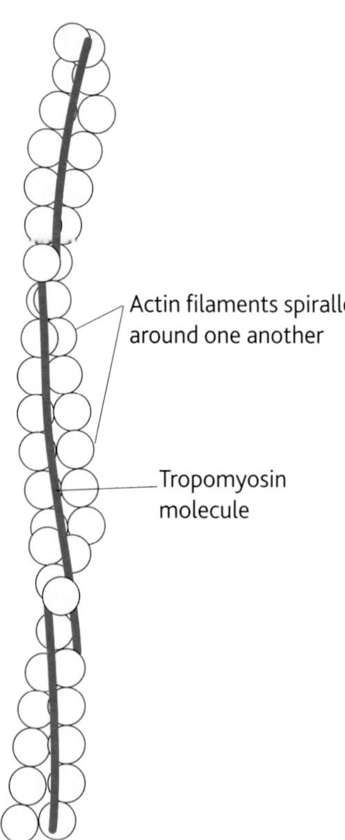

Figure 3 *The relationship of tropomyosin to an actin filament*

The sliding filament mechanism of muscle contraction

The hypothesis that actin and myosin filaments slide past one another during muscle contraction is supported by the changes seen in the band pattern on myofibrils. The next question for the scientists was: by what mechanism do the filaments slide past one another? Clues to the answer lie in the shape of the various proteins involved.

To summarise, the bulbous heads of the myosin filaments form cross-bridges with the actin filaments. They do this by attaching themselves to binding sites on the actin filaments, and then flexing in unison, pulling the actin filaments along the myosin filaments. They then become detached and, using ATP as a source of energy, return to their original angle and re-attach themselves further along the actin filaments. This process is repeated up to 100 times a second. The action is similar to the way a ratchet operates.

The following account describes the sliding filament mechanism of muscle contraction in detail. The process is continuous but, for ease of understanding, has been divided into stimulation, contraction and relaxation.

Hint

The action of the myosin heads is similar to the rowing action of oarsmen in a boat. The oars (myosin heads) are dipped into the water, flexed as the oarsmen pull on them, removed from the water and then dipped back into the water further along. The oarsmen work in unison and the boat and water move relative to one another.

Muscle stimulation

- An **action potential** reaches many **neuromuscular junctions** simultaneously, causing calcium ion channels to open and calcium **ions** to move into the synaptic knob.
- The calcium ions cause the synaptic vesicles to fuse with the presynaptic membrane and release their **acetylcholine** into the synaptic cleft.
- Acetylcholine diffuses across the synaptic cleft and binds with receptors on the postsynaptic membrane, causing it to depolarise.

> **Hint**
>
> An action potential is the result of depolarisation of part of a membrane. The spread of the action potential across the muscle is therefore often referred to as a 'wave of depolarisation'.

Muscle contraction

- The action potential travels deep into the fibre through a system of tubules (T-tubules) that branch throughout the cytoplasm of the muscle (sarcoplasm).
- The tubules are in contact with the endoplasmic reticulum of the muscle (sarcoplasmic reticulum) which has actively absorbed calcium ions from the cytoplasm of the muscle.
- The action potential opens the calcium ion channels on the endoplasmic reticulum and calcium ions flood into the muscle cytoplasm down a diffusion gradient.
- The calcium ions cause the tropomyosin molecules that were blocking the binding sites on the actin filament to pull away (Figure 4, stages 1 and 2).
- The ADP molecule attached to the myosin heads means they are now in a state to bind to the actin filament and form a cross-bridge (Figure 4, stage 3).
- Once attached to the actin filament, the myosin heads change their angle, pulling the actin filament along as they do so and releasing a molecule of ADP (Figure 4, stage 4).
- An ATP molecule attaches to each myosin head, causing it to become detached from the actin filament (Figure 4, stage 5).
- The calcium ions then activate the enzyme ATPase, which hydrolyses the ATP to ADP. The hydrolysis of ATP to ADP provides the energy for the myosin head to return to its original position (Figure 4, stage 6).
- The myosin head, once more with an attached ADP molecule, then reattaches itself further along the actin filament and the cycle is repeated as long as nervous stimulation of the muscle continues (Figure 4, stage 7).

> **Hint**
>
> The hydrolysis of ATP releases energy and produces ADP and inorganic phosphate (P_i). For simplicity, this account just refers to ADP as the product.

Muscle relaxation

- When nervous stimulation ceases, calcium ions are actively transported back into the endoplasmic reticulum using energy from the **hydrolysis** of ATP.
- This reabsorption of the calcium ions allows tropomyosin to block the actin filament again.
- Myosin heads are now unable to bind to actin filaments and contraction ceases, i.e. the muscle relaxes.

1 Tropomyosin molecule prevents myosin head from attaching to the binding site on the actin molecule.

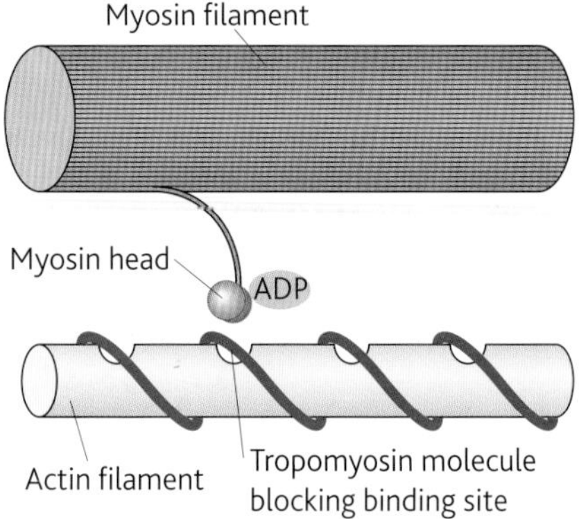

2 Calcium ions released from the endoplasmic reticulum cause the tropomyosin molecule to pull away from the binding sites on the actin molecule.

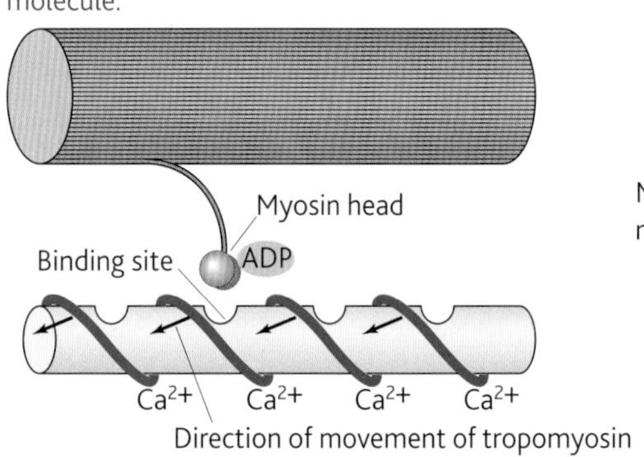

3 Myosin head now attaches to the binding site on the actin filament.

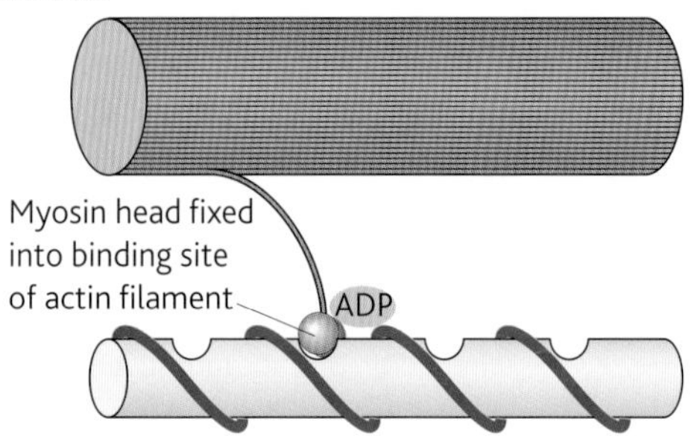

4 Head of myosin changes angle, moving the actin filament along as it does so. The ADP molecule is released.

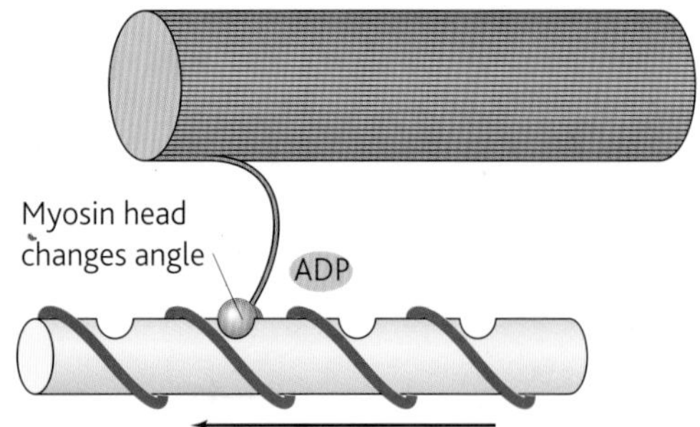

5 ATP molecule fixes to myosin head, causing it to detach from the actin filament.

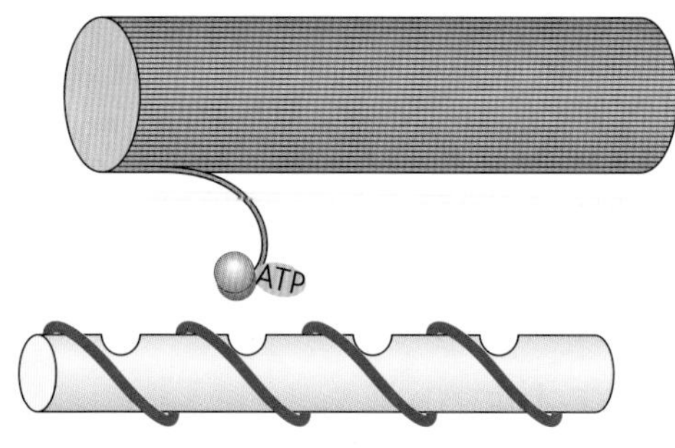

6 Hydrolysis of ATP to ADP by ATPase provides the energy for the myosin head to resume its normal position.

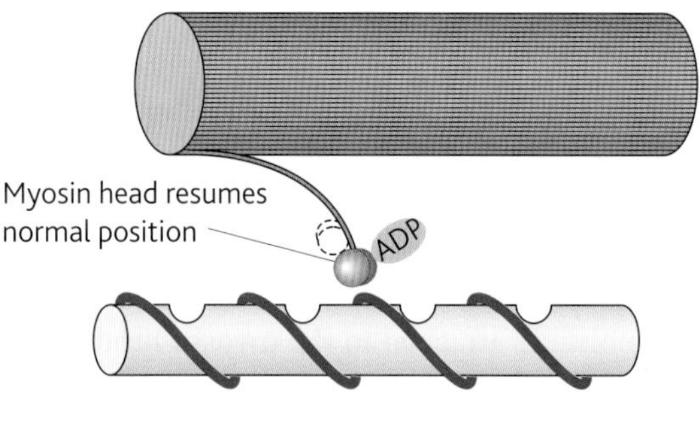

7 Head of myosin reattaches to a binding site further along the actin filament and the cycle is repeated.

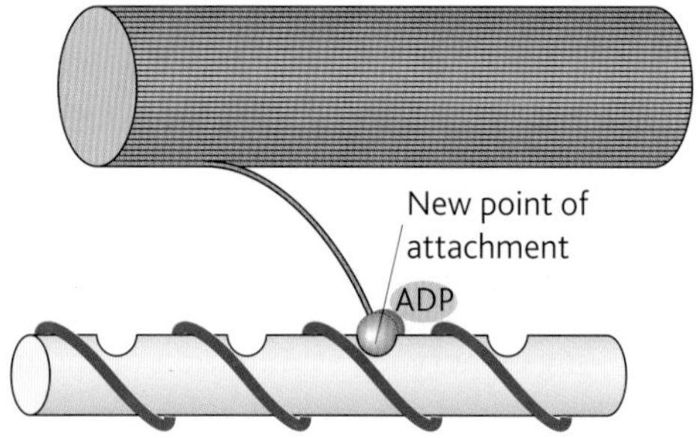

Figure 4 *Sliding filament mechanism of muscle contraction (showing only one myosin head throughout)*

Energy supply during muscle contraction

Muscle contraction requires considerable energy. This is supplied by the hydrolysis of ATP to ADP and inorganic phosphate (P_i). The energy released is needed for:

- the movement of the myosin heads
- the reabsorption of calcium ions into the endoplasmic reticulum by active transport.

In an active muscle, there is clearly a great demand for ATP. In some circumstances, for example, escaping from danger, the ability of muscles to work intensely can be life-saving. Most ATP is regenerated from ADP during the respiration of pyruvate in the mitochondria, which are particularly plentiful in the muscle. However, this process requires oxygen. In a very active muscle oxygen is rapidly used up and it takes time for the blood supply to replenish it. Therefore a means of rapidly generating ATP **anaerobically** is also required. This is achieved using a chemical called **phosphocreatine.**

Figure 5 *Marathon runners undergoing strenuous exercise*

Phosphocreatine cannot supply energy directly to the muscle, so instead it regenerates ATP, which can. Phosphocreatine is stored in muscle and acts as a reserve supply of phosphate, which is available immediately to combine with ADP and so re-form ATP. The phosphocreatine store is replenished using phosphate from ATP when the muscle is relaxed.

Summary questions

1. How is the shape of the myosin molecule adapted to its role in muscle contraction?
2. Trained sprinters have high levels of phosphocreatine in the muscles. Explain the advantage of this.
3. During the contraction of a muscle sarcomere, a single actin filament moves 0.8 µm. If the hydrolysis of a single ATP molecule provides enough energy to move an actin filament 40 nm, how many ATP molecules are needed to move the actin filament 0.8 µm? Show your working.
4. Dead cells can no longer produce ATP. Soon after death, muscles contract, making the body stiff – a state known as *rigor mortis.* From your knowledge of muscle contraction, explain the reasons why *rigor mortis* occurs after death.

AQA Examination-style questions

1 **Figure 1** shows a diagram of part of a muscle myofibril.

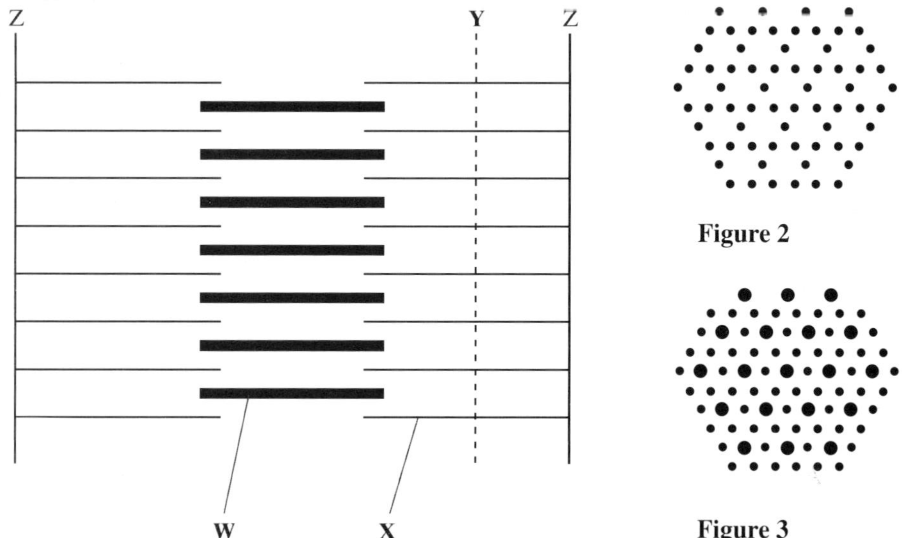

Figure 2

Figure 3

Figure 1

(a) Name the protein present in the filaments labelled **W** and **X**. *(1 mark)*

(b) **Figure 2** shows the cut ends of the protein filaments when the myofibril was cut at position **Y**. **Figure 3** shows the protein filaments when the myofibril was cut at the same distance from a Z-line at a different stage of contraction. Explain why the pattern of protein filaments differs in **Figure 2** and **Figure 3**. *(2 marks)*

(c) Describe the role of calcium ions in the contraction of a myofibril. *(4 marks)*

AQA, 2004

2 **Figure 4** shows the stages in one cycle that results in movement of an actin filament in a muscle myofibril.

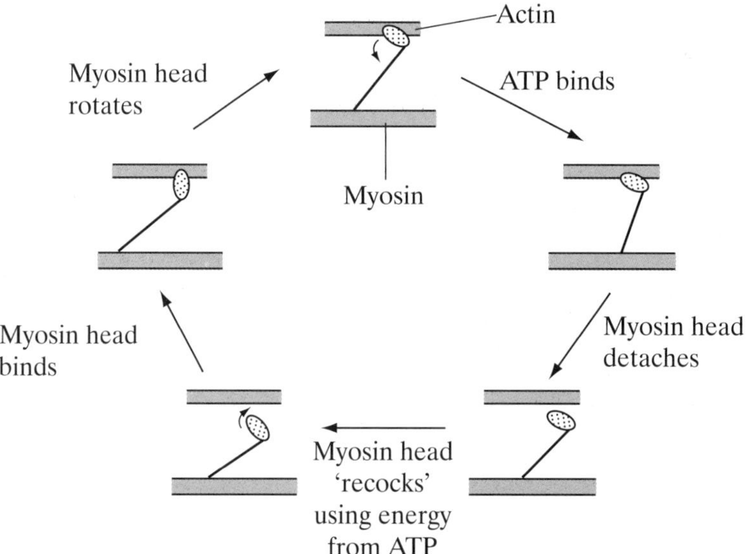

Figure 4

(a) Describe how stimulation of a muscle by a nerve impulse starts the cycle shown in **Figure 4**. *(3 marks)*

(b) Each cycle requires hydrolysis of one molecule of ATP and moves one actin filament 40 nm. During contraction of a muscle sarcomere, a single actin filament moves 0.6 μm. Calculate how many molecules of ATP are required to produce this movement. *(2 marks)*

(c) After death, cross bridges between actin and myosin remain firmly bound resulting in rigor mortis. Using information in the diagram, explain what causes the cross bridges to remain firmly bound. *(2 marks)*

AQA, 2005

3 (a) **Figure 5** shows the banding pattern observed in part of a relaxed muscle fibril.

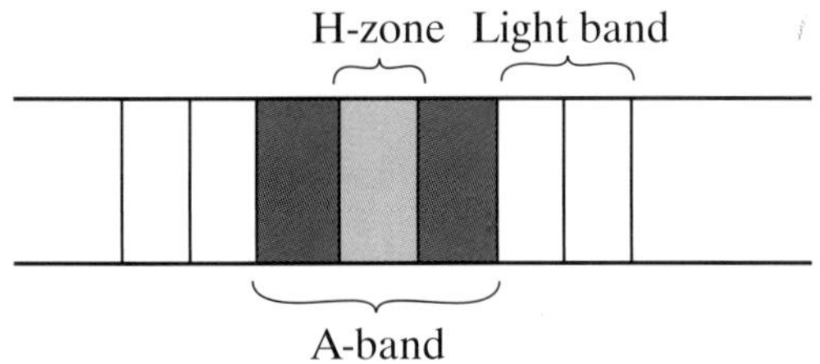

Figure 5

(i) Describe what causes the different bands seen in the muscle myofibril.

(ii) Describe how the banding pattern will be different when the muscle fibril is contracted. *(4 marks)*

(b) There is an increase in the activity of the enzyme ATPase during muscle contraction. An investigation into muscle contraction involved measuring the activity of ATPase in solutions containing ATP, myosin and different muscle components. The table shows the results.

Solution	Contents	ATPase activity / arbitrary units
A	ATP, myosin and actin	1.97
B	ATP, myosin, actin and tropomyosin	0.54
C	ATP, myosin, actin, tropomyosin and calcium ions	3.85

(i) Explain the importance of ATPase during muscle contraction.

(ii) Using your knowledge of muscle contraction, explain the difference in the results between:
A and **B**;
B and **C**. *(6 marks)*

AQA, 2006

4 Skeletal muscle is made of bundles of fibres.

(a) Describe the role of ATP and phosphocreatine in producing contraction of a muscle fibre. *(3 marks)*

(b) The table shows some properties of slow and fast muscle fibres.

Endurance athletes, such as marathon runners, nearly always have a high proportion of slow fibres in their muscles. Explain the benefit of this. *(6 marks)*

Property of fibre	Slow fibres	Fast fibres
Number of mitochondria per fibre	Many	Few
Concentration of enzymes regulating glycolysis	Moderate	High
Resistance to fatigue	High	Low

AQA, 2002

12 Homeostasis

12.1 Principles of homeostasis

Learning objectives:

- What is homeostasis?
- What is the importance of homeostasis?
- How do control mechanisms work?
- How are control mechanisms coordinated?

Specification reference: 3.5.4

In the previous chapter we looked at how organisms develop systems to control and coordinate their activities as they become more complex. In particular we considered the way in which organisms respond rapidly to environmental changes using their nervous system. Another feature of an increase in complexity is the ability of organisms to control their internal environment. By maintaining a relatively constant internal environment for their cells, organisms can limit the external changes these cells experience. This maintenance of a constant internal environment is called **homeostasis**. In this chapter we shall learn about homeostasis and the role of the other coordination system, hormonal coordination, in an organism's physiological control.

The internal environment is made up of **tissue fluids** that bathe each cell, supplying nutrients and removing wastes. Maintaining the features of this fluid at the optimum levels protects the cells from changes in the external environment, thereby giving the organism a degree of independence.

What is homeostasis?

Homeostasis is the maintenance of a constant internal environment in organisms. It involves maintaining the chemical make-up, volume and other features of blood and tissue fluid within restricted limits. Homeostasis ensures that the cells of the body are in an environment that meets their needs and allows them to function normally despite external changes. This does not mean that there are no changes. On the contrary, there are continuous fluctuations brought about by variations in internal and external conditions, such as changes in temperature, pH and water potential. These changes, however, occur around a set point. Homeostasis is the ability to return to that set point and so maintain organisms in a balanced equilibrium.

The importance of homeostasis

Homeostasis is essential for the proper functioning of organisms for the following reasons:

- The enzymes that control the biochemical reactions within cells, and other proteins, such as channel proteins, are sensitive to changes in pH and temperature. Any change to these factors reduces the efficiency of enzymes or may even prevent them working altogether, for example, by denaturing them. Even small fluctuations in temperature or pH can impair the ability of enzymes to carry out their roles effectively. Maintaining a constant internal environment means that reactions take place at a constant and predictable rate.
- Changes to the **water potential** of the blood and tissue fluids may cause cells to shrink and expand (even to bursting point) as a result of water leaving or entering by osmosis. In both instances the cells cannot operate normally. The maintenance of a constant blood glucose concentration is essential in ensuring a constant water

Link

The importance of temperature and pH in relation to enzyme activity is covered in Topic 2.7 of the *AS Biology* book.

Link

Water potential and its importance to cells is covered in Topic 3.7 of the *AS Biology* book.

potential. A constant blood glucose concentration also ensures a reliable source of glucose for respiration by cells.

- Organisms with the ability to maintain a constant internal environment are more independent of the external environment. They have a wider geographical range and therefore have a greater chance of finding food, shelter, etc. Mammals, for example, with their ability to maintain a constant temperature, are found in most habitats, ranging from hot arid deserts to cold, frozen polar regions.

Figure 1 *Homeostasis allows animals such as these camels in the desert (top) and these penguins in the Antarctic (bottom) to survive in extreme environments*

Control mechanisms

The control of any self-regulating system involves a series of stages that feature:

- the **set point**, which is the desired level, or norm, at which the system operates. This is monitored by a ...
- **receptor**, which detects any deviation from the set point and informs the ...
- **controller**, which coordinates information from various receptors and sends instructions to an appropriate ...
- **effector**, which brings about the changes needed to return the system to the set point. This return to normality creates a ...
- **feedback loop**, which informs the receptor of the changes to the system brought about by the effector.

Figure 2 illustrates the relationship between these stages using the everyday example of controlling a central heating system.

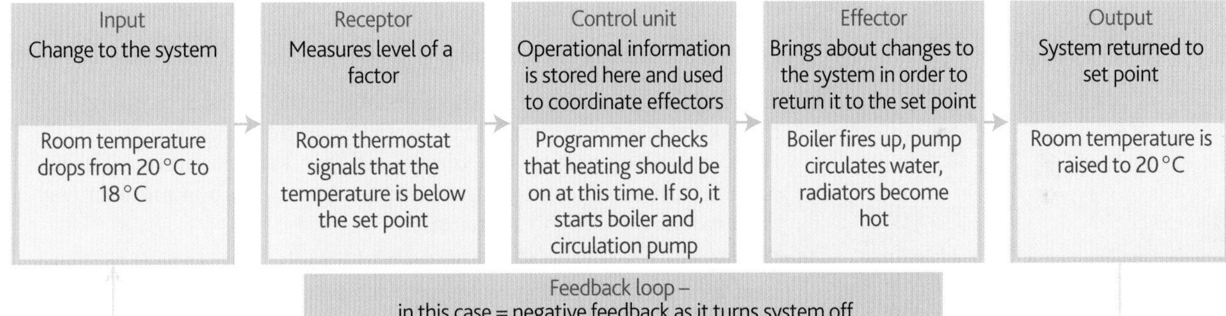

Figure 2 *Components of a typical control system*

Coordination of control mechanisms

Systems normally have many receptors and effectors. It is important to ensure that the information provided by receptors is analysed by the control centre before action is taken. Receiving information from a number of sources allows a better degree of control. For example, temperature receptors in the skin may signal that the skin itself is cold and that body temperature should be raised. However, information from the temperature centre in the brain may indicate that blood temperature is already above normal. This situation could arise during strenuous exercise when blood temperature rises but sweating cools the skin. By analysing the information from all the detectors, the brain can decide the best course of action – in this case not to raise the body temperature further. In the same way, the control centre must coordinate the action of the effectors so that they operate harmoniously. For example, sweating would be less effective in cooling the body if it were not accompanied by **vasodilation**.

Summary questions

1. What is homeostasis?
2. Explain why maintaining a constant temperature is important in mammals.
3. Explain why maintaining a constant blood glucose concentration is important in mammals.

12.2 Regulation of body temperature

Learning objectives:

- What are the main ways in which heat is gained and lost in organisms?
- How is body temperature regulated in ectotherms?
- How is body temperature regulated in endotherms?
- How is body temperature regulated in mammals?

Specification reference: 3.5.4

If an organism's body temperature is too low, the rate at which enzyme-controlled reactions take place may be too slow for the organism to function properly. Equally, if the body temperature is too high, enzymes may be denatured and the organism may cease to function altogether. Therefore, in order to survive, many animals need to regulate their body temperature – a process called thermoregulation.

Mechanisms of heat loss and gain

All animals can control their body temperature to a greater or lesser degree. To do so, they must control heat gain and heat loss by one or more of the following means.

Methods of gaining heat include:

- **production of heat** by the metabolism of food during respiration
- **gain of heat from the environment** by conduction (e.g. from the ground), convection (e.g. from surrounding air or water) and radiation.

Methods of losing heat include:

- **evaporation of water** (e.g. during sweating)
- **loss of heat to the environment** by conduction (e.g. to the ground), convection (e.g. to the surrounding air or water) and radiation.

It is worth explaining these terms:

- **Conduction** occurs mainly in solids and is the transfer of energy through matter from particle to particle (i.e. from atom to atom). Heat causes the particles to vibrate and gain **kinetic energy**. These particles cause adjacent particles to vibrate and so the kinetic energy is transferred through the material.
- **Convection** occurs in fluids (gases and liquids) and is the transfer of heat as a result of the movement of the warmed matter itself. The heat causes the fluid to expand and move, carrying with it the heat that it has absorbed.
- **Radiation** is very different from conduction and convection. The energy is not transferred by the movement of particles but by electromagnetic waves. When these waves hit an object they normally heat it up.

Animals such as birds and mammals derive most of their heat from the metabolic activities that take place inside their bodies. They are therefore known as **endotherms** ('inside heat'). All other animals obtain a large proportion of their heat from sources outside their bodies, namely the environment. They are therefore known as **ectotherms** ('outside heat').

Hint

If you think of heat as a message that needs to be transferred between two people some distance apart, then **conduction** is like passing the message on a piece of paper along a line of stationary people standing close to one another, **convection** is like a person carrying the piece of paper from the sender to the receiver, and **radiation** is like transmitting the message by radio waves rather than by transferring it on a piece of paper.

Regulation of body temperature in ectotherms

Ectotherms gain most of their heat from the environment, so their body temperature fluctuates with that of the environment. They therefore control their body temperature by adapting their behaviour to changes in the external temperature. Reptiles, such as lizards, are ectotherms. Lizards cannot warm up by exercising because, if their body temperature is low, they cannot respire fast enough to provide the energy for rapid movement. Instead, they control their body temperature by:

- **exposing themselves to the Sun**. In order to gain heat lizards orientate themselves so that the maximum surface area of their body is exposed to the warming rays of the Sun.
- **taking shelter**. Lizards will shelter in the shade to prevent over-heating when the Sun's radiation is at its peak. At night they retreat into burrows in order to reduce heat loss when the external temperature is low.
- **gaining warmth from the ground**. Lizards will press their bodies against areas of hot ground to warm themselves up. When the required temperature is reached, they raise themselves off the ground on their legs.
- **generating metabolic heat**. Although not the main source of heat, respiration still provides a proportion of a lizard's body heat.
- **colour variations**. Darker colours absorb more heat while lighter colours reflect more heat. Lizards in colder environments are generally darker in colour than those in warmer areas.

Figure 1 *A lizard showing thermoregulatory behaviour by gaining heat both from the sun and the warm rock*

Regulation of body temperature in endotherms

Endotherms gain most of their heat from internal metabolic activities. Their body temperature remains relatively constant despite fluctuations in the external temperature. The core body temperature of endotherms is usually in the range 35–44 °C. This range is a compromise between having a higher temperature at which enzymes work more rapidly and the amount of energy (and hence food) needed to maintain that higher temperature. Like ectotherms, endothermic animals use behaviour to maintain a constant body temperature. They will shelter from cold wind or hot sunshine, curl up when it is cold and spread out more when it is hot. Unlike ectotherms, however, they also use a wide range of physiological mechanisms to regulate their temperature.

Conserving and gaining heat in response to a cold environment

Mammals and birds that live in cold climates have evolved a number of genetic adaptations in order to survive in these environments. One of the most important is having a body with a small surface area to volume ratio. It is from within the 'volume' that heat is produced and from the 'surface area' that heat is lost. It follows that the smaller the surface area to volume ratio the easier it is to maintain a high body temperature in a cold environment. Mammals and birds in cold climates therefore tend to be relatively large, for example, the polar bear and penguin. When compared to animals in warmer climates they also have smaller extremities, such as ears, and thick fur, feathers or fat layers to insulate the body.

While some of these features can be varied over a period of time, for example, more fat can be laid down in preparation for the colder seasons of the year, the response is not rapid enough to adapt to daily temperature variations. To make rapid changes, mammals rely on one or more of the following mechanisms:

- **vasoconstriction**. The diameter of the arterioles near the surface of the skin is made smaller. This reduces the volume of blood reaching the skin surface through the capillaries. Most of the blood entering the skin therefore passes beneath the insulating layer of fat and so loses little heat to the environment (Figure 3).

Figure 2 *The penguin (top) and the polar bear both have large compact bodies with a small surface area to volume ratio. This helps them to conserve heat in the cold environments of the South and North Poles where they live*

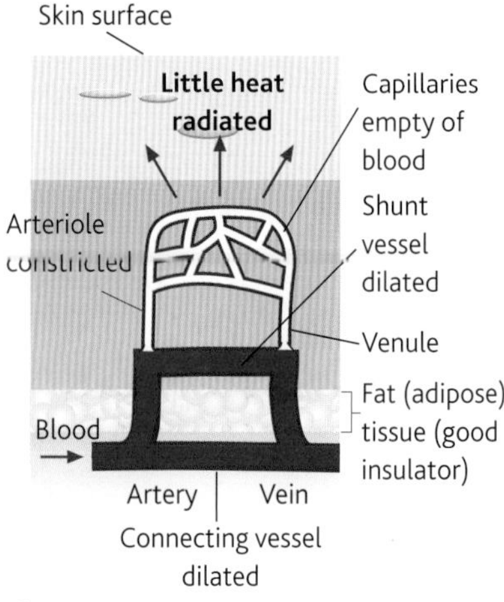

Figure 3 *Vasoconstriction*

- **shivering**. The muscles of the body undergo involuntary rhythmic contractions that produce metabolic heat.
- **raising of hair**. The hair erector muscles in the skin contract, thus raising the hairs on the body. This enables a thicker layer of still air, which is a good insulator, to be trapped next to the skin, thus improving insulation and conserving heat in mammals with thick fur.
- **increased metabolic rate**. In cold conditions more of the hormones that increase metabolic rate are produced. As a result metabolic activity, including respiration, is increased and so more heat is generated.
- **decrease in sweating**. Sweating is reduced, or ceases altogether, in cold conditions.
- **behavioural mechanisms**. Sheltering from the wind, basking in the sun and huddling together all help animals to maintain their core body temperature.

Losing heat in response to a warm environment

Long-term adaptations to life in a warm climate include having a large surface area to volume ratio (often due to large extremities, such as ears) and lighter coloured fur to reflect heat. Rapid responses that enable heat to be lost when the environmental temperature is high include:

- **vasodilation**. The diameter of the arterioles near the surface of the skin becomes larger. This allows warm blood to pass close to the skin surface through the capillaries. The heat from this blood is then radiated away from the body (Figure 4).
- **increased sweating**. To evaporate water from the skin surface requires energy in the form of heat. In relatively hairless mammals, such as humans, sweating is a highly effective means of losing heat. In mammals with fur, cooling is achieved by the evaporation of water from the mouth and tongue, which occurs when air is rapidly passed over these surfaces during panting.
- **lowering of body hair**. The hair erector muscles in the skin relax and the elasticity of the skin causes them to flatten against the body. This reduces the thickness of the insulating layer and allows more heat to be lost to the environment when the internal temperature is higher than the external temperature.
- **behavioural mechanisms**. Avoiding the heat of the day by sheltering in burrows and seeking out shade help to prevent the body temperature from rising.

Much heat radiated
Capillaries full of blood
Shunt vessel constricted
Venule
Arteriole dilated
Blood
Artery
Vein
Connecting vessel constricted

Figure 4 *Vasodilation*

Figure 5 *The Arctic fox (left) lives in cold climates and has short ears to reduce the surface area of the body and thus conserve heat. The fennec fox, by contrast, lives in hot climates and has long ears to increase the surface area of the body and thus increase heat loss*

Hint

As humans have very little hair compared to other mammals, hair erection is ineffective at conserving heat. It takes place all the same and can be observed by the raised areas of skin around each hair, known as 'goose pimples'.

Control of body temperature

Regulation of core body temperature in endotherms is an example of **homeostasis**. In this case, the stimulus (a change in body temperature) is detected by receptors (thermorcceptors), which pass the information to a coordinator (the **hypothalamus**) in the brain, which then causes an effector (the skin) to produce the appropriate response (an increase or decrease in core temperature). Within the hypothalamus there is a thermoregulatory centre consisting of two parts:

- **a heat gain centre**, which is activated by a fall in blood temperature. This controls the mechanisms that increase body temperature.
- **a heat loss centre**, which is activated by a rise in blood temperature. This controls the mechanisms that decrease body temperature.

The hypothalamus monitors the temperature of the blood passing through it. In addition, the thermoreceptors in the skin measure skin temperature. These thermoreceptors send impulses along the **autonomic nervous system** to the hypothalamus. They provide information on the environmental temperature and so give advanced warning of potential changes in the core body temperature. The animal can therefore take measures to conserve or lose heat as appropriate, before the core temperature is affected. The two sets of thermoreceptors (in the hypothalamus and the skin) interact to control temperature as shown in Figure 6. Of the two, it is the core temperature, as measured in the blood passing through the hypothalamus, which is most important.

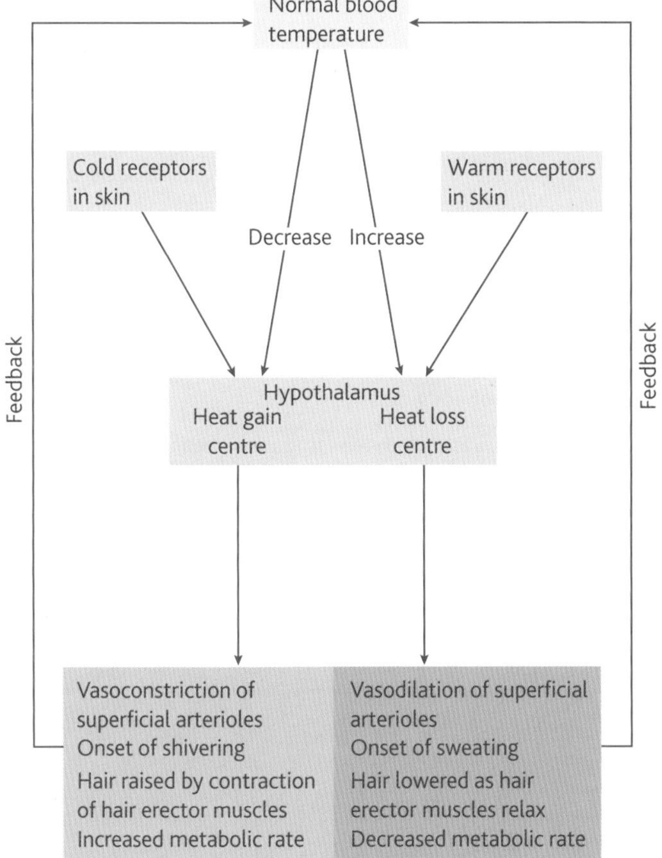

Figure 6 *Summary of body temperature control by the hypothalamus and autonomic nervous system*

Summary questions

1. State the **three** ways by which heat is lost to, and gained from, the environment.
2. Name **four** structures in the skin that are involved with thermoregulation in mammals.
3. Explain why a large, compact mammal will probably require less food per gram of body mass to maintain its body temperature in a cold climate than a small, less compact mammal.
4. Explain why the skin of humans often appears paler on a cold day than on a hot day. What is the significance of this difference?

Application

Comparing heat gain and loss in ectotherms and endotherms

The graphs shown in Figure 7 compare the rates of metabolic heat generation and evaporative heat loss in a mammal and a reptile as the environmental temperature changes.

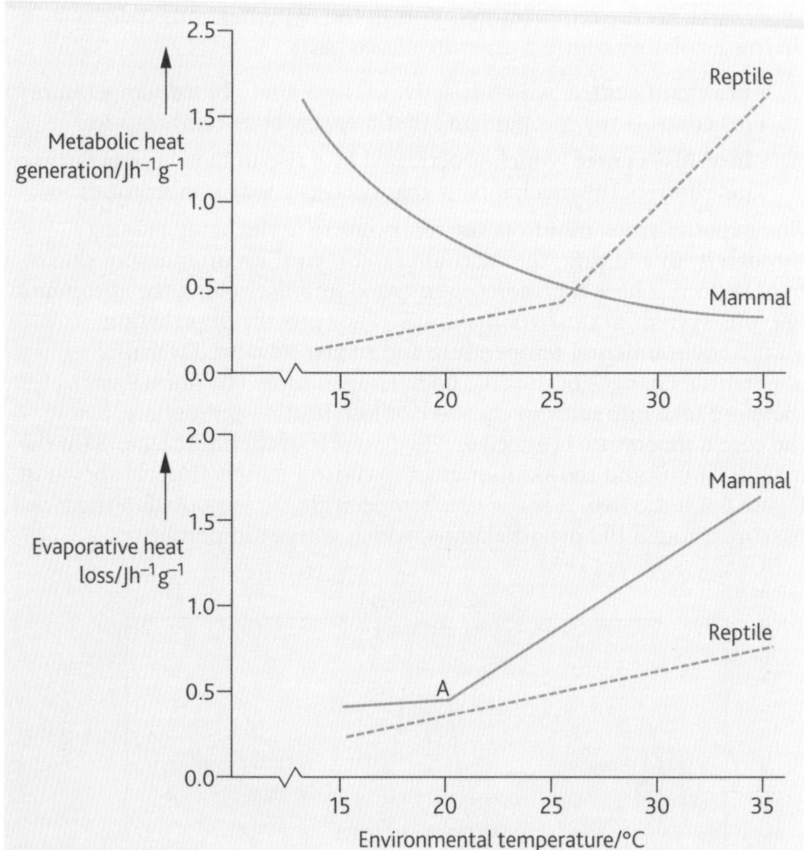

Figure 7

1. Give a reason why the values for heat generation and heat loss are measured per gram of body mass.
2. a Describe the relationship between metabolic heat generation and evaporative heat loss in a reptile.
   b How does this relationship differ in a mammal?
3. Reptiles frequently seek shade when the environmental temperature rises above 25° C. Use the graphs to explain this type of behaviour.
4. Suggest a reason for the change in the evaporative heat loss in the mammal at point A on the graph.

Application

Role of the hypothalamus in temperature regulation

In an experiment to determine the role of the hypothalamus, a ground squirrel had its hypothalamus alternately cooled and warmed. At the same time the metabolic rate and body temperature were measured. The results are shown in the graphs in Figure 8.

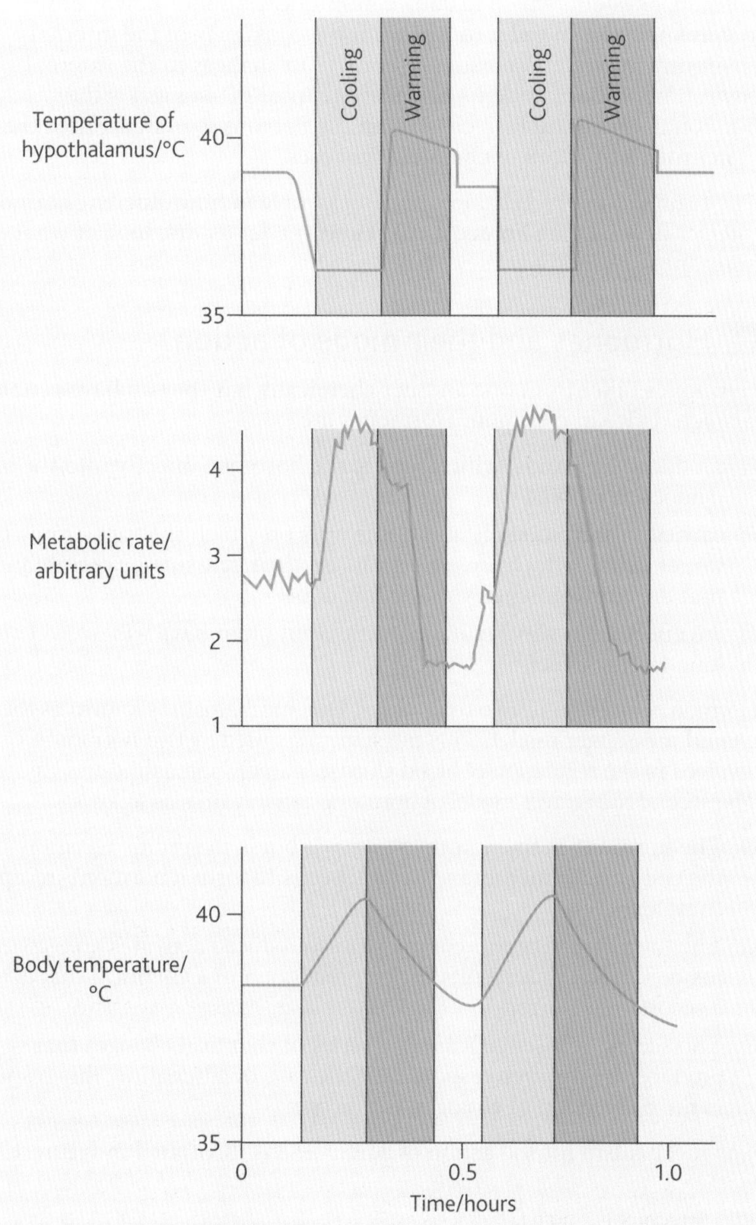

Figure 8

Figure 9 *A ground squirrel*

1 Describe the relationships between the temperature of the hypothalamus, the metabolic rate and the body temperature.

2 Explain the relationships that you have just described.

12.3 Hormones and the regulation of blood glucose

Learning objectives:

- How do hormones work?
- What is the role of the pancreas in regulating blood glucose?
- What factors influence blood glucose concentration?
- What are the roles of insulin, glucagon and adrenaline in regulating blood glucose?

Specification reference: 3.5.4

We saw in Topic 10.1 that animals possess two principal coordinating systems: the nervous system, which communicates rapidly, and the hormonal system, which usually communicates more slowly. Both systems interact in order to maintain the constancy of the internal environment while also being responsive to changes in the external environment. Both systems also use chemical messengers – the hormonal system exclusively so, and the nervous system through the use of **neurotransmitters** in chemical **synapses**.

The regulation of blood glucose is an example of how different hormones interact in achieving **homeostasis**. However, let us first look at what hormones are and how they work.

Hormones and their mode of action

Hormones differ from one another chemically but they all have certain characteristics in common. Hormones are:

- produced by glands, which secrete the hormone directly into the blood (endocrine glands)
- carried in the blood plasma to the cells on which they act – known as **target cells** – which have receptors on their cell-surface membranes that are complementary to the hormone
- are effective in very small quantities, but often have widespread and long-lasting effects.

Hormones function in two main ways. One mechanism is known as the **second messenger model**. This mechanism is used by two hormones involved in the regulation of blood glucose, namely adrenaline and glucagon. The second messenger model of hormone action works as follows:

- The hormone is the first messenger. It binds to specific receptors on the cell-surface membrane of target cells to form a hormone–receptor complex.
- The hormone–receptor complex thus produced activates an enzyme inside the cell that results in the production of a chemical that acts as a second messenger.
- The second messenger causes a series of chemical changes that produce the required response. In the case of adrenaline, this response is the conversion of glycogen to glucose.

The mechanism for the action of adrenaline is illustrated in Figure 1.

The role of the pancreas in regulating blood glucose

The pancreas is a large, pale-coloured gland that is situated in the upper abdomen, behind the stomach. It produces enzymes (protease, amylase and lipase) for digestion and hormones (insulin and glucagon) for regulating blood glucose.

When examined microscopically, the pancreas is made up largely of the cells that produce its digestive enzymes. Scattered throughout these cells

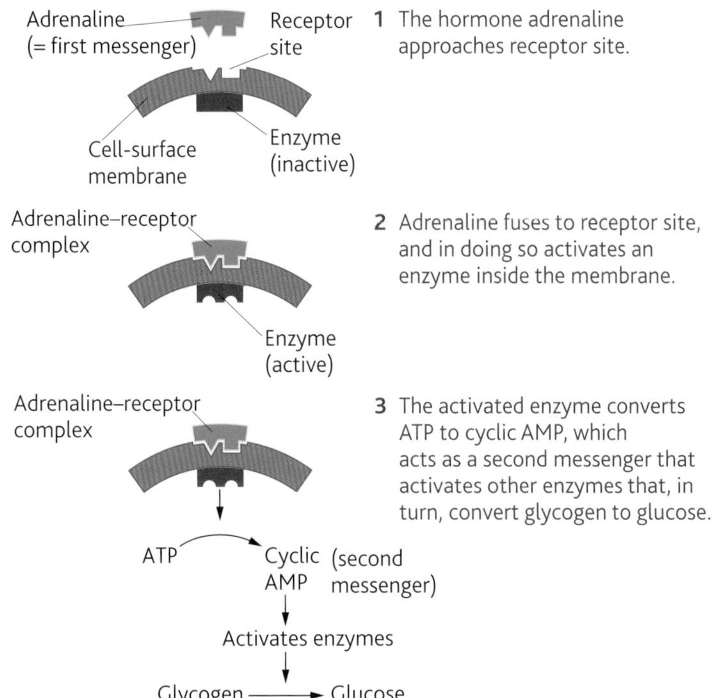

Figure 1 *Second messenger model of hormone action as illustrated by the action of adrenaline in regulating blood sugar*

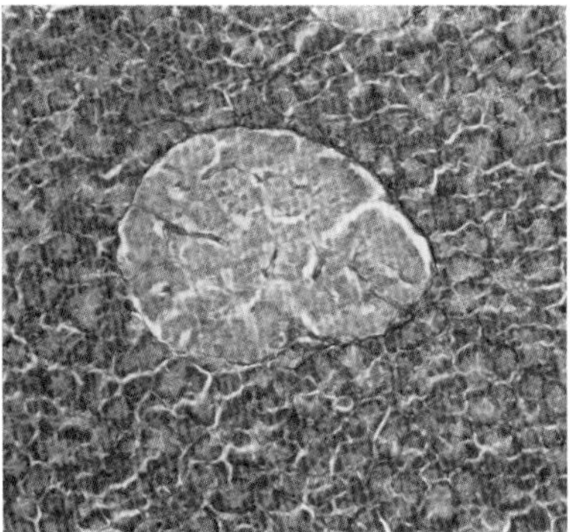

Figure 2 *LM of the pancreas showing an islet of Langerhans (centre) containing α cells and β cells. The meshworks of blue and white in the islet are blood capillaries. Around the islet are the enzyme-producing pancreatic cells*

are groups of hormone-producing cells known as **islets of Langerhans**. The cells of the islets of Langerhans are of two types:

- **α cells**, which are the larger and produce the hormone **glucagon**
- **β cells**, which are smaller and produce the hormone **insulin**.

Regulation of blood glucose

Glucose is the main substrate for respiration, providing the source of energy for almost all organisms. It is broken down during glycolysis (see Topic 4.1) and, if oxygen is present, the Krebs cycle (see Topic 4.2) and electron transport system (see Topic 4.3), to provide ATP – the energy currency of cells (see Topic 2.1). It is therefore essential that the blood of mammals contains a relatively constant level of glucose for respiration. If the level falls too low, cells will be deprived of energy and die – brain cells are especially sensitive in this respect because they can only respire glucose. If the level rises too high, it lowers the **water potential** of the blood and creates osmotic problems that can cause dehydration and be equally dangerous. Homeostatic control (see Topic 12.1) of blood glucose is therefore essential.

Blood glucose and variations in its level

The normal level of blood glucose is 90 mg in each 100 cm^3 of blood. Blood glucose comes from three sources:

- **directly from the diet** in the form of glucose resulting from the breakdown of other carbohydrates such as starch, maltose, lactose and sucrose
- **from the breakdown of glycogen (glycogenolysis)** stored in the liver and muscle cells. A normal liver contains 75–100 g of glycogen, produced by converting excess glucose from the diet in a process called glycogenesis

- **from gluconeogenesis**, which is the production of new glucose, that is, glucose from sources other than carbohydrate. The liver, for example, can make glucose from glycerol and amino acids.

As animals do not eat continuously and their diet varies, their intake of glucose fluctuates. Likewise, glucose is used up at different rates depending on the level of mental and physical activity. It is against these changes in supply and demand that the three main hormones, **insulin**, **glucagon** and **adrenaline**, operate to maintain a constant blood glucose level.

Hint

You will find it easier to understand the terms used in this topic if you remember the following:

'gluco' / 'glyco' = glucose

'glycogen' = glycogen

'neo' = new

'lysis' = splitting

'genesis' = birth / origin

Therefore:

glycogen – o – lysis = splitting of glycogen

gluco – neo – genesis = formation of new glucose

Insulin and the β cells of the pancreas

The β cells of the islets of Langerhans in the pancreas detect a rise in blood glucose level and respond by secreting the hormone insulin directly into the blood plasma. Insulin is a globular protein made up of 51 amino acids.

Almost all body cells (red blood cells being a notable exception) have glycoprotein receptors on their cell-surface membranes that bind with insulin molecules. When it combines with the receptors, insulin brings about:

- a change in the tertiary structure of the glucose transport protein channels, causing them to change shape and open, allowing more glucose into the cells
- an increase in the number of carrier molecules in the cell-surface membrane
- activation of the enzymes that convert glucose to glycogen and fat.

As a result, the blood glucose level is lowered in one or more of the following ways:

- by increasing the rate of absorption of glucose into the cells, especially in muscle cells
- by increasing the respiratory rate of the cells, which therefore use up more glucose, thus increasing their uptake of glucose from the blood
- by increasing the rate of conversion of glucose into glycogen (glycogenesis) in the cells of the liver and muscles
- by increasing the rate of conversion of glucose to fat.

The effect of these processes is to remove glucose from the blood and so return its level to normal. This lowering of the blood glucose level causes the β cells to reduce their secretion of insulin (= negative feedback).

Glucagon and the α cells of the pancreas

The α cells of the islets of Langerhans detect a fall in blood glucose and respond by secreting the hormone glucagon directly into the blood plasma. Only the cells of the liver have receptors that bind to glucagon, so only liver cells respond. They do this by:

- activating an enzyme that converts glycogen to glucose
- increasing the conversion of amino acids and glycerol into glucose (= gluconeogenesis).

The overall effect is therefore to increase the amount of glucose in the blood and return it to its normal level. This raising of the blood glucose level causes the α cells to reduce the secretion of glucagon (= negative feedback).

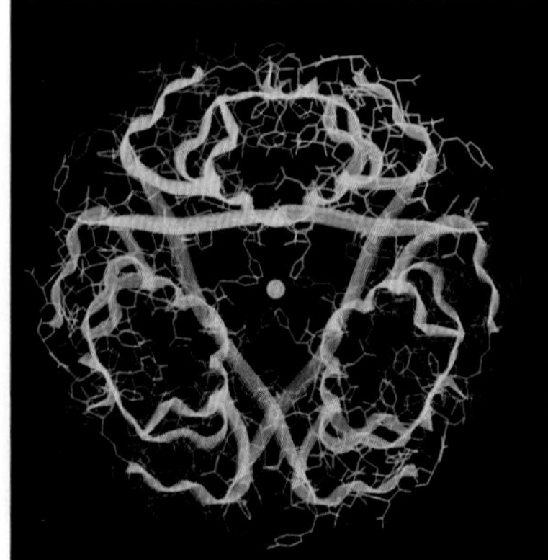

Figure 4 *Molecular graphic of an insulin molecule. Insulin is made up of 51 amino acids arranged in two chains (shown here as yellow and green ribbons)*

Role of adrenaline in regulating the blood glucose level

There are at least four other hormones apart from glucagon that can increase blood glucose level. The best known of these is adrenaline. At times of excitement or stress, adrenaline is produced by the adrenal glands that lie above the kidneys. Adrenaline raises the blood glucose level by:

- activating an enzyme that causes the breakdown of glycogen to glucose in the liver
- inactivating an enzyme that synthesises glycogen from glucose.

Hormone interaction in regulating blood glucose

The two hormones insulin and glucagon act in opposite directions. Insulin lowers the blood glucose level, whereas glucagon increases it. The two hormones are said to act antagonistically. The system is self-regulating in that it is the level of glucose in the blood that determines the quantity of insulin and glucagon produced. In this way the interaction of these two hormones allows highly sensitive control of the blood glucose level. The level of glucose is not, however, constant, but fluctuates around a set point. This is because of the way negative feedback mechanisms work. Only when the blood glucose level falls below the set point is insulin secretion reduced (negative feedback), leading to a rise in blood glucose. In the same way, only when the level exceeds the set point is glucagon secretion reduced (negative feedback), causing a fall in the blood glucose level.

The control of blood glucose level is summarised in Figure 3.

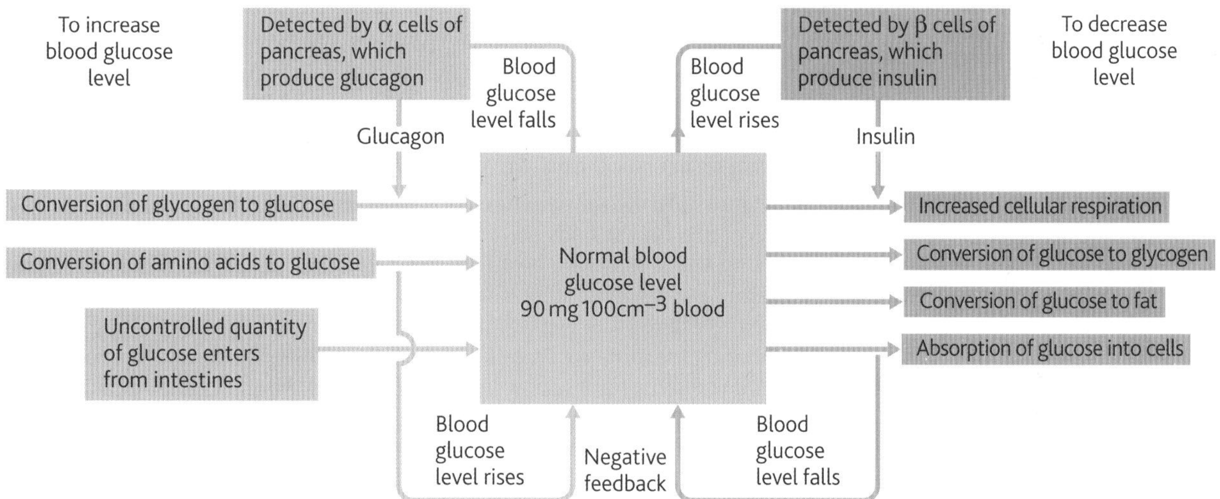

Figure 3 *Summary of regulation of blood glucose*

Summary questions

In the following passage, state the most suitable word to replace the numbers in brackets.

The chemical energy in glucose is released by cells during the process known as (1). It is therefore important that the blood glucose level is maintained at a constant level because if it falls too low cells are deprived of energy, and (2) cells are especially sensitive in this respect. If it gets too high (3) problems occur that may cause dehydration. Blood glucose is formed directly from (4) in the diet or from the breakdown of (5), which is stored in the cells of the liver and (6). The liver can also increase blood glucose levels by making glucose from other sources, such as glycerol and (7), in a process known as (8). Blood glucose is used up when it is absorbed into cells, converted into fat or (9) for storage, or used up during (10) by cells. In order to maintain a constant level of blood glucose the pancreas produces two hormones from clusters of cells within it called (11). The β cells produce the hormone (12), which causes the blood glucose level to fall. The α cells produce the hormone (13), which has the opposite effect. Another hormone, called (14), can also raise blood glucose levels.

12.4 Diabetes and its control

Learning objectives:

- What are the two main types of diabetes and how do they differ?
- How can each type of diabetes be controlled?

Specification reference: 3.5.4

Diabetes is a chronic disease in which a person is unable to metabolise carbohydrate, especially glucose, properly. There are over 100 million people worldwide with diabetes, 1.4 million of whom are in the UK. In addition, a further 1 million people in the UK are thought to have the disease but are currently unaware of it. There are two distinct forms, of which diabetes mellitus, or 'sugar diabetes', is much more common.

Types of sugar diabetes

Diabetes is a metabolic disorder caused by an inability to control blood glucose levels due to a lack of the hormone insulin or a loss of responsiveness to insulin.

There are two forms of diabetes:

- **Type I (insulin dependent)** is due to the body being unable to produce insulin. It normally begins in childhood. It may be the result of an autoimmune response whereby the body's immune system attacks its own cells, in this case the β cells of the islets of Langerhans. Type I diabetes develops quickly, usually over a few weeks, and the symptoms (see 'Hint') are normally obvious.
- **Type II (insulin independent)** is normally due to the **glycoprotein** receptors on the body cells losing their responsiveness to insulin. However, it may also be due to an inadequate supply of insulin from the pancreas. Type II diabetes usually develops in people over the age of 40 years. There is, however, an increasing number of cases of obesity and poor diet leading to Type II diabetes in adolescents. It develops slowly, and the symptoms are normally less severe and may go unnoticed. People who are overweight are particularly likely to develop Type II diabetes. Over 75 per cent of people with diabetes have Type II.

Figure 2 illustrates the differences in blood glucose levels between diabetic and non-diabetic persons who have swallowed a glucose solution.

Control of diabetes

Although diabetes cannot be cured, it can be treated very successfully. Treatment varies depending on the type of diabetes.

- **Type I diabetes** is controlled by injections of insulin. This cannot be taken by mouth because, being a protein, it would be digested in the alimentary canal. It is therefore injected, typically either two or four times a day. The dose of insulin must be matched exactly to the glucose intake. If a diabetic takes too much insulin, he or she will experience a low blood glucose level that can result in unconsciousness. To ensure the correct dose, blood glucose levels are monitored using biosensors. By injecting insulin and managing their carbohydrate intake and exercise carefully, diabetics can lead normal lives.
- **Type II diabetes** is controlled by regulating the intake of carbohydrate in the diet and matching this to the amount of exercise taken. In some cases, this may be supplemented by injections of insulin or by the use of drugs that stimulate insulin production. Other drugs slow down the rate at which the body absorbs glucose from the intestine.

Hint

Symptoms of diabetes

- high blood glucose level
- presence of glucose in the urine
- increased thirst and hunger
- need to urinate excessively
- genital itching or regular episodes of thrush
- tiredness
- weight loss
- blurred vision.

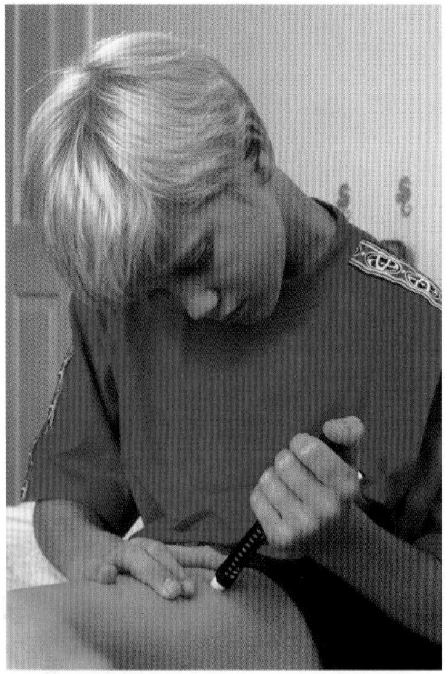

Figure 1 *A diabetic person injecting insulin*

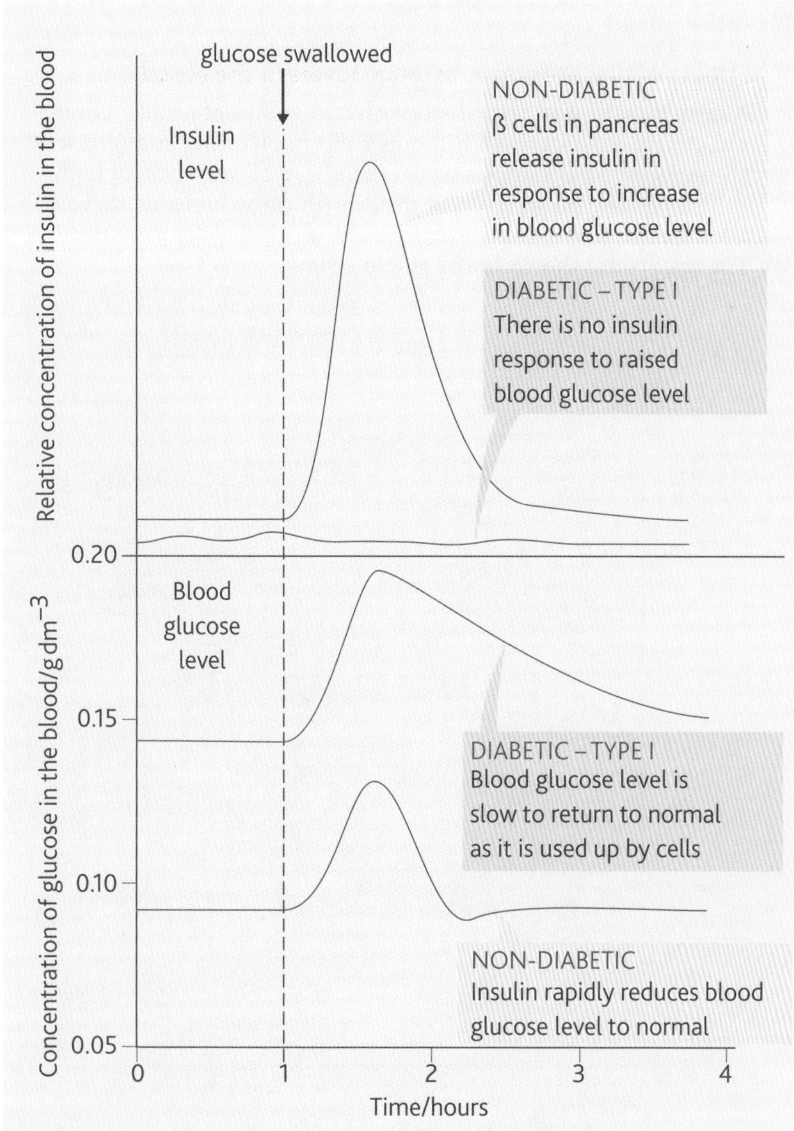

Figure 2 *Comparison of blood glucose and insulin levels in a person with Type I diabetes and a non-diabetic person after each has swallowed a glucose solution*

Hint

Blood glucose concentration can be controlled by changing the uptake of glucose from the gut (diet) and by changing the rate at which glucose is removed from the blood (exercise and insulin).

Summary questions

1. State **one** difference between the causes of Type I and Type II diabetes.
2. State **one** difference between the main ways of controlling Type I and Type II diabetes.
3. Suggest an explanation for why tiredness is a symptom of diabetes.
4. What lifestyle advice might you give someone in order to help them avoid developing Type II diabetes?

Application

Effects of diabetes on substance levels in the blood

An experiment was carried out with two groups of people. Group X had Type I diabetes while group Y did not (control group). Every 15 minutes blood samples were taken from all members of both groups and the mean levels of insulin, glucagon and glucose were calculated. After an hour, every person was given a glucose drink. The results are shown in the graphs below.

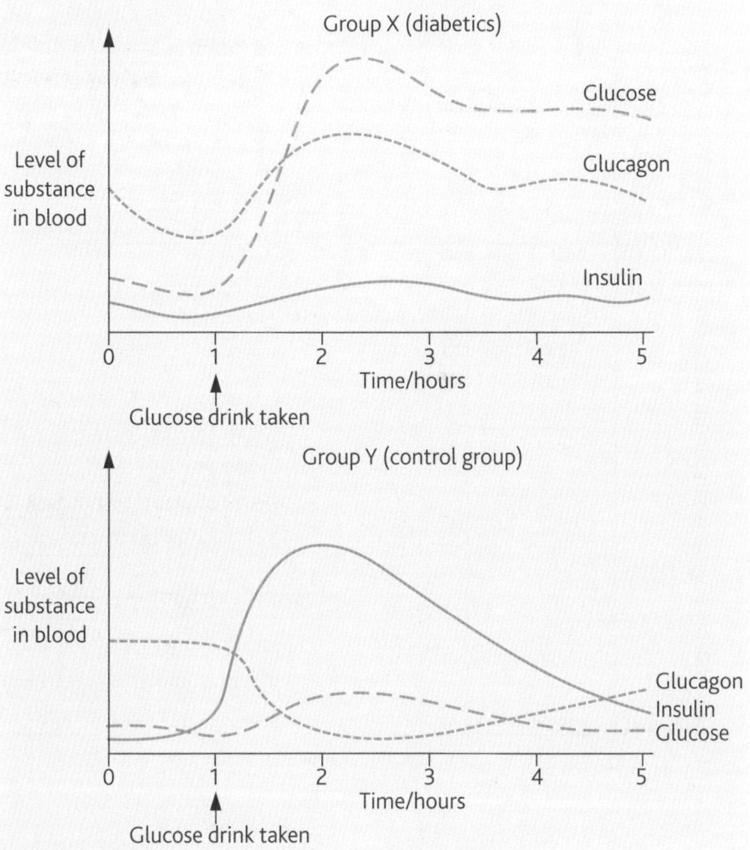

Figure 3

1. Name a hormone other than insulin and glucagon that is involved in regulating blood glucose levels.
2. State **two** differences between groups X and Y in the way insulin secretion responds to the drinking of glucose.
3. Suggest a reason why the glucose level falls in both groups during the first hour.
4. Using information from the graphs, explain the changes in the blood glucose level in group Y after drinking the glucose.
5. Explain the difference in blood glucose level of group X compared to group Y.
6. Suggest what might happen to the blood glucose level of group X if they have no food over the next 24 hours.

Examination-style questions

1 (a) (i) What is meant by homeostasis?

(ii) Giving **one** example, explain why homeostasis is important in mammals. *(3 marks)*

(b) (i) Cross-channel swimmers experience a large decrease in external temperature when they enter the water. Describe the processes involved in thermoregulation in response to this large decrease in external temperature.

(ii) A person swimming in cold water may not be able to maintain their core body temperature and will begin to suffer from hypothermia. Explain why a tall, thin swimmer is more likely to suffer from hypothermia than a short, stout swimmer of the same body mass. *(9 marks)*

AQA, 2006

2 Mammals are endotherms; reptiles are ectotherms.

(a) Explain **two** advantages of endothermy over ectothermy. *(2 marks)*

(b) **Figure 1** shows how the rates of metabolic heat generation and evaporative heat loss in a reptile change with environmental temperature. Each plot is the mean of several values. The vertical bars on the graphs represent the standard deviation about the mean.

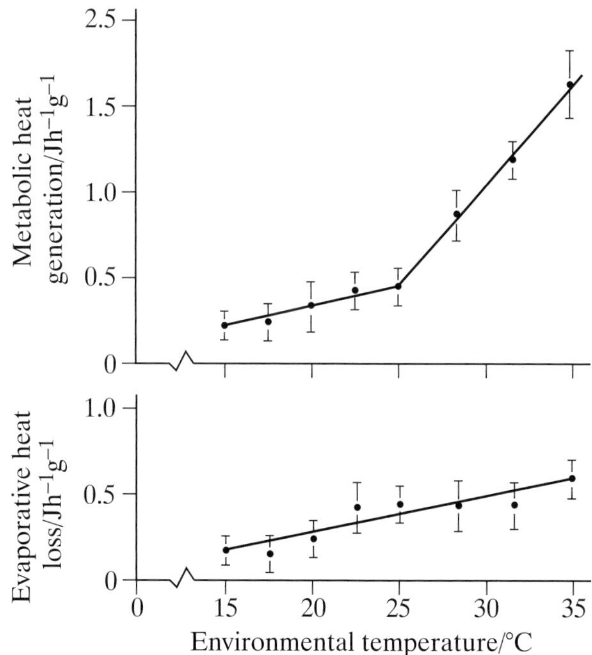

Figure 1

(i) Explain why it is more useful to show the standard deviation rather than the range of values.

(ii) Explain why the values for metabolic heat generation are given per gram of body mass.

(iii) Describe the relationship between metabolic heat generation and evaporative heat loss shown in **Figure 1**.

(iv) Use **Figure 1** to explain why these reptiles often seek shade when the environmental temperature rises above 25 °C. *(8 marks)*

(c) **Figure 2** shows the relationship between metabolic heat generation and evaporative heat loss in a small mammal.

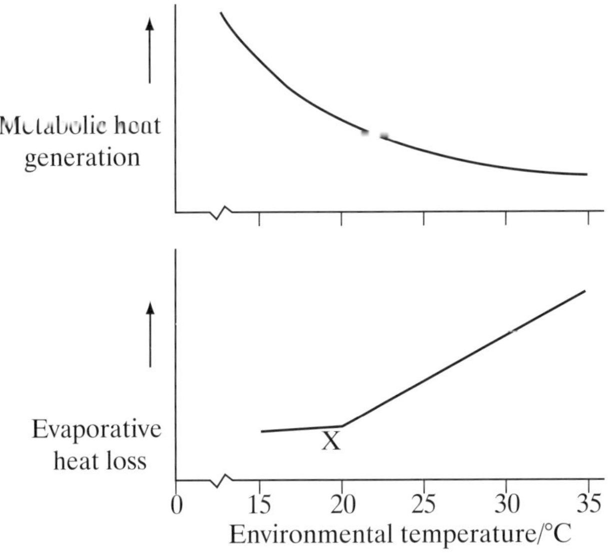

Figure 2

(i) How is the relationship between heat generation and evaporative heat loss in a mammal different from that in a reptile?

(ii) Suggest an explanation for the change in the slope of the graph for evaporative heat loss at the point marked **X**.

(iii) Explain how the change in metabolic heat generation in a small mammal is bought about as environmental temperature rises. *(5 marks)*

AQA, 2003

3 (a) Describe the role of insulin in the control of blood glucose concentration. *(4 marks)*

Figure 3 shows the pathway by which glycogen is broken down in liver and muscle cells.

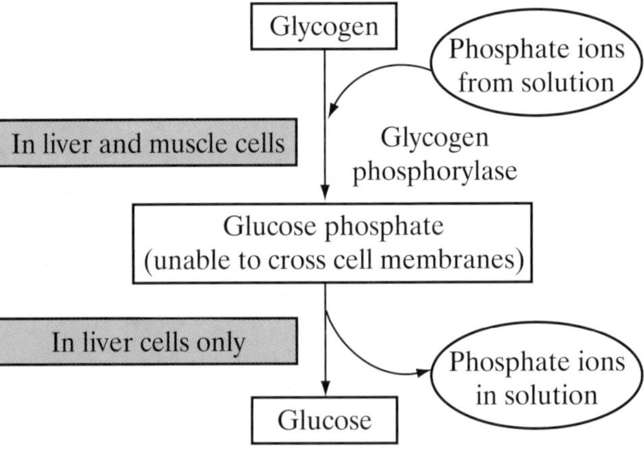

Figure 3

(b) Suggest why it is important that muscle cells do not convert glucose phosphate to glucose. *(2 marks)*

(c) The production of glycogen phosphorylase from an inactive form of the enzyme is triggered by the hormones glucagon and adrenaline, and by calcium ions. Adrenaline is a hormone released when an animal senses danger. This is controlled by the sympathetic nervous system. **Figure 4** shows the receptors for glucagon and adrenaline on liver and muscle cells.

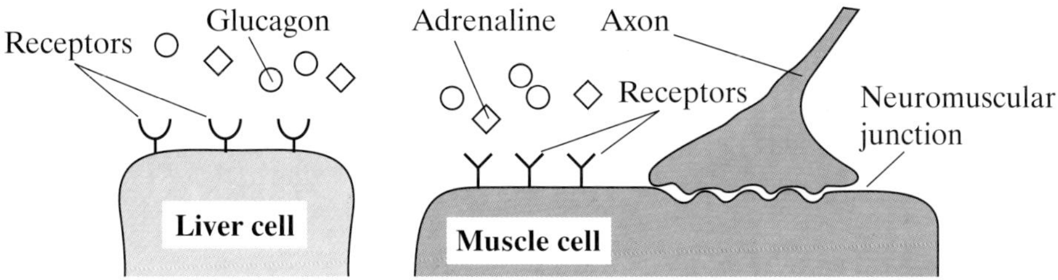

Figure 4

Use the information in **Figures 3** and **4** to suggest how glycogen breakdown in liver and muscle cells is increased when an animal runs away from a predator. *(6 marks)*

AQA, 2003

4 Some people produce no insulin. As a result they have a condition called diabetes. In an investigation, a man with diabetes drank a glucose solution. The concentration of glucose in his blood was measured at regular intervals. The results are shown in **Figure 5**.

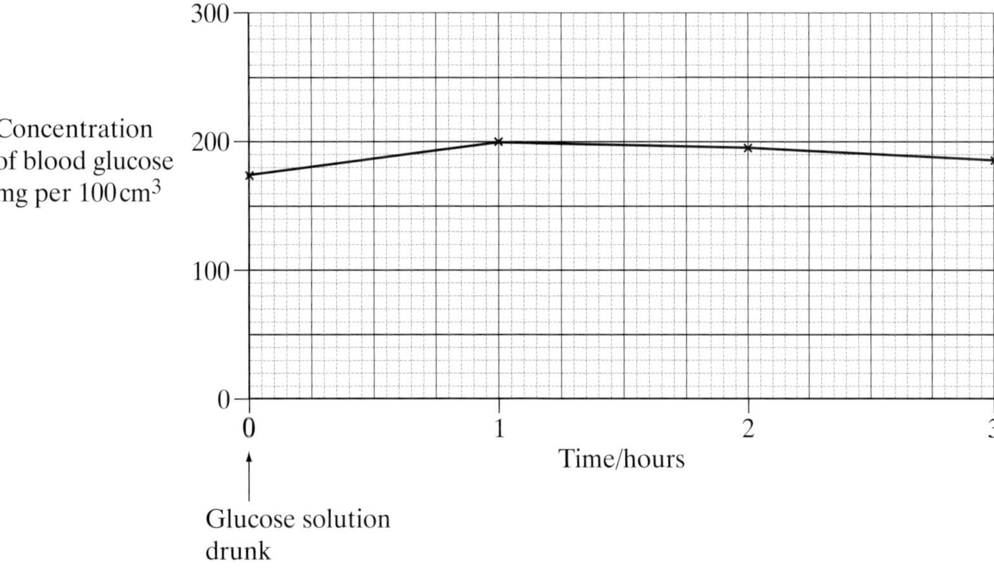

Figure 5

(a) Suggest **two** reasons why the concentration of glucose decreased after 1 hour even though this man's blood contained no insulin. *(2 marks)*

(b) The investigation was repeated on a man who did not have diabetes. The concentration of glucose in his blood before drinking the glucose solution was 80 mg per 100 cm^3. Sketch a curve on the graph to show the results you would expect. *(1 mark)*

(c) The diabetic man adopted a daily routine to stabilise his blood glucose concentration within narrow limits. He ate three meals a day: breakfast, a midday meal and an evening meal. He injected insulin once before breakfast and once before the evening meal.

The injection he used before breakfast was a mixture of two types of insulin. The mixture contained slow-acting insulin and fast-acting insulin.

(i) Explain the advantage of injecting both types of insulin before breakfast.

(ii) One day, the man did not eat a midday meal. Suggest one reason why his blood glucose concentration did not fall dangerously low even though he had injected himself with the mixture of insulin before breakfast. *(3 marks)*

AQA, 2004

13 Feedback mechanisms

13.1 The principles of feedback mechanisms

Learning objectives:

- What is negative feedback?
- How does it help to control homeostatic processes?
- How does it differ from positive feedback?

Specification reference: 3.5.5

We saw in Topic 12.1 that the **homeostatic** control of any system involves a series of stages featuring:

- **the set point**, or desired level (norm), at which the system operates
- **a receptor**, which detects any deviation from the set point (norm)
- **a controller**, which coordinates information from various sources
- **an effector**, which brings about the corrective measures needed to return the system to the set point (norm)
- **a feedback loop**, which informs the receptor of the changes to the system brought about by the effector.

Let us now look in more detail at the last stage in the list – the feedback loop. When an effector has corrected any deviation and returned the system to the set point, it is important that this information is fed back to the receptor. If the information is not fed back, the receptor will continue to stimulate the effector, leading to an over-correction and causing a deviation in the opposite direction. There are two types of feedback: negative feedback and positive feedback.

Negative feedback

Negative feedback occurs when the feedback causes the corrective measures to be turned off. In doing so it returns the system to its original (normal) level.

Let us take the example of temperature regulation that we looked at in Topic 12.2. If the temperature of the blood increases, thermoreceptors in the **hypothalamus** send nerve impulses to the heat loss centre, which is also in the hypothalamus. This in turn sends impulses to the skin (effector organ). **Vasodilation**, sweating and lowering of body hairs all lead to a reduction in blood temperature. If the fact that blood temperature has returned to normal is not fed back to the hypothalamus, it will continue to stimulate the skin to lose body heat. Blood temperature will then fall below normal and may continue to do so causing **hypothermia** and the death of the organism.

What happens in practice is that the cooler blood returning from the skin passes through the hypothalamus. The thermoreceptors detect that blood temperature is at its normal set point again and so they cease to send impulses to the heat loss centre. This in turn stops sending impulses to the skin and so vasodilation, sweating, etc. cease and blood temperature remains at its normal level rather than continuing to fall. The blood, having been cooled to its normal temperature, has turned *off* the effector (the skin) that was correcting the rise in temperature. This is therefore negative feedback and is illustrated in Figure 1.

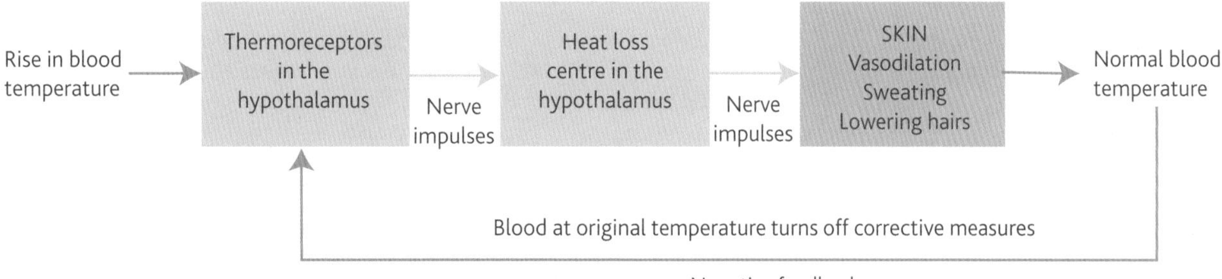

Figure 1 *Negative feedback in the control of body temperature*

There are separate negative feedback mechanisms to regulate departures from the norm in each direction. In temperature regulation for example, if the blood temperature falls, rather than rises, the heat gain centre will cause **vasoconstriction**, the raising of hair and reduced sweating. There will again be negative feedback to turn off these mechanisms when the blood returns to its normal temperature.

Another example is in the control of blood glucose that we considered in Topic 12.3. If there is a fall in the concentration of glucose in the blood, the α cells in the **islets of Langerhans** in the pancreas produce the hormone **glucagon**. Glucagon causes the conversion of glycogen to glucose and **gluconeogenesis** in the liver. As a result the blood glucose concentration rises to normal. As this blood circulates back to the pancreas, the α cells detect the change and stop producing glucagon (= negative feedback). These events are illustrated in Figure 3.

Figure 2 *Control of body temperature involves negative feedback mechanisms*

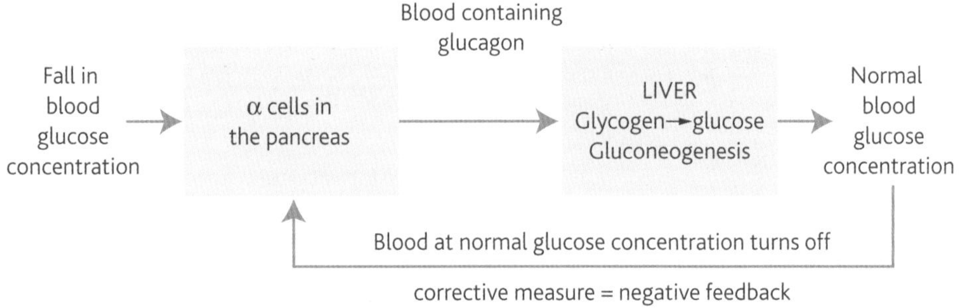

Figure 3 *Negative feedback in the control of blood glucose levels*

In the same way, if the blood glucose level rises, rather than falls, insulin will be produced from the β cells in the pancreas. **Insulin** increases the uptake of glucose by cells and its conversion to glycogen and fat. The fall in blood glucose concentration that results will turn off insulin production once blood glucose levels return to normal (= negative feedback).

Having separate negative feedback mechanisms that control departures from the norm in either direction gives a greater degree of homeostatic control.

AQA Examiner's tip

If you are answering a question about negative feedback, e.g. control of blood glucose, make certain that you write about negative feedback. Do not just give a description of how insulin and glucagon work.

Summary questions

1 Distinguish between positive and negative feedback.

2 Why is negative feedback important in maintaining a system at a set point?

3 What is the advantage of having separate negative feedback mechanisms to control deviations away from normal?

Positive feedback

Positive feedback occurs when the feedback causes the corrective measures to remain turned on. In doing so it causes the system to deviate even more from the original (normal) level. Examples are rare, but one occurs in neurones when a stimulus causes a small influx of sodium ions (10.3). This influx increases the permeability of the neurone to sodium ions so more ions enter, causing a further increase in permeability and even more rapid entry of ions. This results in a very rapid build-up of an action potential that allows an equally rapid response to a stimulus.

Positive feedback occurs more often when there is a breakdown of control systems. In certain diseases, for example typhoid fever, there is a breakdown of temperature regulation resulting in a rise in body temperature leading to **hyperthermia**. In the same way, when the body gets too cold (hypothermia) the temperature control system tends to break down, leading to positive feedback resulting in the body temperature dropping even lower.

Application

Control of blood water potential

Figure 4 shows some of the changes that occur as a result of water being lost from the blood due to sweating.

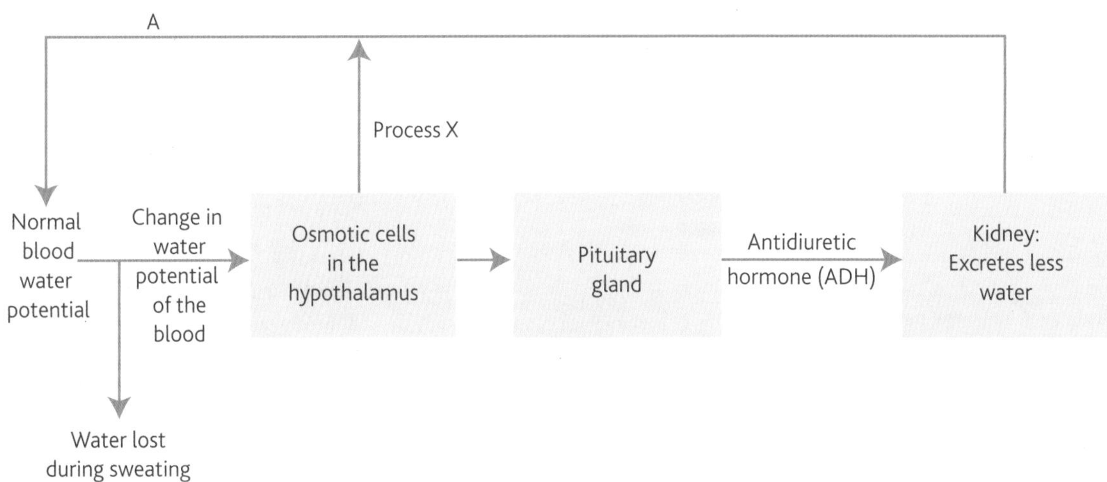

Figure 4

1 Describe the change in water potential that occurs in the blood as a result of sweating.

2 Which of the structures shown in Figure 4 acts as:

a a receptor

b an effector?

3 Describe how ADH gets from the pituitary gland to the kidney.

4 The kidney conserves the water that is already in the blood. Given that the water potential of the blood returns to its normal level prior to sweating, suggest what is happening in process X.

5 State as concisely as possible what mechanism is shown by the line labelled A.

13.2 Control of the oestrous cycle

Learning objectives:

- Which hormones are involved in the control of oestrous cycles?
- How do these hormones interact in the control of the human menstrual cycle?
- How are different forms of feedback loop involved in this control?

Specification reference: 3.5.5

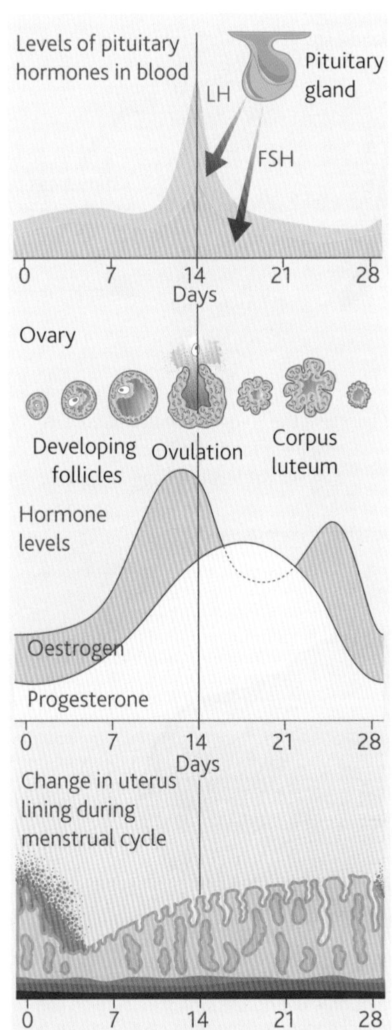

Figure 1 *Summary of changes during the menstrual cycle*

After looking at the principles of negative feedback in Topic 13.1, let us now examine its role in the oestrous cycle of mammals. The oestrous cycle is the regular pattern of changes that takes place in the reproductive system of female mammals. In humans it begins at puberty and continues until the menopause at 45–50 years of age. It is controlled through the interaction of a number of hormones. These hormones circulate in the blood plasma and so reach all parts of the body. However, only cells with the appropriate receptors can respond to a particular hormone. In humans, and some other primates, the lining of the uterus is shed along with some blood between each cycle. The name given to this particular type of oestrous cycle is the **menstrual cycle**.

Hormonal control of the menstrual cycle

There are four main hormones that control the menstrual cycle, each interacting with the other in a way that ensures a regular cycle of events. Two hormones are released from the pituitary gland (which lies at the base of the brain):

- **Follicle-stimulating hormone (FSH)** stimulates the development of follicles in the ovary, which contain eggs, and stimulates the follicles in the ovaries to produce oestrogen.
- **Luteinising hormone (LH)** causes ovulation to occur, and stimulates the ovary to produce progesterone from the corpus luteum.

The remaining two hormones are produced by the ovaries:

- **Oestrogen** causes the rebuilding of the uterus lining after menstruation and stimulates the pituitary gland to produce LH.
- **Progesterone** maintains the lining of the uterus in readiness to receive the fertilised egg and inhibits the production of FSH from the pituitary gland.

The menstrual cycle is controlled by the interaction of these four hormones and their negative and positive feedback loops. This process is described below and illustrated in Figure 1.

- The menstrual cycle begins when the uterus lining is shed, along with some blood (days 1–5).
- From day 1, the pituitary gland releases FSH into the blood which stimulates follicles in the ovary to grow and mature. Each follicle contains an egg.
- The growing follicles secrete small amounts of oestrogen into the blood. This low level of oestrogen causes the uterus lining to build up again and also inhibits the release of FSH and LH from the pituitary gland (= negative feedback).
- As the follicles grow, more oestrogen is produced. The level of oestrogen increases until, at around day 10, it reaches a critical point where it stimulates the pituitary gland to release more FSH and LH (= positive feedback).
- There is a surge in FSH and LH production. The surge in LH causes one of the follicles in the ovary to release its egg. This is called ovulation and occurs on day 14.

Hint

The hormones released during the oestrous cycle are carried to all parts of the body in the blood plasma. However, cells can only respond to a hormone if they have the appropriate receptors, i.e. receptors that are complementary to the hormone in question. Therefore only certain cells respond to each hormone.

- After ovulation LH stimulates the empty follicle to develop into a structure called the corpus luteum, which secretes progesterone (and smaller amounts of oestrogen).
- Progesterone maintains the thick lining of the uterus and also inhibits the release of FSH and LH by the pituitary gland (= negative feedback).
- If the egg is not fertilised, the corpus luteum degenerates and so no longer produces progesterone.
- With less progesterone, the lining of the uterus is no longer maintained and so breaks down (menstruation). Less progesterone also means that FSH release is no longer inhibited.
- FSH release therefore resumes and the cycle repeats itself.

The alternate switching on and off of these hormones is responsible for the regular sequence of events in the menstrual cycle. The interaction of these hormones is shown in Figure 2.

AQA Examiner's tip

Look carefully at any question about oestrous cycles. Does it say that the example is human? If not, the timings may be different and menstruation is unlikely to occur.

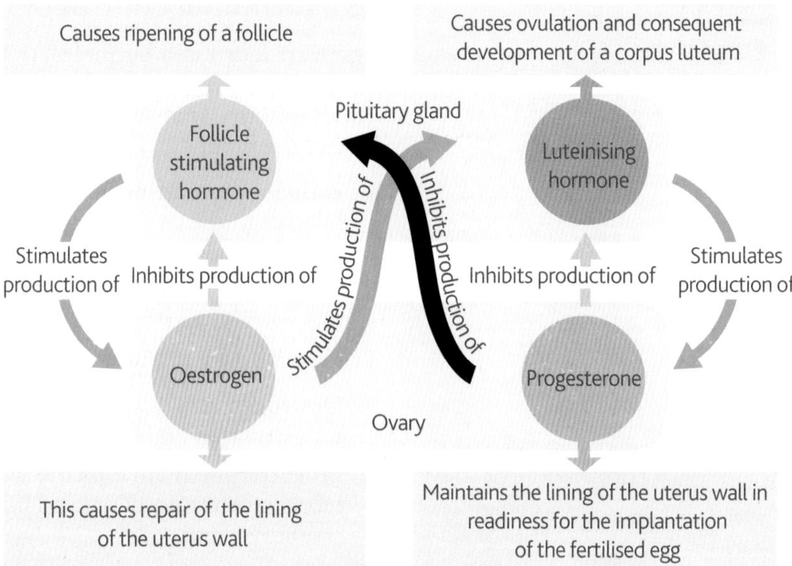

Figure 2 *Summary of main hormone interactions in the menstrual cycle*

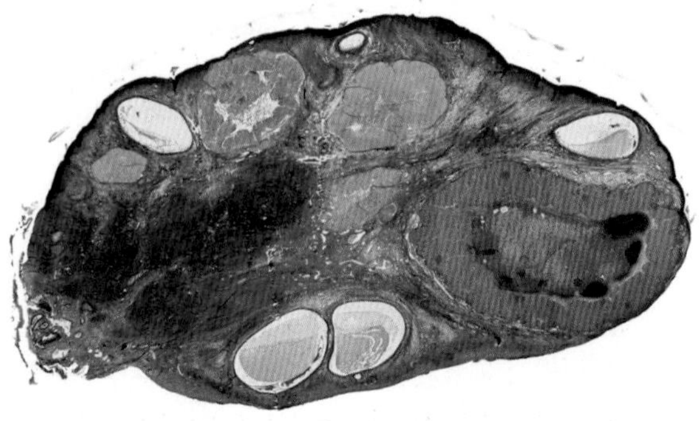

Figure 3 *Light micrograph of a section through a human ovary. The pink structure on the right is the corpus luteum that secretes the hormone progesterone to prepare the uterus for the implantation of a fertilised egg. The white ovals are the follicles in which eggs develop*

Summary questions

1 In each of the following cases name as precisely as possible the structure that produces the hormone:

a LH

b progesterone

c oestrogen

2 Using the information in Figure 2, describe an example of control by negative feedback.

3 When a female human reaches the menopause, she has very few follicles left in her ovaries. Suggest a reason why levels of FSH in the blood rise in women reaching the menopause.

4 Some female farm animals were given progesterone in their diet. When the progesterone was withdrawn from the diet, they all ovulated (produced eggs) a few days later. Suggest an explanation for this.

Application

The oestrous cycle in pigs

The graph in Figure 4 shows the concentration of four hormones during part of the oestrous cycle of a female pig.

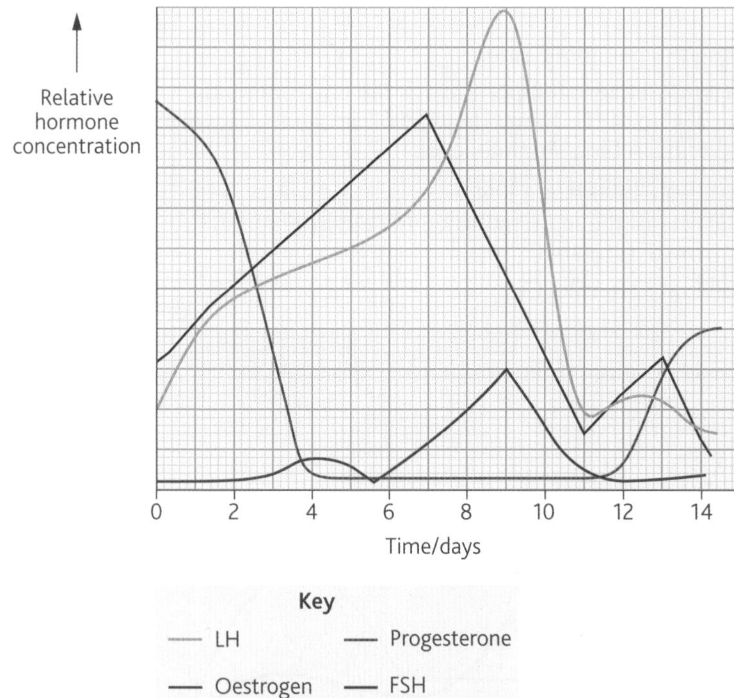

Figure 4

1 From your knowledge of the different hormones, suggest on which day ovulation is most likely to occur. Give a reason for your answer.

2 Using the graph and your knowledge of the two hormones, explain the effects of oestrogen on the production of FSH from day 0 to day 12.

3 Explain why pigs injected with progesterone may not ovulate.

AQA Examination-style questions

1 **Figure 1** shows the effect of increasing the environmental temperature on the metabolic rate of a small mammal.

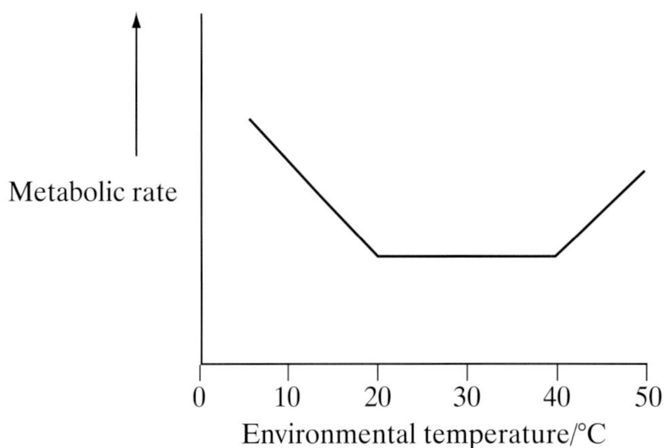

Figure 1

(a) Suggest **one** way of measuring the metabolic rate. *(1 mark)*

(b) The mammal maintained a constant core temperature as the environmental temperature increased from 5 °C to 40 °C. Name the type of mechanism involved in restoring physiological systems to their original level. *(1 mark)*

(c) Use your knowledge of thermoregulation to explain:

(i) the change in metabolic rate of the mammal when the environmental temperature increases from 5 °C to 40 °C;

(ii) the increase in metabolic rate after 40 °C. *(5 marks)*

AQA, 2006

2 **Figure 2** shows the changes in concentration of the hormones responsible for controlling the menstrual cycle.

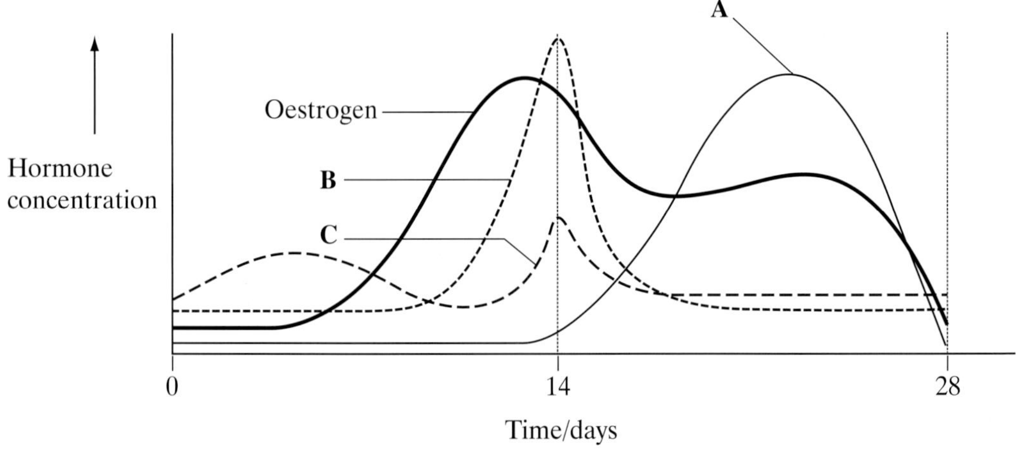

Figure 2

(a) Which curve, **A**, **B**, **C** or **D** shows changes in the concentration of:

(i) FSH;

(ii) LH. *(2 marks)*

(b) Explain how the release of FSH is controlled by negative feedback. *(2 marks)*

(c) Using **Figure 2**, give one example of how positive feedback is involved in the secretion of hormone **C**. *(2 marks)*

AQA, 2003

3 The menstrual cycle in humans is controlled by hormones. The equivalent cycle in cows is controlled by the same hormones. **Figure 3** shows changes in the concentration of some of these hormones in a cow.

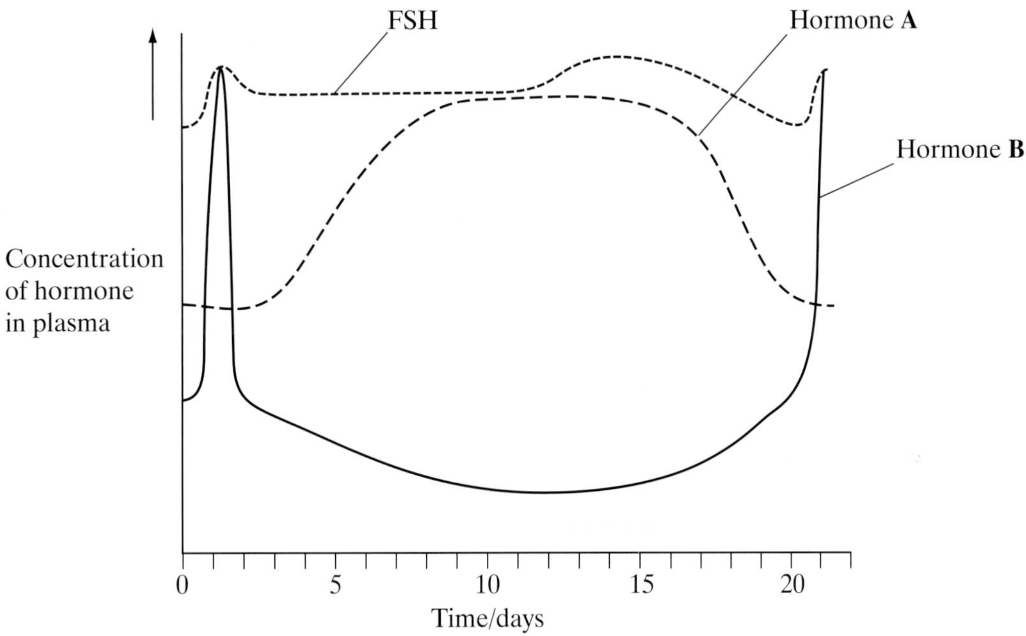

Figure 3

(a) (i) Name hormone **A**. Explain your answer.

(ii) Name hormone **B**. Explain your answer. *(2 marks)*

(b) Scientists can clone cows. To do this they need to collect eggs from a cow. Using the graph, suggest **two** time periods in which the scientist should attempt to collect eggs. Give an explanation for your answer. *(2 marks)*

AQA, 2007

14 Genetic control of protein structure and function

14.1 Structure of ribonucleic acid

Learning objectives:

- What is the genetic code and what are its main features?
- What is the structure of ribonucleic acid (RNA)?
- What are the structure and the role of messenger RNA (mRNA)?
- What are the structure and the role of transfer RNA (tRNA)?

Specification reference: 3.5.6

In your AS Biology course you will have learned about the importance of DNA and how it acts as a code for the sequence of amino acids that make up a protein. In this chapter, we turn our attention to exactly how these DNA triplets are used to make the proteins that they code for.

The genetic code

We know that the sequence of nucleotides in DNA forms a code that determines the sequence of amino acids in the proteins of an organism. In eukaryotic cells DNA is largely confined to the nucleus. However, the synthesis of proteins takes place in the cytoplasm. So how is the code on the DNA in the nucleus transferred to the cytoplasm where it is translated into proteins? The answer is that sections of the DNA code are transcribed onto a single-stranded molecule called ribonucleic acid (RNA).

There are a number of types of RNA. The one that transfers the DNA code from the nucleus to the cytoplasm acts as a type of messenger and is hence given the name **messenger RNA**, or **mRNA** for short. This mRNA is small enough to leave the nucleus through the nuclear pores and to enter the cytoplasm, where the code that it contains is used to determine the sequence of amino acids in the proteins which are synthesised there.

It is the sequence of nucleotide bases on mRNA that is referred to as the genetic code. As we shall see later, although the mRNA code is derived from the DNA code, it is not identical to it. The mRNA code is complementary to the DNA code. The term **codon** refers to the sequence of three bases (triplet) on mRNA that codes for a single amino acid.

The main features of the genetic code are as follows:

- Each amino acid in a protein is coded for by a sequence of three nucleotide bases on mRNA (i.e. a codon).
- A few amino acids have only a single codon.
- The code is a **degenerate code**. This means that most amino acids have more than one codon. For example, the amino acid leucine has six different codons (UUA, UUG, CUU, CUC, CUA and CUG – see Table 1 for the names of the amino acids to which these codons refer).
- Three codons do not code for any amino acid. These are called **stop codons** and mark the end of a polypeptide chain. They act in much the same way as a full stop at the end of a sentence.
- The code is **non-overlapping**, that is, each base in the sequence is read only once. Thus six bases numbered 123456 are read as codons 123 and 456, rather than as codons 123, 234, 345 and 456.
- It is a **universal code**, that is, the same codon codes for the same amino acid in all organisms (with a few minor exceptions).

In the AS Biology course we learned that DNA is composed of two nucleotide chains wound around each other (double helix). We shall now look at the structure of a related molecule that is usually made up of a single nucleotide chain: **ribonucleic acid** (**RNA**).

Link

Before starting work on the following topics, you should revise all the information on DNA and meiosis in Chapter 8 of the *AS Biology* book.

AQA Examiner's tip

The term 'genetic code' is often used very loosely and is applied to a range of things, including the whole genome of an organism or the sequence of nucleotide bases on DNA. For the purposes of this specification, it should only be used to mean the sequence of nucleotide bases on mRNA, which codes for amino acids.

Table 1 *The genetic code. The base sequences shown are those on mRNA*

First position	Second position				Third position
	U	C	A	G	
U	Phe	Ser	Tyr	Cys	U
	Phe	Ser	Tyr	Cys	C
	Leu	Ser	Stop	Stop	A
	Leu	Ser	Stop	Trp	G
C	Leu	Pro	His	Arg	U
	Leu	Pro	His	Arg	C
	Leu	Pro	Gln	Arg	A
	Leu	Pro	Gln	Arg	G
A	Ile	Thr	Asn	Ser	U
	Ile	Thr	Asn	Ser	C
	Ile	Thr	Lys	Arg	A
	Met	Thr	Lys	Arg	G
G	Val	Ala	Asp	Gly	U
	Val	Ala	Asp	Gly	C
	Val	Ala	Glu	Gly	A
	Val	Ala	Glu	Gly	G

The genetic code

A codon is made up of three nucleotide bases read in the sequence shown. For example, UGC codes for the amino acid Cys (cysteine). The first letter (U) is in the 'first position' column, the second letter (G) is in the 'second position' column and the third letter (C) is in the 'third position' column.

Ribonucleic acid (RNA) structure

Ribonucleic acid (RNA) is a **polymer** made up of repeating mononucleotide sub-units (Figure 1). It forms a single strand in which each nucleotide is made up of:

- the **pentose sugar ribose**
- one of the organic bases adenine (A), guanine (G), cytosine (C) and uracil (U)
- a phosphate group.

The two types of RNA that are important in protein synthesis are:

- messenger RNA (mRNA)
- transfer RNA (tRNA).

Messenger RNA (mRNA)

Consisting of thousands of mononucleotides, mRNA is a long strand that is arranged in a single helix. Because it is manufactured when DNA forms a mirror copy of part of one of its two strands, there is a great variety of different types of mRNA. Once formed, mRNA leaves the nucleus via pores in the nuclear envelope and enters the cytoplasm, where it associates with the ribosomes. There it acts as a template upon which proteins are built. Its structure is suited to this function because it possesses the correct sequences of the many triplets of organic bases that code for specific polypeptides. It is also easily broken down and therefore exists only while it is needed to manufacture a given protein.

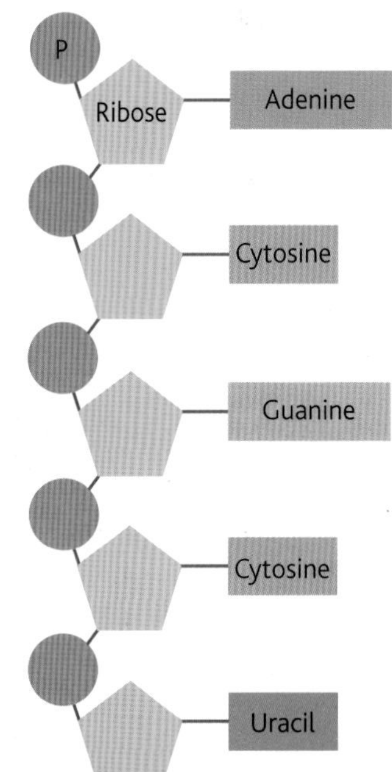

Figure 1 *Section of ribonucleic acid (RNA) molecule*

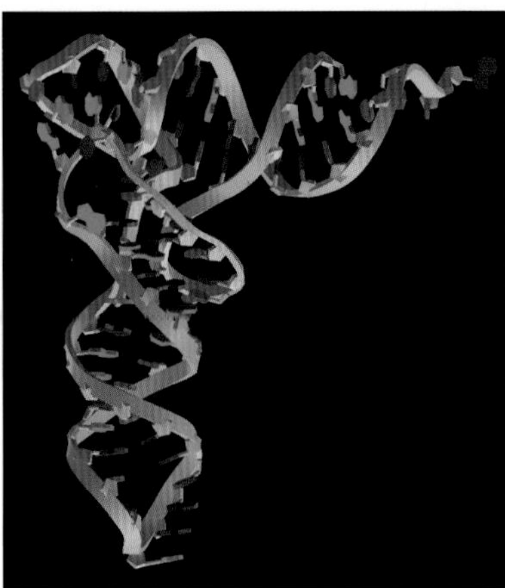

Figure 2 *Computer artwork of a tRNA molecule*

Transfer RNA (tRNA)

Transfer RNA (tRNA) is a relatively small molecule that is made up of around 80 nucleotides. It is a single-stranded chain folded into a clover-leaf shape, with one end of the chain extending beyond the other. This is the part of the tRNA molecule to which amino acids can easily attach. There are several types of tRNA, each able to carry a single amino acid. At the opposite end of the tRNA molecule is a sequence of three other organic bases, known as the **anticodon**. For each amino acid there is a different sequence of organic bases on the anticodon.

You will recall from the AS Biology course that the organic bases in DNA pair up in a precise way, for example, guanine pairs with cytosine, and adenine pairs with thymine. These are known as complementary base pairs. In RNA, however, the base thymine is always replaced by a similar base called uracil. RNA can link with both DNA and other RNA molecules. The complementary base pairings that RNA forms are therefore:

- guanine with cytosine
- adenine with uracil (in RNA) or thymine (in DNA).

During protein synthesis, the anticodon pairs with the three complementary organic bases that make up the triplet of bases (codon) on mRNA. The tRNA structure (Figure 3), with its end chain for attaching amino acids and its anticodon for pairing with the codon of the mRNA, is structurally suited to its role of lining up amino acids on the mRNA template during protein synthesis.

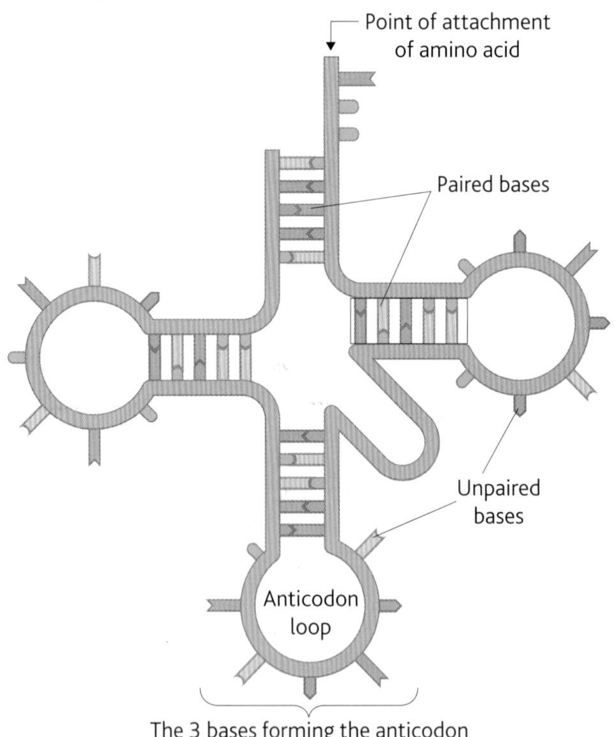

Figure 3 *Clover-leaf structure of tRNA*

Summary questions

1 Explain why the genetic code is described as:

a universal;

b degenerate;

c non-overlapping.

2 State **three** ways in which the molecular structure of RNA differs from DNA.

3 Distinguish between a codon and an anticodon.

Application

Comparison of DNA, messenger RNA and transfer RNA

Table 2 compares the structure, function and composition of DNA, mRNA and tRNA.

Table 2 *Comparison of DNA, mRNA and tRNA*

DNA	Messenger RNA	Transfer RNA
Double polynucleotide chain	Single polynucleotide chain	Single polynucleotide chain
Largest molecule of the three	Molecule is smaller than DNA but larger than tRNA	Smallest molecule of the three
Double-helix molecule	Single-helix molecule (except in a few viruses)	Clover-shaped molecule
Pentose sugar is deoxyribose	Pentose sugar is ribose	Pentose sugar is ribose
Organic bases are adenine, guanine, cytosine and thymine	Organic bases are adenine, guanine, cytosine and uracil	Organic bases are adenine, guanine, cytosine and uracil
Found mostly in the nucleus	Manufactured in the nucleus but found throughout the cell	Manufactured in the nucleus but found throughout the cell
Quantity is constant for all cells of a species (except gametes)	Quantity varies from cell to cell and with level of metabolic activity	Quantity varies from cell to cell and with level of metabolic activity
Chemically very stable	Chemically unstable – easily broken down	Chemically more stable than mRNA but less stable than DNA

1 Table 2 states that, for DNA, the 'quantity is constant for all cells of a species (except gametes)'.
   a How does the quantity in a gamete differ from that in a body cell?
   b What is the significance of the difference you have described?

2 Explain why:
   a DNA needs to be chemically very stable;
   b mRNA needs to be easily broken down (chemically unstable).

14.2 Polypeptide synthesis – transcription and splicing

Learning objectives:

- How is pre-messenger RNA produced from DNA in the process called transcription?
- How is pre-messenger RNA modified to form messenger RNA?

Specification reference: 3.5.6

We saw in the AS Biology course that proteins are made up of polypeptides. Proteins, especially enzymes, are essential to all aspects of life. Every organism needs to make its own, sometimes unique, proteins. The biochemical machinery in the cytoplasm of each cell has the capacity to make every protein from just 20 amino acids. Exactly which proteins it manufactures depends upon the instructions that are provided, at any given time, by the DNA in the cell's nucleus. The basic process is as follows.

- DNA provides the instructions in the form of a long sequence of **nucleotides** and the bases they possess.
- A complementary section of part of this sequence is made in the form of a molecule called pre-mRNA – a process called **transcription**.
- The pre-mRNA is spliced to form mRNA.
- The mRNA is used as a template to which complementary tRNA molecules attach and the amino acids they carry are linked to form a polypeptide – a process called **translation**.

The process can be likened to a bakery, where the basic equipment and ovens (cell organelles) can manufacture any variety of cake (protein) from relatively few basic ingredients (amino acids). Which particular variety of cake is made depends on the recipe (genetic code) that the baker uses on any particular day. By choosing different recipes at different times, rather than making everything all the time, the baker can meet seasonal demands, adapt to changing customer needs and avoid waste.

DNA replication can be likened to the publication of many copies of a recipe book (genome); making a photocopy of a recipe to use in the bakery is therefore transcription. Making the cakes, using the photocopied recipe, is translation. If the book is not removed from the library, many copies of the recipe can be made, and the same cakes can be produced in many places at the same time or over many years.

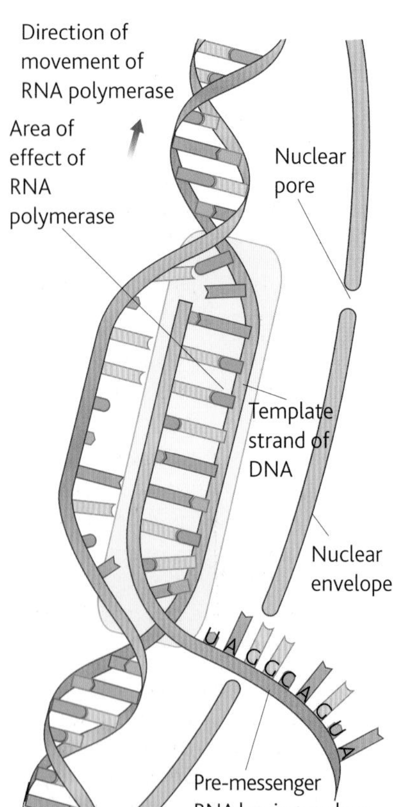

Figure 1 *Summary of transcription*

Transcription

Transcription is the process of making pre-mRNA using part of the DNA as a template. The process, which is illustrated in Figure 1, is as follows.

- The enzyme **DNA helicase** acts on a specific region of the DNA molecule to break the **hydrogen bonds** between the bases, causing the two strands to separate and expose the nucleotide bases in that region.
- The enzyme RNA polymerase moves along one of the two DNA strands, known as the **template strand**, causing the nucleotides on this strand to join with the individual complementary nucleotides from the pool which is present in the nucleus.
- In this way an exposed guanine base on the DNA is linked to the cytosine base of a free nucleotide. Similarly, cytosine links to guanine, and thymine joins to adenine. The exception is adenine, which links to uracil rather than thymine.
- As the RNA polymerase adds the nucleotides one at a time to build a strand of pre-mRNA, the DNA strands rejoin behind it. As a result, only about 12 base pairs on the DNA are exposed at any one time.

- When the RNA polymerase reaches a particular sequence of bases on the DNA that it recognises as a 'stop' triplet code, it detaches, and the production of pre-mRNA is then complete.

Splicing of pre-mRNA

DNA is made up of sections called exons that code for proteins and sections called introns that do not. These intervening introns would interfere with the synthesis of a polypeptide. In the pre-mRNA of **eukaryotic cells**, these intervening non-functional introns are removed and the functional exons are joined together in a process called **splicing.** This process is shown in Figure 2.

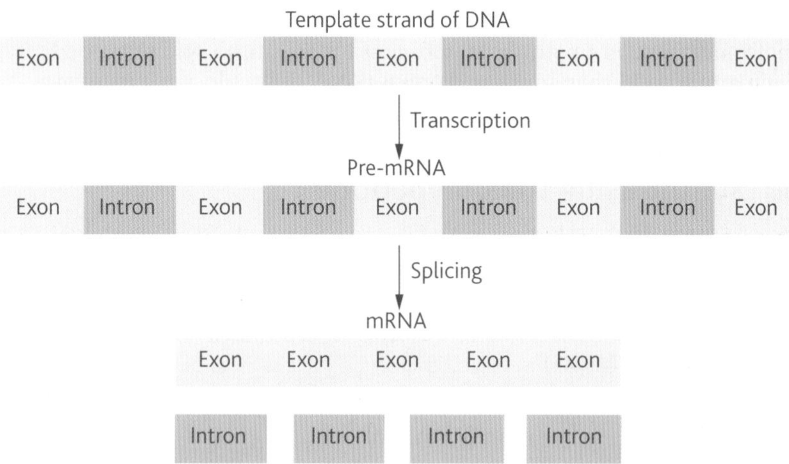

Figure 2 *Splicing of pre-mRNA to form mRNA*

Once the introns have been removed, the remaining exon sections can be rejoined in a variety of different combinations. This means that a single section of DNA (gene) can code for up to a dozen different proteins, depending on the order in which the exons are recombined. Mutations can affect the splicing of pre-mRNA. Certain disorders such as Alzheimer's disease, are the result of splicing failures that lead to non-functional polypeptides being made.

A summary of the complete process of polypeptide synthesis is therefore:

- DNA provides the instructions in the form of a long sequence of nucleotides and the bases they possess.
- A complementary section of part of this sequence is made in the form of a molecule called pre-mRNA – a process called **transcription**.
- The pre-mRNA is modified to mRNA by removing the base sequences copied from non-functional DNA (introns) – a process called **splicing**.
- The mRNA is used as a template to which complementary tRNA molecules attach and the amino acids they carry are linked to form a polypeptide – a process called **translation**.

The mRNA molecules are too large to diffuse out of the nucleus and so, once they have been spliced, they leave via a nuclear pore. Outside the nucleus, the mRNA is attracted to the ribosomes to which it becomes attached, ready for the next stage of the process: translation.

Summary questions

1. Describe the role of RNA polymerase in transcription.
2. Which other enzyme is involved in transcription and what is its role?
3. Why is splicing of pre-mRNA necessary?
4. A sequence of bases along the template strand of DNA is ATGCAAGTCCAG.
   - a What is the sequence of bases on a pre-messenger RNA molecule that has been transcribed from this part of the DNA molecule?
   - b How many amino acids does the sequence code for?
5. A gene is made up of 756 base pairs. The mRNA that is transcribed from this gene is only 524 nucleotides long. Explain why there is this difference.

14.3 Polypeptide synthesis – translation

Learning objectives:

- How is a polypeptide synthesised during the process of translation?
- What are the roles of messenger RNA and transfer RNA in translation?

Specification reference: 3.5.6

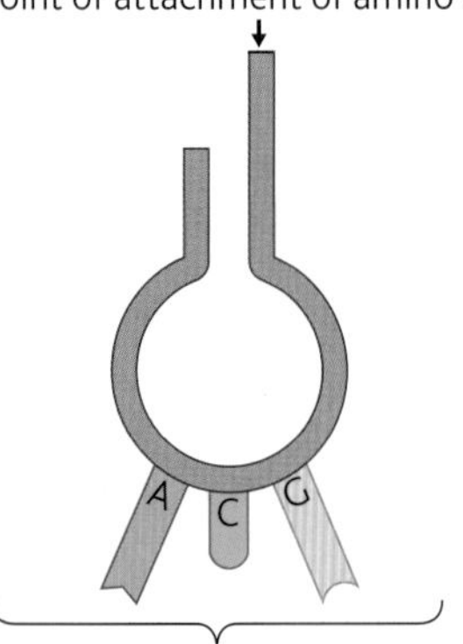

Anticodon – this sequence of ACG means that the amino acid cysteine will attach to the other end of this tRNA molecule. This anticodon will combine with the codon UGC on a mRNA molecule during the formation of a polypeptide. The mRNA codon UGC therefore translates into the amino acid cysteine.

Figure 1 *Simplified structure of one type of tRNA*

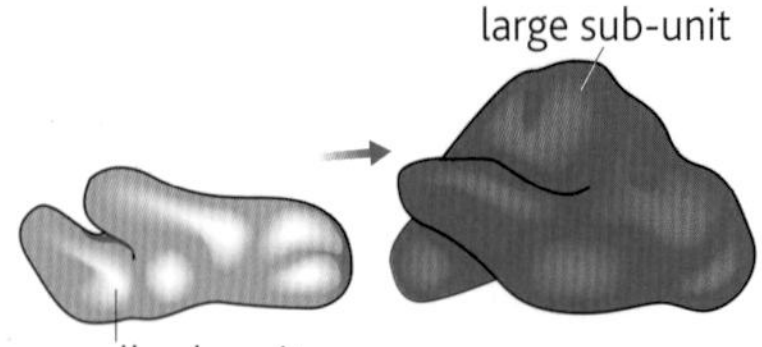

Figure 2 *Structure of a ribosome. The smaller sub-unit fits into a depression on the surface of the larger one*

Hint

Remember that there is no thymine in any RNA molecule. It is uracil in RNA that pairs with adenine.

In Topic 14.2 we looked at how the triplet code of DNA is transcribed into a sequence of **codons** (genetic code) on messenger RNA (mRNA). The next stage is to translate the codons on the mRNA into a sequence of amino acids that make up a polypeptide.

Although the basic structure of transfer RNA (tRNA) is always the same (see Topic 14.1), the sequence of three bases on the **anticodon** loop varies. There are at least 60 variants, each of which corresponds to a codon of three bases on the mRNA. At the other end of the tRNA molecule, there is a point of attachment for an amino acid (Figure 1). Each amino acid therefore has its own tRNA molecule, with its own anticodon of bases.

Synthesising the polypeptide

Once mRNA has passed out of the nuclear pore it determines the synthesis of a polypeptide. The following explanation of how a polypeptide is made is illustrated in Figures 3 and 4. (The information given in brackets below is only to help you follow the process and does not need to be learned.)

- A ribosome (Figure 4, part 1) becomes attached to the starting codon (AUG) at one end of the mRNA molecule.
- The tRNA molecule with the complementary anticodon sequence (UAC) moves to the ribosome and pairs up with the sequence on the mRNA. This tRNA carries an amino acid (methionine).
- A tRNA molecule with a complementary anticodon (UGC) pairs with the next codon on the mRNA (ACG). This tRNA molecule carries another amino acid (threonine).
- The ribosome moves along the mRNA, bringing together two tRNA molecules at any one time, each pairing up with the corresponding two codons on the mRNA.
- By means of an enzyme and ATP, the two amino acids (methonine and threonine) on the tRNA are joined by a **peptide bond**.
- The ribosome moves on to the third codon (GAU) in the sequence on the mRNA, thereby linking the amino acids (threonine and aspartic acid) on the second and third tRNA molecules (Figure 4, part 2).
- As this happens, the first tRNA is released from its amino acid (methionine) and is free to collect another amino acid (methionine) from the amino acid pool in the cell.
- The process continues in this way, with up to 15 amino acids being linked each second, until a complete polypeptide chain is built up (Figure 4, part 3).
- Up to 50 ribosomes can pass immediately behind the first, so that many identical polypeptides can be assembled simultaneously (Figure 3).
- The synthesis of a polypeptide continues until a ribosome reaches a stop codon. At this point, the ribosome, mRNA and the last tRNA molecule all separate and the polypeptide chain is complete.

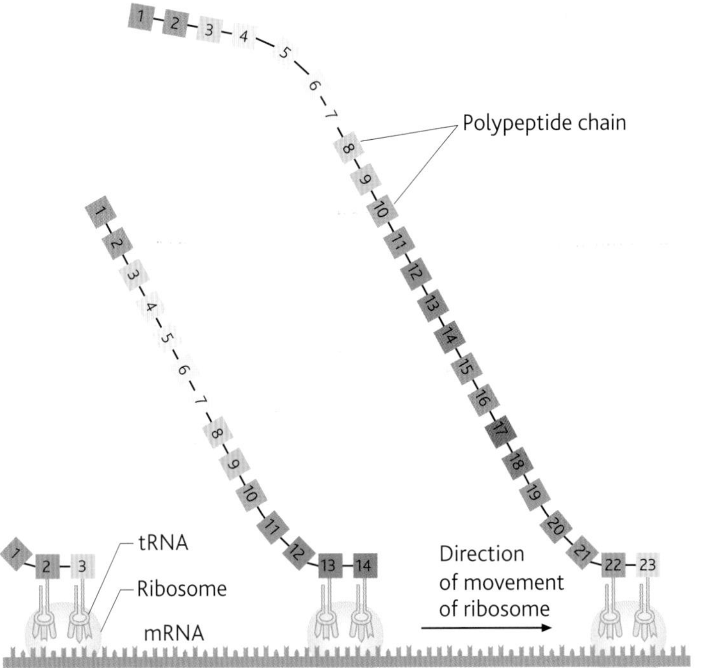

Figure 3 *Polypeptide formation*

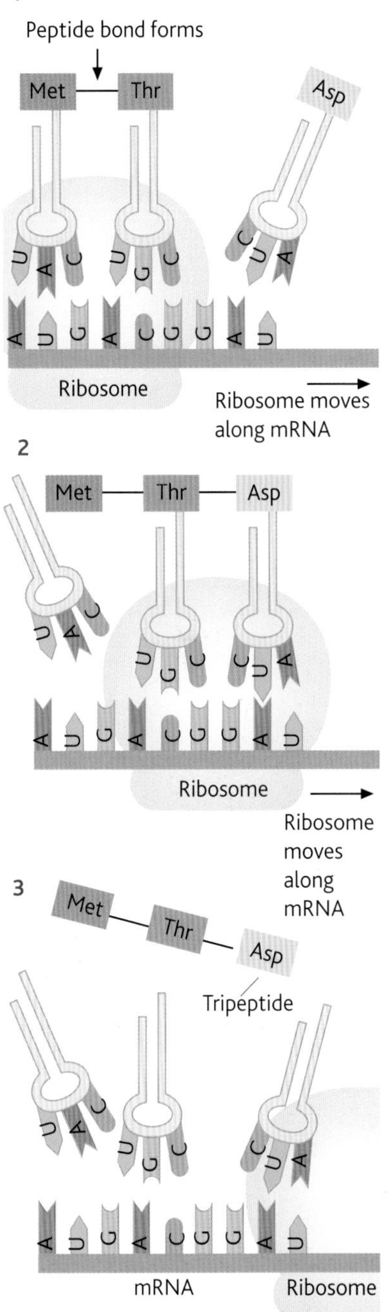

Figure 4 *Translation*

In summary, the DNA triplets that make up a gene determine the codons on mRNA. The codons on mRNA determine the order in which the tRNA molecules line up. They, in turn, determine the sequence of amino acids in the polypeptide. In this way genes precisely determine which proteins a cell manufactures. As many of these proteins are enzymes, genes effectively control the activities of cells.

Assembling a protein

Sometimes a single polypeptide chain is a functional protein. Often, a number of polypeptides are linked together to give a functional protein (quaternary structure). What happens to the polypeptide next depends upon the protein being made, but usually involves the following:

- The polypeptide is coiled or folded, producing its secondary structure.
- The secondary structure is folded, producing the tertiary structure.
- Different polypeptide chains, along with any non-protein groups, are linked to form the quaternary structure.

Summary questions

1 Name the cell organelle involved in translation.

2 A codon found on a section of mRNA has the sequence of bases AUC. List the sequence of bases found on:

a the tRNA anticodon that attaches to this codon;

b the template strand of DNA that formed the mRNA codon.

3 Describe the role of tRNA in the process of translation.

4 A strand of mRNA has 64 codons but the protein produced from it has only 63 amino acids. Suggest a reason for this difference.

Link

Details of protein structure and the formation of peptide bonds is given in Topic 2.5 of the *AS Biology* book. Revision of this topic will help you understand how polypeptides and proteins are assembled.

Table 1

Amino acid	Codon
Tyrosine	UAC
Serine	AGU
Aspartic acid	GAC
Glutamic acid	GAG
Histidine	CAU
Leucine	CUA
Alanine	GCA
Lysine	AAA
Proline	CCU
Glycine	GGC

Application

Interpreting the genetic code

Table 1 lists a number of amino acids and their corresponding codons on mRNA. The strand of DNA against which mRNA is transcribed is called the template strand.

1. Using Table 1, state:
   a. the tRNA anticodon for histidine;
   b. the triplet on the DNA template strand that codes for serine.
2. Name the amino acid coded by the tRNA anticodon GAU.
3. The sequence of bases on a template strand of DNA is CTCCGTGGAATGCGT. List the sequence of amino acids that would appear in a polypeptide coded for by this DNA.
4. The sequence of amino acids in a section of polypeptide is histidine, proline, aspartic acid and leucine. List the sequence of bases on the DNA template strand that codes for this polypeptide section.

Application and How science works

Cracking the code

How exactly did scientists decipher which amino acid was coded for by which codon? Nirenberg and others did so by making synthetic mRNA and using this to make polypeptides. This is an example of How science works (HSW: B, E & I).

The basic stages of the experiments were as follows:

- The enzyme DNase was used to destroy the DNA of living cells.
- Synthetic mRNA was added to the cells without DNA and all 20 amino acids attached to their appropriate tRNA.
- One amino acid was radioactively labelled with carbon 14 (^{14}C) while the remaining 19 had normal, non-radioactive carbon 12 (^{12}C).
- The cells were incubated and the polypeptide produced was later extracted.
- The radioactivity level of the polypeptide produced in each case was measured.

1. Suggest a reason why the DNA of the cells was destroyed.

- In one experiment the radioactive amino acid was phenylalanine and four mixtures, differing only in their mRNA, were set up as follows:
  - mRNA made up of a chain of nucleotides containing only the base adenine = poly A
  - mRNA made up of a chain of nucleotides containing only the base uracil = poly U
  - mRNA made up of a chain of nucleotides containing only the base cytosine = poly C
  - no mRNA was present.

The results are shown in Table 2.

Table 2 *Results of experiment using radioactively labelled phenylalanine*

Type of synthetic mRNA	Radioactivity / counts min^{-1}
Poly A	50
Poly U	39800
Poly C	38
None	44

2 State one codon for the amino acid phenylalanine that is suggested by the results of this experiment. Explain your answer.

3 Why was a mixture without any synthetic RNA used?

The formation of a polypeptide in the mixture with poly U had not been predicted in the original hypothesis. As is often the case in science, an anomalous or unpredicted result can lead to new discoveries. In this case it helped to decipher the genetic code.

Using this method Nirenberg deciphered 47 of the 64 possible codons in the genetic code. The remaining 17 codons, however, gave ambiguous results. This led another scientist, called Khorana, to devise a different technique. He formed very long mRNA molecules that had a repeating sequence of nucleotide bases, such as GUGUGUGUGUGUG. This sequence has two alternating codons: GUG and UGU. The polypeptide produced by this mRNA was made up of alternating cysteine and valine amino acids. The question was, which codon related to which amino acid?

4 Suggest a reason why it is not possible to say which codon relates to which amino acid.

From Nirenberg's earlier experiments, Khorana knew that UGU was a codon for cysteine. This meant that GUG was a codon for valine. By analysing the results of similar experiments using specific sequences of mRNA he was able to decipher the complete genetic code and to show that the code was degenerate. Further experiments by Nirenberg verified these findings.

5 What is meant by 'the code was degenerate'?

6 Despite all these experiments, it was still not possible to find the amino acids coded for by three codons. Why not?

Cracking the genetic code has led to substantial developments, especially in our understanding of many medical conditions. These developments have improved the length and quality of life for many people.

14.4 Gene mutation

Learning objectives:

- What is a gene mutation?
- How do deletion and substitution of bases result in different amino acid sequences in polypeptides?
- Why do some mutations not result in a changed amino acid sequence?
- What are the causes of gene mutations?
- How is cell division genetically controlled?

Specification reference: 3.5.6

Any change to the quantity or the structure of the DNA of an organism is known as a **mutation**. Mutations arising in body cells are not passed on to the next generation. Mutations occurring during the formation of gametes may be inherited, often producing sudden and distinct differences between individuals. They are therefore the basis of **discontinuous variation**. Any change to one or more nucleotide bases, or any rearrangement of the bases, in DNA is known as a **gene mutation**.

We have seen that a sequence of triplets on DNA is transcribed into mRNA and is then translated into a sequence of amino acids that make up a polypeptide. It follows that any changes to one or more bases in the DNA triplets could result in a change in the amino acid sequence of the polypeptide. There are a number of ways in which the DNA bases can change, two examples being substitution and deletion of bases.

Substitution of bases

The type of gene mutation in which a nucleotide in a DNA molecule is replaced by another nucleotide that has a different base is known as a substitution. Depending on which new base is substituted for the original base, there are three possible consequences. As an example, let us take the DNA triplet of bases, guanine–thymine–cytosine (GTC) that codes for the amino acid glutamine. A change to a single base could result in one of the following.

- **A nonsense mutation** occurs if the base change results in the formation of one of the three stop **codons** that mark the end of a polypeptide chain. For example, if the first base, guanine, is replaced by adenine, then GTC becomes ATC. The triplet ATC is transcribed as UAG in mRNA. UAG is one of the three stop codons. As a result the production of the polypeptide would be stopped prematurely. The final protein would almost certainly be significantly different and the protein could not perform its normal function.
- **A mis-sense mutation** arises when the base change results in a different amino acid being coded for. In our example, if the final base, cytosine, is replaced by guanine, then GTC becomes GTG. GTG is one of the DNA triplet codes for the amino acid histidine and this then replaces the original amino acid: glutamine. The polypeptide produced will differ in a single amino acid. The significance of this difference will depend upon the precise role of the original amino acid. If it is important in forming bonds that determine the tertiary structure of the final protein, then the replacement amino acid may not form the same bonds. The protein may then be a different shape and therefore not function properly. For example, if the protein is an enzyme, its active site may no longer fit the substrate and it will no longer catalyse the reaction.
- **A silent mutation** occurs when the substituted base, although different, still codes for the same amino acid as before. This is due to the degenerate nature of the genetic code, in which most amino acids have more than one codon. For instance, if the third base in our example is replaced by thymine, then GTC becomes GTT. However, as both DNA triplets code for glutamine, there is no change in the polypeptide produced and so the mutation will have no effect.

Link

Discontinuous variation is due to genetic factors and is covered in Topic 7.2 of the *AS Biology* book.

Hint

The various gene mutations are illustrated by specific examples that name bases and amino acids. These are only to illustrate the points being made and do not need to be remembered.

Examples of all these types of substitution mutation are shown in Table 1.

Table 1 *Types of substitution mutation.*

	Usual triplet of DNA bases	Nonsense mutation	Mis-sense mutation	Silent mutation
Sequence of bases in DNA	GTC	ATC	GTG	GTT
Sequence of bases in mRNA	CAG	UAG	CAC	CAA
Amino acid in polypeptide	Glutamine	None (stop code)	Histidine	Glutamine

Link

You may recall from Topic 8.2 of the *AS Biology* book that amino acids often have more than one triplet code. This explains why histidine is shown here as being coded for by GTG, but in the answer to a question in Topic 14.3 , it was shown as being coded for by GTA. Both triplets code for histidine.

Hint

Consider the following sentence, which consists only of three-letter words: THE RED HEN ATE HER TEA. If we delete the first letter 'T' but continue to divide the sentence into three-letter words, it becomes HER EDH ENA TEH ERT EA and is incomprehensible. If we delete the final 'T,' this leaves the strange but mostly readable sentence: THE RED HEN ATE HER EA.

Deletion of bases

A gene mutation by deletion arises when a nucleotide is lost from the normal DNA sequence. The loss of a single nucleotide from the thousands in a typical gene may seem a minor change but the consequences can be considerable. Usually the amino acid sequence of the polypeptide is entirely different. This is because the genetic code is read in units of three bases (triplet). One deleted nucleotide creates what is known as a 'frame-shift' because the reading frame that contains each three letters of the code has been shifted to the left by one letter. The gene is now read in the wrong three-base groups and the genetic message is altered. One deleted base at the very start of a sequence could alter every triplet in the sequence. A deleted base near the end of the sequence is likely to have a smaller impact but can still have consequences (see 'Hint'). An example of the effect of a deletion mutation is shown in Figure 1.

Figure 1 *Effects of the deletion of a DNA nucleotide on the amino acid sequence in the final polypeptide*

Causes of mutations

Gene mutations can arise spontaneously during DNA replication. Spontaneous mutations are permanent changes in DNA that occur without any outside influence. Despite being random occurrences, mutations occur with a set frequency. The natural mutation rate varies from species to species, but is typically around one or two mutations per 100 000 genes per generation. This basic mutation rate is increased by outside factors known as **mutagenic agents** or mutagens. These include the following:

- high-energy radiation that can disrupt the DNA molecule
- chemicals that alter the DNA structure or interfere with transcription.

Mutations have both costs and benefits. On the one hand they produce the genetic diversity necessary for natural selection and speciation (see Topics 8.6 and 8.7). On the other hand they often produce an organism that is less well suited to its environment. Additionally, mutations that occur in body cells rather than in gametes can disrupt normal cellular activities, such as cell division.

Figure 2 *This albino hedgehog is the result of a mutation that prevents the production of the pigment melanin*

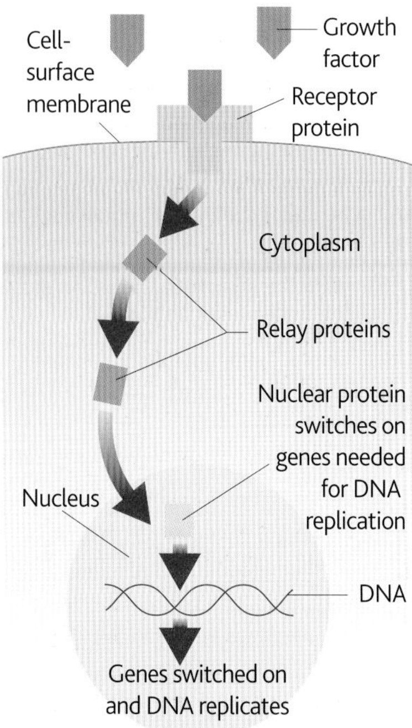

Figure 3 *Control of cell division in a normal cell*

Hint

The interaction of the two genes that control cell division can be likened to the controls of a car. The proto-oncogene is like the accelerator pedal that increases the speed of the car. The tumour suppressor gene is like the brake pedal that reduces the speed of the car.

Genetic control of cell division

Cell division is controlled by genes. Most cells divide at a fairly constant rate to ensure that dead or worn out cells are replaced. In normal cells, this rate is tightly controlled by two genes.

- **proto-oncogenes** that stimulate cell division
- **tumour suppressor genes** that slow cell division.

Role of proto-oncogenes

Proto-oncogenes stimulate cell division. In a normal cell, growth factors attach to a receptor protein on the cell-surface membrane and, via relay proteins in the cytoplasm, 'switch on' the genes necessary for DNA replication. This process is shown in Figure 3. A gene mutation can cause proto-oncogenes to mutate into oncogenes. These oncogenes can affect cell division in two ways:

- The receptor protein on the cell-surface membrane can be permanently activated, so that cell division is switched on even in the absence of growth factors.
- The oncogene may code for a growth factor that is then produced in excessive amounts, again stimulating excessive cell division.

The result is that cells divide too rapidly and a tumour or cancer, develops.

Role of tumour suppressor genes

Research into hereditary forms of cancer led to the discovery of tumour suppressor genes. These genes have the opposite role to proto-oncogenes in the cell, that is, they inhibit cell division. A normal tumour suppressor gene will therefore maintain normal rates of cell division and prevent the formation of tumours – hence its name. If a tumour suppressor gene becomes mutated it is inactivated. In other words it stops inhibiting cell division, which therefore increases. The mutant cells so formed are usually structurally and functionally different from normal cells. Most mutated cells die. However, any that survive are capable of making clones of themselves and forming tumours. Not all tumours are harmful (malignant); some are harmless (benign).

Application and How science works

Mutagenic agents

This is an example of How science works (HSW: H, K & L).

Mutations can be induced by external influences called mutagenic agents. These cause damage in a number of ways.

- **Certain chemicals can remove groups from nucleotide bases.** Nitrous acid can remove an $-NH_2$ group from cytosine in DNA, changing it into uracil.

1 Suggest what might be the result of this change on the codons on a mRNA molecule that is transcribed from a section of DNA with the triplets GCA CTC ATC.

- **Other chemicals can add groups to nucleotides**. Benzpyrene is a chemical found in tobacco smoke. It adds a large group to guanine that makes it unable to pair with cytosine. When DNA polymerase reaches the affected guanine it inserts any of the other bases.

2 What type of mutation is caused by benzpyrene?

- **Ionising radiation**, such as X-rays, can produce highly reactive agents, called free radicals, in cells. These free radicals can alter the shape of bases in DNA so that DNA polymerase can no longer act on them.

3 Explain why DNA polymerase cannot act on DNA that has been damaged by X-rays.

4 State **one** genetic effect of DNA polymerase being unable to act on DNA.

Ultraviolet radiation from the Sun or tanning lamps affects thymine in DNA, causing it to form bonds with the nucleotides on either side of it. This seriously disrupts DNA replication.

Scientific research and experimentation has enabled us to identify potentially dangerous mutagenic agents. The effects of such agents are complex and the amount of harm they cause is often a matter of debate. Commercial organisations, such as the tobacco industry, manufacturers of sunbeds, and producers and retailers of sun-block lotions all have an interest in the research that is undertaken. They are more likely to fund research that may benefit their business than research that may harm it. It is therefore important that the results of any research are subjected to the scrutiny of other scientists from a wide variety of backgrounds, views, interests and organisations, in a process known as peer review.

This is usually achieved by publishing research findings in reputable scientific journals that have an extensive global readership. The conclusions and claims made by researchers and their sponsors can then be debated and the scientific community at large can test the claims by further experimentation. These claims then become accepted, modified or rejected, depending on the outcome of this further research.

Armed with all this scientific information, decision-makers such as governments and heads of business can take appropriate action that benefits society. Governments, for example, can introduce legislation that controls cigarette sales and smoking, and the use of sunbeds and sets a minimum age at which cigarettes or tanning treatments can be bought. The decisions are often not clear-cut however. X-rays, for example, can be harmful on one hand but are an invaluable diagnostic tool, with countless health benefits, on the other.

5 Leaders in business and government have to make decisions about the use of scientific discoveries. Who else, apart from research scientists, might influence the advice that these leaders give to the public on the use of sunbeds?

Summary questions

1 The following is a sequence of 12 nucleotides within a much longer mRNA molecule: AUGCAUGUUACU. Following a gene mutation the same 12-nucleotide portion of the mRNA molecule is AUGCUGUUACUG. What type of gene mutation has occurred? Show your reasoning.

2 Explain why a deletion gene mutation is more likely to result in a change to an organism than a substitution gene mutation.

3 Explain why a mutation that is transcribed on to mRNA may not result in any change to the polypeptide that it codes for.

4 Errors in transcription occur about 100 000 times more often than errors in DNA replication. Explain why errors in DNA replication can be far more damaging than errors in transcription.

5 Which two types of gene control cell division in normal cells. What is the role of each?

Figure 4 *Ultraviolet radiation from sunbeds has the potential to disrupt DNA replication*

Examination-style questions

1 The table shows the sequence of bases on part of the template strand of DNA.

Base sequence on template strand of DNA	C	G	T	T	A	C
Base sequence of mRNA						

(a) Copy and complete the table to show the base sequence of the mRNA transcribed from this DNA strand. *(2 marks)*

(b) A piece of mRNA is 660 nucleotides long but the DNA template strand from which it was transcribed is 870 nucleotides long.

(i) Explain this difference in numbers of nucleotides.

(ii) What is the maximum number of amino acids in the protein translated from this piece of mRNA? Explain your answer. *(3 marks)*

(c) Draw a table to give **two** differences between the structure of mRNA and the structure of tRNA. *(2 marks)*

AQA, 2006

2 (a) **Figure 1** shows the exposed bases (anticodons) of two tRNA molecules involved in the synthesis of a protein. Complete the boxes to show the sequence of bases found along the corresponding section of the coding DNA strand. *(2 marks)*

AGC		**UUC**

Figure 1

TACAAGGTCGTCTTTGTCAAG

Figure 2

(b) Describe the role of tRNA in the process of translation. *(3 marks)*

(c) **Figure 2** shows the sequence of bases in a section of DNA coding for a polypeptide of seven amino acids. The polypeptide was hydrolysed. It contained four different amino acids. The number of each type obtained is shown in the table.

Amino acid	**Number present**
Phe	2
Met	1
Lys	1
Gln	3

Use the base sequence shown in **Figure 2** to work out the order of amino acids in the polypeptide. *(2 marks)*

AQA, 2006

3 Read the following passage.

The sequence of bases in a molecule of DNA codes for proteins. Different sequences of bases code for different proteins. The genetic code, however, is degenerate. Although the base sequence for AGT codes for serine, other sequences may also code for this same amino acid. There are four base sequences which code for amino acid glycine. These are CCA, CCC, CCG and CCT. There are also four base sequences coding for the amino acid proline. These are GGA, GGC, GGG, and GGT.

Pieces of DNA which have a sequence where the same base is repeated many times are called 'slippery' When 'slippery' DNA is copied during replication, errors may occur in copying. Individual bases may be copied more than once. This may give rise to differences in the protein which is produced by the piece of DNA containing the errors.

Use information in the passage and your own knowledge to answer the following questions.

(a) Different sequences of bases code for different proteins (lines 1–2). Explain how. *(2 marks)*

(b) The base sequence AGT codes for serine (lines 2–3). Give the mRNA codon transcribed from this base sequence. *(2 marks)*

(c) Glycine-proline-proline is a series of amino acids found in a particular protein. Give the sequence of DNA bases for these three amino acids which contain the longest 'slippery' sequence. *(2 marks)*

(d) Explain how copying bases more than once may give rise to differences in the protein (lines 9–10). *(2 marks)*

(e) Starting with mRNA in the nucleus of a cell, describe how a molecule of protein is synthesised. *(6 marks)*

AQA, 2005

4 (a) (i) What is the role of RNA polymerase in transcription?

(ii) Name the organelle involved in translation. *(2 marks)*

(b) **Figure 3** shows some molecules involved in protein synthesis.

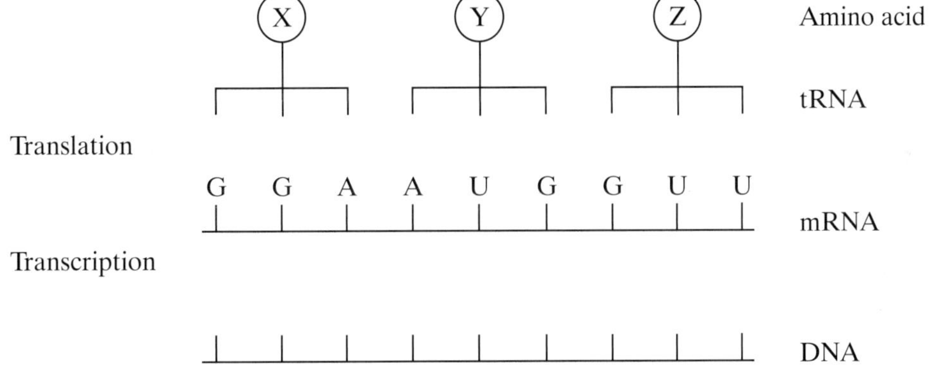

Figure 3

Copy and complete **Figure 3** to show:

(i) the bases on the DNA strand from which the mRNA was transcribed;

(ii) the bases forming the anticodons of the tRNA molecules. *(2 marks)*

Figure 4 shows the effects of two different mutations of the DNA on the base sequence of the mRNA codons for three amino acids.

Original mRNA: G C A A U G G U U

Mutation 1: G C U A U G G U U

Mutation 2: G C A A U G G C U

Amino acid	mRNA codon
methionine	AUG
valine	GUC GUC
alanine	GCA GCC GCU

Figure 4

(c) Name the type of mutation represented by mutation 1. *(1 mark)*

(d) Use the information in the table to:

(i) identify amino acid **X** in **Figure 3**;

(ii) explain how each mutation may affect the polypeptide for which this section of DNA is part of the code. *(5 marks)*

AQA, 2006

15 Control of gene expression

15.1 Totipotency and cell specialisation

Learning objectives:

- What are totipotent cells?
- Which types of cells are totipotent in plants and animals?
- How do cells lose their totipotency and become specialised?
- How can totipotent stem cells be used to treat human disorders?

Specification reference: 3.5.7

All cells contain the same genes. Every cell is therefore capable of making everything that the body can produce. A cell in the lining of the small intestine has the gene coding for **insulin** just as a β cell of the pancreas has the gene coding for maltase. So why do the cells of the small intestine produce maltase rather than insulin and β cells of the pancreas produce insulin rather than maltase? The answer is that, although all cells contain all genes, only certain genes are expressed (switched on) in any one cell at any one time.

Some genes are permanently expressed (switched on) in all cells. For example, the genes that code for essential chemicals, such as the enzymes involved in respiration, are expressed in all cells. Other genes are permanently not expressed (switched off), for example, the gene for insulin in cells lining the small intestine. Further genes are switched on and off as and when they are needed. In this chapter we shall look at how the expression of genes is controlled.

Differentiated cells differ from each other, often visibly so. This is mainly because they each produce different proteins. The proteins that a cell produces are coded for by the genes it possesses or, more accurately, by the genes that are expressed (switched on).

An organism develops from a single fertilised egg. A fertilised egg clearly has the ability to give rise to all types of cells. Cells such as fertilised eggs, which can mature into any body cell, are known as **totipotent cells**. The early cells that are derived from the fertilised egg are also totipotent. These later differentiate and become specialised for a particular function. For example, **mesophyll** cells become specialised for photosynthesis and muscle cells become specialised for contraction. This is because, during the process of cell specialisation, only some of the genes are expressed. This means that only part of the DNA of a cell is translated into proteins. The cell therefore only makes those proteins that it requires to carry out its specialised function. Although it is still capable of making all the other proteins, these are not needed and so it would be wasteful to produce them. Therefore, the genes for these other proteins are not expressed. The ways in which genes are prevented from expressing themselves include:

- preventing transcription and hence preventing the production of mRNA
- breaking down mRNA before its genetic code can be translated.

Details of these mechanisms are given in Topic 15.2.

If specialised cells still retain all the genes of the organism, can they still develop into any other cell? The answer is: it depends – there are no hard and fast rules. **Xylem vessels**, which transport water in plants, and red blood cells, which carry oxygen in animals, are so specialised that they lose their nuclei once they are mature. As the nucleus contains the genes, then clearly these cells cannot develop into other cells. In fact, specialisation is irreversible in most animal cells. Once cells have matured and specialised they can no longer develop into other cells or, in other words, they lose their totipotency. Only a few totipotent cells exist in mature animals. These are called **adult stem cells**.

Link

As a starting point for this topic, it would be useful to revise Topic 12.1, 'Cell differentiation,' in the *AS Biology* book.

Hint

Differentiation results from differential gene expression.

Stem cells are undifferentiated dividing cells that occur in adult animal tissues and need to be constantly replaced. They are found in the inner lining of the small intestine, in the skin and also in the bone marrow, which produces red and white blood cells. Under certain conditions, stem cells can develop into any other types of cell. As a result they can be used to treat a variety of genetic disorders, such as the blood diseases thalassaemia and sickle cell anaemia.

In addition to the adult stem cells found in mature organisms, stem cells also occur at the earliest stage of the development of an embryo, before the cells have differentiated. These are called **embryonic stem cells**.

The situation in plants is different. Mature plants have many totipotent cells. Under the right conditions, many plant cells can develop into any other cell. For example, if we take a cell from the root of a carrot, place it in a suitable nutrient medium and give it certain chemical stimuli at the right time, we can develop a complete new carrot plant. Growing cells outside of a living organism in this way is called *in vitro* development. Since this new carrot plant is genetically identical to the one from which the single root cell came, it is a **clone**. Cells from most plant species can be used to clone new plants in this way.

Summary questions

1 What are totipotent cells?

2 How does the distribution of totipotent cells in animals differ from that in plants?

3 All cells possess the same genes and yet a skin cell can produce the protein keratin but not the protein myosin, while a muscle cell can produce myosin but not keratin. Explain why.

Application

Growth of plant tissue cultures

There are many factors that influence the growth of plant tissue cultures from totipotent cells. One group consists of plant growth factors, which are chemicals involved in the growth and development of plant tissues. Plant growth factors have a number of features:

- They have a wide range of effects on plant tissues.
- The effects on a particular tissue depend upon the concentration of the growth factor.
- The same concentration affects different tissues in different ways.
- The effect of one growth factor can be modified by the presence of another.

An experiment was carried out to investigate the effects on the development of a plant tissue culture of three growth factors: cytokinin, IAA and 2,4-D. Samples of totipotent plant cells were grown on a basic growth medium in a series of test tubes. Each test tube contained a mixture of the three growth factors in different concentrations. After 2 weeks, the tubes were observed to see the effects of the growth factor mixtures on shoot and root growth. The results are shown in Table 1.

Table 1 *Effect of growth factors on shoot and root development*

Tube no.	Relative concentration of growth factors			Shoot development	Root development
	Cytokinin	IAA	2,4-D		
1	None	Low	None	Moderate	Little
2	Low	High	None	Extensive	Little
3	High	Low	None	Little	Moderate
4	None	High	High	Extensive	Extensive
5	None	None	None	Very little	Very little

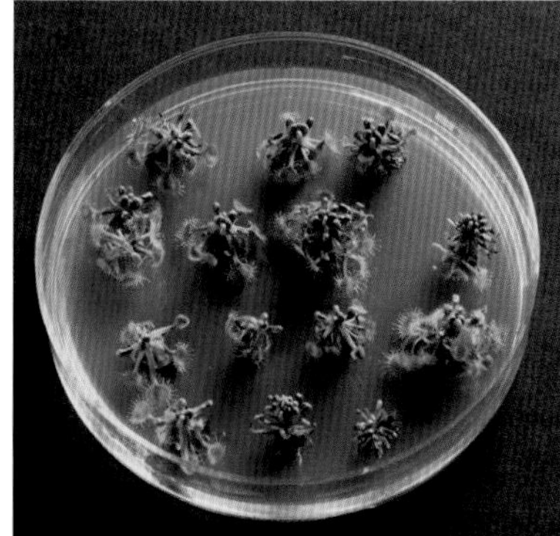

Figure 1 *Plants growing from tissue cultures in a Petri dish*

1 Name the process by which the totipotent cells of the plant tissue culture change in appearance and develop into shoot or root cells.

2 From Table 1, state which **two** growth factors together produced the greatest development of both shoots and roots.

3 Describe one piece of evidence from Table 1 that supports the view that 'the effects of one growth factor can be modified by another'.

Application and How science works

Human embryonic stem cells and the treatment of disease

This is an example of How science works (HSW: I & J).

Although there are a number of types of stem cell in the human body, it is the first few cells from the division of the fertilised egg that have the greatest potential to treat human diseases. As they come from the early stages of an embryo, they are called human embryonic stem cells. These cells can be grown *in vitro* and then induced to develop into a wide range of different human tissues. The process is illustrated in Figure 2.

There are many possible uses of cells grown in this manner. The cells can be used to regrow tissues that have been damaged in some way, either by accident (e.g. skin grafts for serious burn damage) or as a result of disease (e.g. neuro-degenerative diseases, such as Parkinson's disease). Table 2 lists some of the potential uses of human cells produced from stem cells.

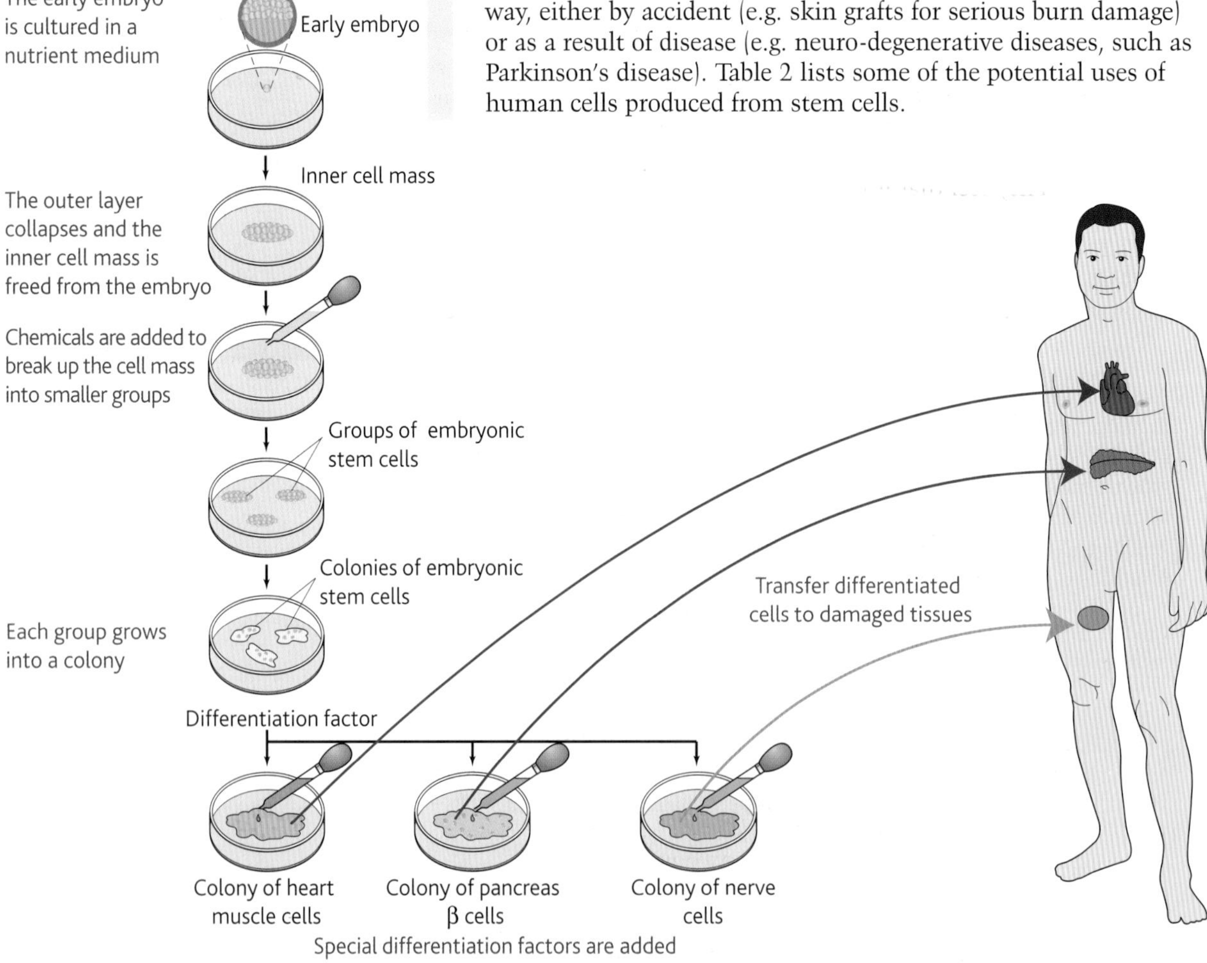

Figure 2 In vitro *culturing of human embryonic stem cells*

Table 2 *Potential uses of human cells produced from stem cells*

Type of cell	Disease that could be treated
Heart muscle cells	Heart damage, e.g. as a result of a heart attack
Skeletal muscle cells	Muscular dystrophy
β cells of the pancreas	Type I diabetes
Nerve cells	Parkinson's disease, multiple sclerosis, strokes, Alzheimer's disease, paralysis due to spinal injury
Blood cells	Leukaemia, inherited blood diseases
Skin cells	Burns and wounds
Bone cells	Osteoporosis
Cartilage cells	Osteoarthritis
Retina cells of the eye	Macular degeneration

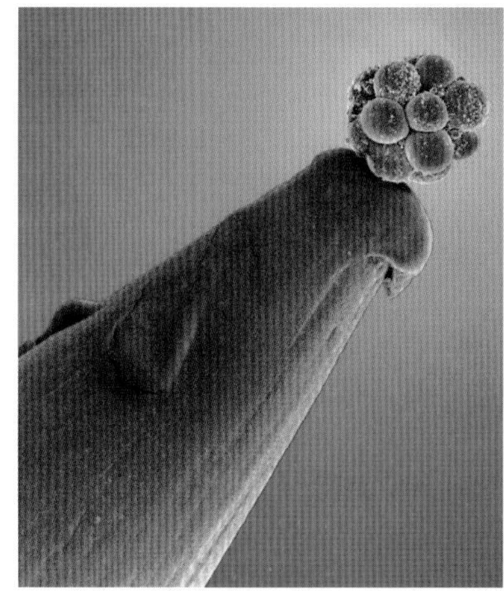

Figure 3 *SEM of a 3-day-old human embryo at the 16-cell stage on the tip of a pin*

At present, embryonic stem cell research is only allowed in the UK under licence and specified conditions. These conditions include its use as a means of increasing knowledge about embryo development and serious diseases, including their treatment. Embryos used in this type of research are obtained from *in vitro* fertilisation. The process nevertheless presents a number of ethical issues.

One issue surrounds the argument as to whether a human embryo less than 14 days old should be afforded the same respect as a fetus or an adult person. Some people feel that using embryos in this way undermines our respect for human life and could progress to the use of fetuses, and even newborn babies, for research or the treatment of disease. They feel that it is a further move towards reproductive cloning and, even if this remains illegal in the UK, the information gained could be used to clone humans elsewhere. Others disagree, arguing that an embryo at such an early stage of development is just a ball of identical, undifferentiated cells, bearing no resemblance to a human being. They feel that the laws prohibiting cloning, in the UK and elsewhere, provide sufficient protection.

Supporters of human embryonic stem cell research contend that it is wrong to allow human suffering to continue when there is a possibility of alleviating it. They further argue that, since embryos are produced for other purposes, e.g. fertility treatments, it makes no sense to destroy superfluous embryos that could be used in research. Opponents of embryonic stem cell research contend that it is wrong to use humans, including human embryos, as a means to an end, even if that end is the laudable one of alleviating human suffering.

However, human embryos are not the only source of stem cells. For example, they can be obtained from the bone marrow of adult humans. As long as a person gives consent, this source of stem cells raises no real ethical issues. At present, these cells have far more restricted medical applications but scientists hope, in time, to be able to make them behave more like embryonic stem cells.

1 Write **two** accounts, each of around 200 words, evaluating the case a) for, and b) against, the continued use of embryos for stem cell research.

15.2 Regulation of transcription and translation

Learning objectives:

- How does oestrogen affect gene transcription?
- What is small interfering RNA?
- How does it affect gene expression?

Specification reference: 3.5.7

In Topic 15.1, we saw how cell specialisation is the result of the selective expression of certain **genes** out of the full complement found in every cell. Let us now investigate some ways in which cells control which genes are expressed.

The effect of oestrogen on gene transcription

In Topic 12.3 we learned that there are two mechanisms of hormone action and looked at how protein hormones, such as **insulin**, operate by using a second messenger. Here we will examine the second mechanism, which is used by lipid-soluble hormones such as oestrogen. Before looking at how oestrogen operates, let us consider the general principles involved in preventing of the expression of a gene by preventing transcription.

- For **transcription** to begin the gene needs to be stimulated by specific molecules that move from the cytoplasm into the nucleus. These molecules are called **transcriptional factors**.
- Each transcriptional factor has a site that binds to a specific region of the DNA in the nucleus.
- When it binds, it stimulates this region of DNA to begin the process of transcription.
- Messenger RNA (mRNA) is produced and the genetic code it carries is then translated into a polypeptide.
- When a gene is not being expressed (i.e. is switched off), the site on the transcriptional factor that binds to DNA is blocked by an inhibitor molecule.
- This inhibitor molecule prevents the transcriptional factor binding to DNA and so prevents transcription and polypeptide synthesis.

Hormones like oestrogen can switch on a gene and thus start transcription by combining with a receptor on the transcriptional factor. This releases the inhibitor molecule. The process is illustrated in Figure 1 and operates as follows:

- Oestrogen is a lipid-soluble molecule and therefore diffuses easily through the **phospholipid** portion of cell-surface membranes (Figure 1, stage 1).
- Once inside the cytoplasm of a cell, oestrogen combines with a site on a receptor molecule of the transcriptional factor. The shape of this site and the shape of the oestrogen molecule complement one another (Figure 1, stage 3).
- By combining with the site, the oestrogen changes the shape of the receptor molecule. This change of shape releases the inhibitor molecule from the DNA binding site on the transcriptional factor (Figure 1, stage 4).
- The transcriptional factor can now enter the nucleus through a nuclear pore and combine with DNA (Figure 1, stage 5).
- The combination of the transcriptional factor with DNA stimulates transcription of the gene that makes up the portion of DNA (Figure 1, stage 5).

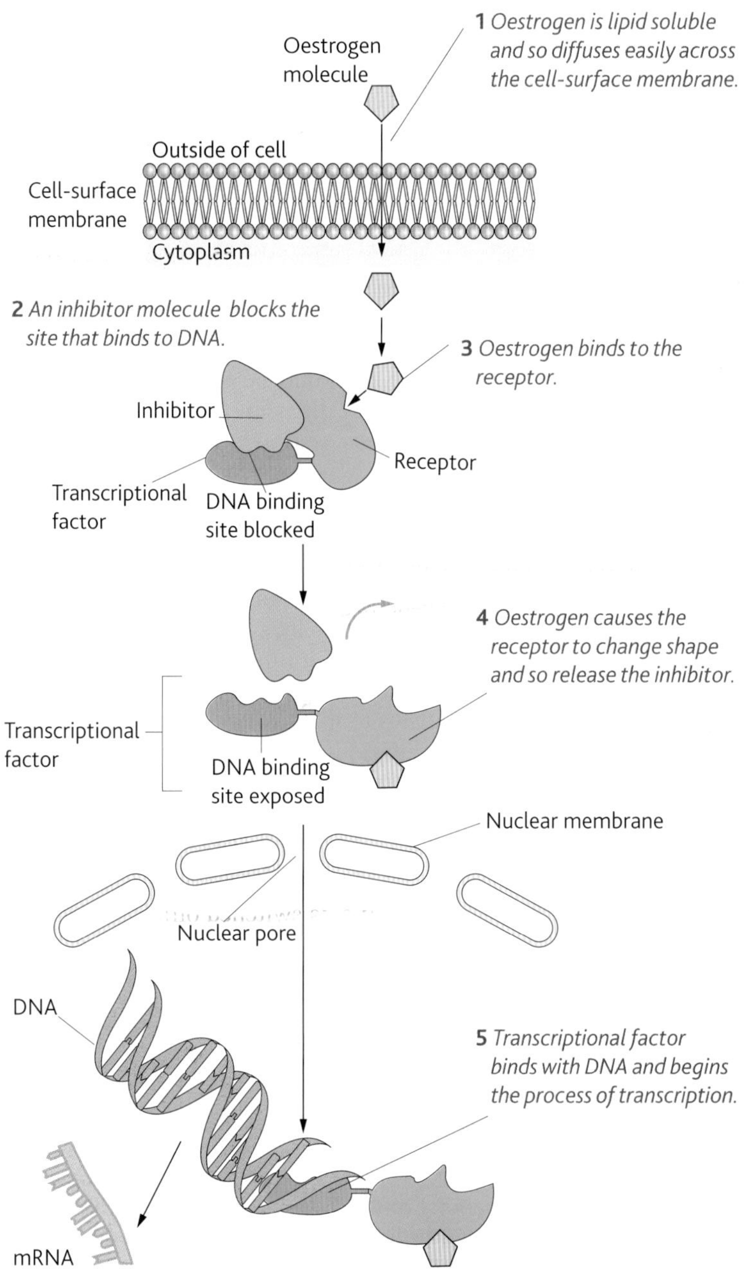

Figure 1 *The effect of oestrogen on gene transcription*

The effect of siRNA on gene expression

Gene expression can be prevented by breaking down messenger RNA before its genetic code can be translated into a polypeptide. Essential to this process are small double-stranded sections of RNA called **small interfering RNA (siRNA)**. The process (see Figure 2) operates as follows.

- An enzyme cuts large double-stranded molecules of RNA into smaller sections called small interfering RNA (siRNA) (Figure 2, stage 1).
- One of the two siRNA strands combines with an enzyme (Figure 2, stage 2).
- The siRNA molecule guides the enzyme to a messenger RNA molecule by pairing up its bases with the complementary ones on a section of the mRNA molecule (Figure 2, stage 3).

Link

The attachment of oestrogen to a receptor causes changes in the shape of the receptor in the same way as the attachment of a non-competitive inhibitor to an enzyme molecule changes its shape and also its active site. This involves the same basic mechanism, which is described in Topic 2.8 of the *AS Biology* book.

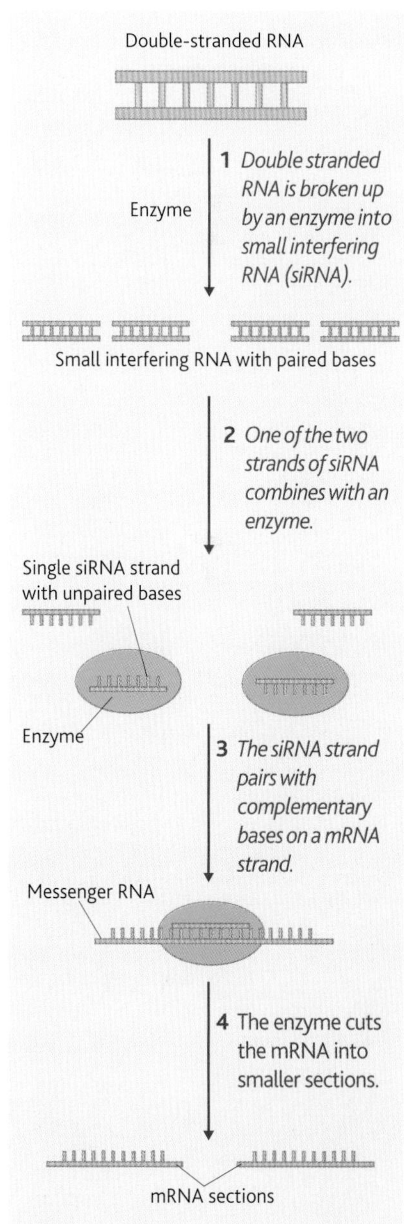

Figure 2 *The effect of siRNA on gene expression*

- Once in position the enzyme cuts the mRNA into smaller sections (Figure 2, stage 4).
- The mRNA is no longer capable of being translated into a polypeptide.
- This means that the gene has not been expressed, that is, it has been blocked.

The siRNA has a number of potential scientific and medical uses:

- It could be used to identify the role of genes in a biological pathway. Some siRNA that blocks a particular gene could be added to cells. By observing the effects (or lack of them) we could determine what the role of the blocked gene is.
- As some diseases are caused by genes, it may be possible to use siRNA to block these genes and so prevent the disease.

Application

Cancer – the 'two hit' hypothesis

We saw in Topic 14.4 that tumours can develop as a result of a mutation of proto-oncogenes that causes cells to divide more rapidly than normal. Tumours can also develop by a mutation of tumour suppressor genes that prevents them from inhibiting cell division. It only takes a single mutated allele to activate proto-oncogenes but it takes a mutation of both alleles to inactivate tumour suppressor genes (two-hits). As natural mutation rates are slow, it takes a considerable time for both tumour suppressor alleles to mutate. This explains why the risk of many cancers increases as one gets older. It is thought that some people are born with one mutated allele. These people are at greater risk of cancer as they need only one further mutation, rather than two, to develop the disease. This explains why certain cancers carry an inherited increased risk.

1. Explain why a doctor may enquire about a patient's family medical history before deciding on using X-ray analysis for a condition other than cancer.
2. Suggest a reason why a single mutant allele of a proto-oncogene can cause cancer, but it requires two mutant alleles of the tumour suppressor gene to do so.
3. One experimental treatment for cancer involves introducing tumour suppressor genes into rapidly dividing cells in order to arrest tumour growth. Explain how this treatment might work.
4. Another experimental treatment is the development of an antibiotic drug that will destroy certain protein receptors on membranes of cancer cells. Explain how this treatment might be effective.

Application

Gene expression in haemoglobin

We saw in the *AS Biology* book that a haemoglobin molecule is made up of four polypeptide chains. Each is known as a globulin. In adult humans two of the polypeptides in a haemoglobin molecule are alpha-globulin and two are beta-globulin. In other words, 50% of the total globulin in all haemoglobin is alpha and 50% is beta. In a human fetus, however, the haemoglobin is different, with much of the beta-globulin being replaced by a third type: gamma-globulin.

Summary questions

1. What is the role of a transcriptional factor?
2. Describe how oestrogen stimulates the expression of a gene.
3. One of the two strands of siRNA combines with an enzyme and guides it to an mRNA molecule which it then cuts. Explain why the mRNA is unlikely to be cut if the other siRNA strand combines with the enzyme.

Fetal haemoglobin has a greater affinity for oxygen than adult haemoglobin. The changes in the production of the three types of globulin during early human development are shown in Figure 3.

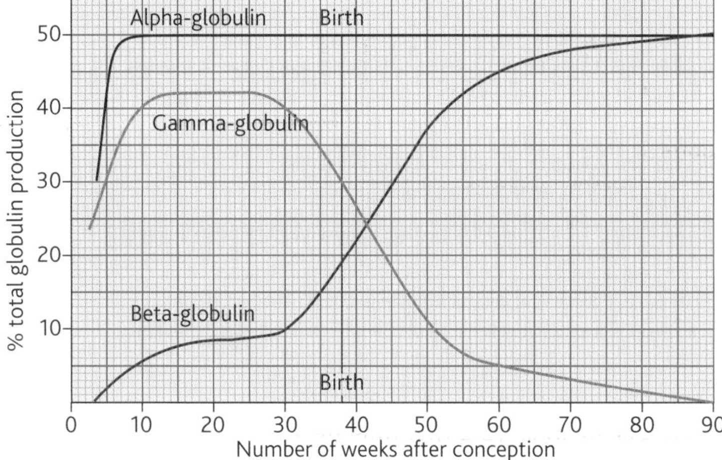

Figure 3 *Percentage total globulin production during early human development*

Humans have genes that code for the production of all three types of globulin. The production of the different haemoglobins depends upon which gene is expressed. The expression of these genes changes at different times during development.

1. Suggest an advantage of fetal haemoglobin having a greater affinity for oxygen than adult haemoglobin.
2. At birth, what percentage of the total globulin production is of each globulin type?
3. Describe the changes in gene expression that occur at 25 weeks.
4. Outline **two** possible explanations for the change in the expression of the gene for gamma-globulin after 25 weeks.
5. Sickle cell disease is the result of a mutant form of haemoglobin. In Saudia Arabia and India, some individuals have high levels of fetal haemoglobin in their blood, even as adults. Where these individuals have sickle cell disease, their symptoms are much reduced. Suggest how controlling the expression of the genes for globulin might provide a therapy for sickle cell disease.

AQA Examination-style questions

1 Scientists carried out a study into the use of stem cells to treat fourteen patients with heart disease. The patients were injected with their own stem cells. The injections into the damaged tissue led to improvements in heart contraction within a few weeks.

(a) What are stem cells? *(1 mark)*

(b) Suggest why the patients were injected with their own stem cells? *(1 mark)*

(c) Suggest and explain how the injection of stem cells led to improvements in heart contraction. *(2 marks)*

(d) A newspaper reported that this study showed heart disease could be treated successfully using stem cells. Give two reasons why the results of this study should be viewed with caution. *(2 marks)*

2 Plant tissue culture is a method used to propagate plants. The flow diagram in **Figure 1** shows one method of plant tissue culture.

Small piece (explant) of tissue is removed from a plant, e.g. bud, shoot, root tissue → Using sterile conditions the explant is transferred to a culture vessel containing nutrients → A mass of unspecialised cells (callus tissue) develops → Callus tissue is transferred to a new culture medium containing nutrients and plant growth regulators → New shoots or roots develop → Developing plants are separated and grown under optimum conditions

Figure 1

(a) Name the type of cell division involved in plant tissue culture. *(1 mark)*

(b) Give **two** advantages of producing plants using this method rather than from seeds. *(2 marks)*

(c) Callus tissue develops into either shoots or roots depending on the relative concentration of the plant growth regulators used. Explain how these unspecialised cells can develop into different types of plant tissue. *(4 marks)*

AQA, 2005

3 Children with severe combined immunodeficiency disorder (SCID) cannot produce the many types of white blood cells that fight infections. This is because they do not have the functional gene to make the enzyme ADA.

Some children with SCID have been treated with stem cells.

The treatment used with the children is described in **Figure 2**.

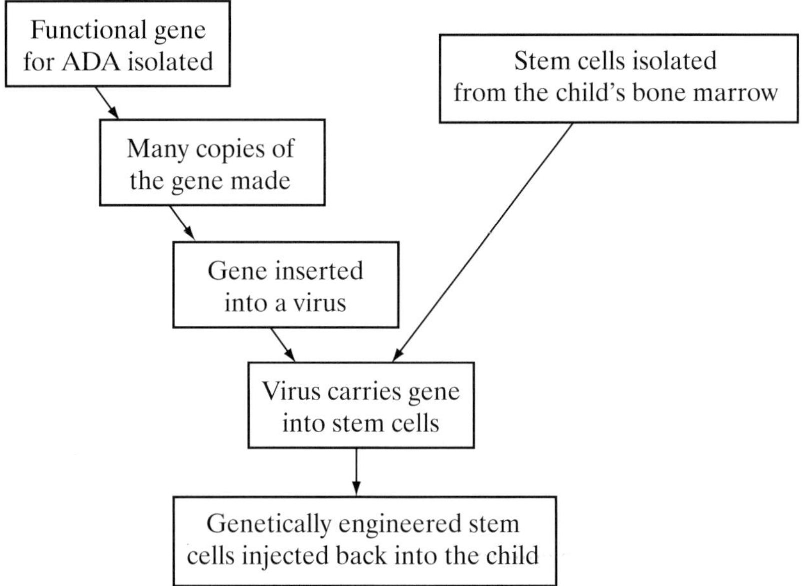

Figure 2

(a) Using the information given, suggest and explain two reasons why stem cells were used in this treatment. *(4 marks)*

(b) A child was treated with genetically engineered stem cells. **Figure 3** shows the number of functioning white blood cells in the child during the year following treatment. Children who do not suffer from SCID have between 5000 and 8000 white blood cells per mm^3 of blood. Describe and explain these results. *(3 marks)*

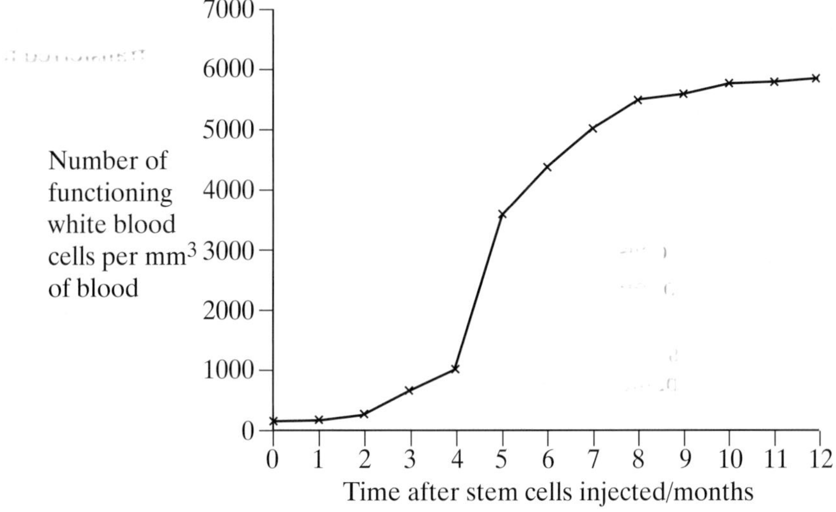

Figure 3

AQA, 2008

4 The hormone oestrogen stimulates some types of breast cancer cells to divide and multiply. Tamoxifen is a drug used in the treatment of breast cancer. It reduces the rate at which breast cancer cells divide.

(a) Tamoxifen acts on cancer cells by binding to oestrogen receptors. Use your knowledge of the role of oestrogen in gene regulation to explain how tamoxifen affects cancer cells. *(4 marks)*

(b) Scientists are investigating the use of siRNA molecules to interfere with gene expression in cancer cells.

(i) Describe the structure of siRNA molecules.

(ii) Explain how siRNA molecules interfere with gene expression. *(6 marks)*

16 DNA technology

16.1 Producing DNA fragments

Learning objectives:

- How is complementary DNA made using reverse transcriptase?
- How are restriction endonucleases used to cut DNA into fragments?

Specification reference: 3.5.8

Perhaps the most significant scientific advance in recent years has been the development of technology that allows genes to be manipulated, altered and transferred from organism to organism – even to transform DNA itself. These techniques have enabled us to understand better how organisms work and to design new industrial processes and medical applications.

A number of human diseases result from individuals being unable to produce for themselves various metabolic chemicals. Many of these chemicals are proteins, such as **insulin**. They are therefore the product of a specific portion of DNA, that is, the product of a gene. Treatment of such deficiencies previously involved extracting the chemical from a human or animal donor and introducing it into the patient. This presents problems such as rejection by the immune system and risk of infection. The cost is also considerable.

It follows that there are advantages in producing large quantities of 'pure' proteins from other sources. As a result, techniques have been developed to isolate genes, **clone** them and transfer them into microorganisms. These microorganisms are then grown to provide a 'factory' for the continuous production of a desired protein. The DNA of two different organisms that has been combined in this way is called **recombinant DNA**. The resulting organism is known as a **genetically modified organism** (**GMO**).

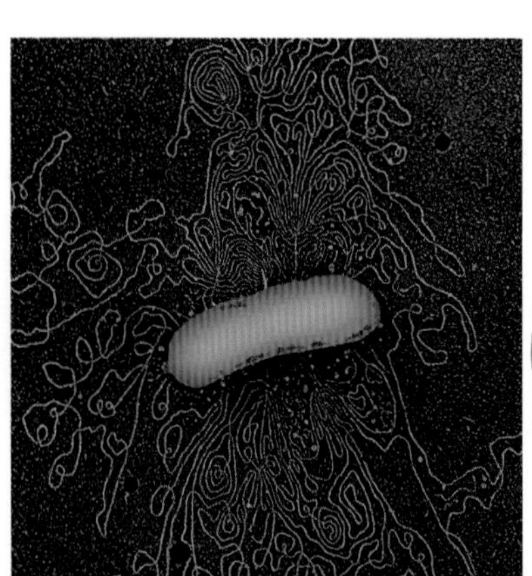

Figure 1 *An* Escherichia coli *bacterial cell that has been treated so that its DNA is ejected*

The process of making a protein using the DNA technology of gene transfer and cloning involves a number of stages:

1. **isolation** of the DNA fragments that have the gene for the desired protein
2. **insertion** of the DNA fragment into a vector
3. **transformation**, that is, the transfer of DNA into suitable host cells
4. **identification** of the host cells that have successfully taken up the gene by use of **gene markers**
5. **growth/cloning** of the population of host cells.

Let us consider each stage in detail.

Before a gene can be transplanted, it must be identified and isolated from the rest of the DNA. Given that the required gene may consist of a sequence of a few hundred bases amongst the many millions in human DNA, this is no small feat! Two of the methods employed use enzymes that have important roles in microorganisms: reverse transcriptase and restriction endonucleases.

Using reverse transcriptase

Retroviruses are a group of viruses of which the best known is human immunodeficiency virus (HIV). The genetic information of retroviruses is in the form of RNA. However, they are able to synthesise DNA from their RNA using an enzyme called reverse transcriptase. It is so-named because it catalyses the production of DNA from RNA, which is the reverse of the more usual transcription of RNA from DNA. The process of using reverse transcriptase to isolate a gene is illustrated in Figure 2 and described below:

- A cell that readily produces the protein is selected (e.g. the β-cells of the islets of Langerhans from the pancreas are used to produce insulin).
- These cells have large quantities of the relevant mRNA, which is therefore extracted.
- Reverse transcriptase is then used to make DNA from RNA. This DNA is known as **complementary DNA (cDNA)** because it is made up of the **nucleotides** that are complementary to the mRNA.
- To make the other strand of DNA, the enzyme DNA polymerase is used to build up the complementary nucleotides on the cDNA template. This double strand of DNA is the required gene.

Link

DNA polymerase acts in the same way when forming the second DNA strand during DNA replication, as described in Topic 11.1 of the *AS Biology* book.

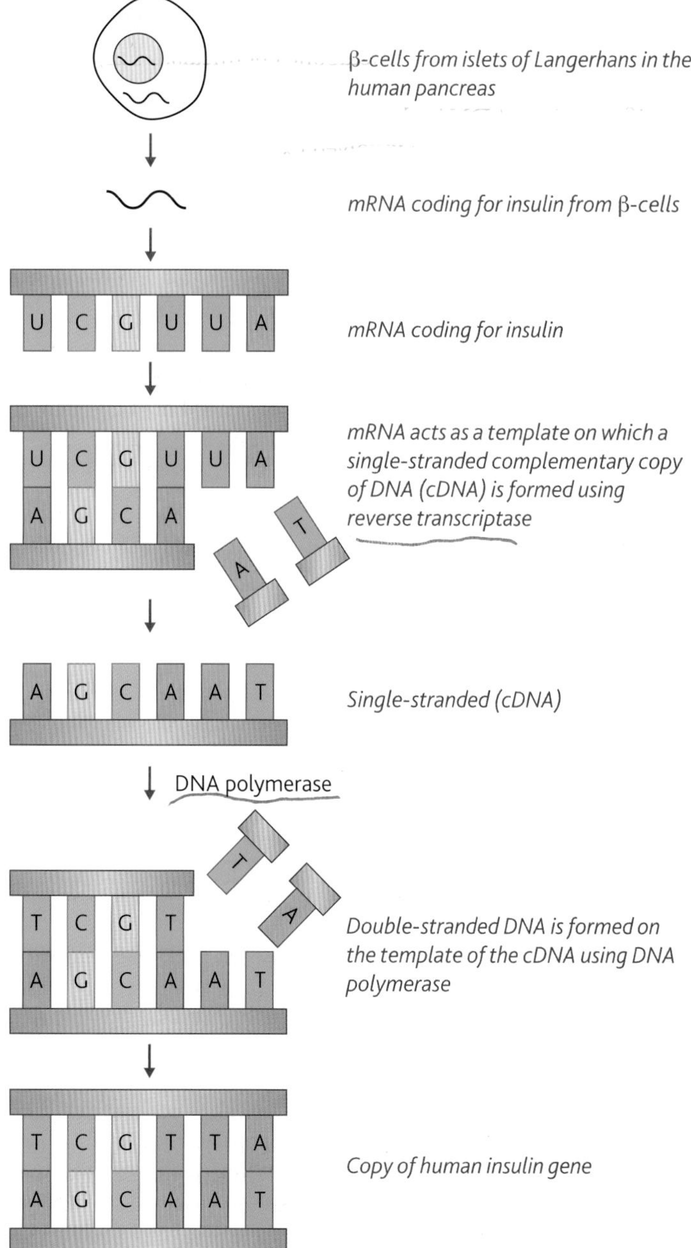

Figure 2 *The use of reverse transcriptase to isolate the gene that codes for insulin*

Hint

Each restriction endonuclease recognises and cuts DNA at a specific sequence of bases. These sequences occur in the DNA of all species of organisms – but not in the same places!

Using restriction endonucleases

All organisms use defensive measures against invaders. Bacteria are frequently invaded by viruses that inject their DNA into them in order to take over the cell. Some bacteria defend themselves by producing enzymes that cut up the viral DNA. These enzymes are called restriction endonucleases.

There are many types of restriction endonucleases. Each one cuts a DNA double strand at a specific sequence of bases called a recognition sequence. Sometimes, this cut occurs between two opposite base pairs. This leaves two straight edges known as blunt ends. For example, one restriction endonuclease cuts in the middle of the base recognition sequence GTTAAC (see Figure 3).

Other restriction endonucleases cut DNA in a staggered fashion. This leaves an uneven cut in which each strand of the DNA has exposed, unpaired bases. An example is a restriction endonuclease that recognises a six-base pair (or 6 bp) AAGCTT, as shown in Figure 3. In this figure, look at the sequence of unpaired bases that remain. If you read both the four unpaired bases at each end from left to right, the two sequences are opposites of one another, that is, they are a **palindrome**. The recognition sequence is therefore referred to as a 6 bp palindromic sequence. This feature is typical of the way restriction endonucleases cut DNA to leave 'sticky ends'. We shall look at the importance of these 'sticky ends' in Topic 16.2.

a HpaI *restriction endonuclease has a recognition site GTTAAC, which produces a straight cut and therefore blunt ends:*

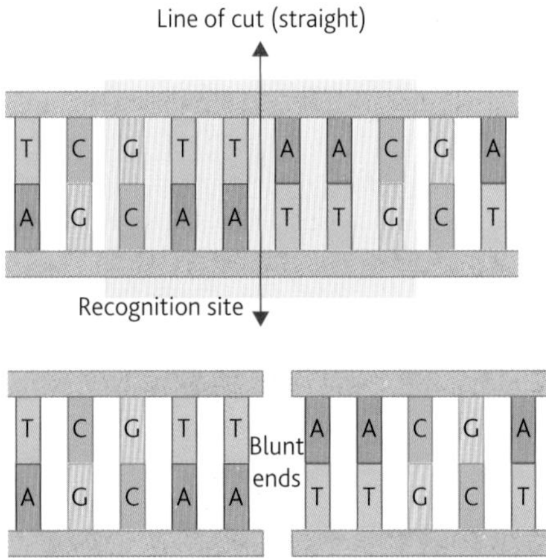

b HindIII *restriction endonuclease has the recognition site AAGCTT, which produces a staggered cut and therefore 'sticky ends':*

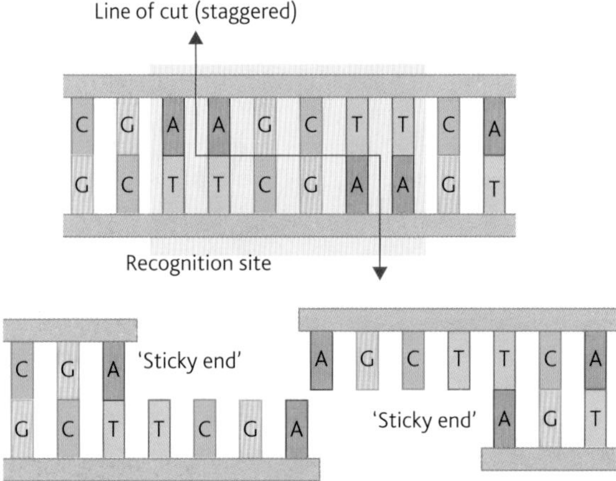

Figure 3 *Action of restriction endonucleases*

Summary questions

In the following passage replace each number with the most appropriate word or words.

Where the DNA of two different organisms is combined, the product is known as (1) DNA. One method of producing DNA fragments is to make DNA from RNA using an enzyme called (2). This enzyme initially forms a single strand of DNA called (3) DNA. To form the other strand requires an enzyme called (4). Another method of producing DNA fragments is to use enzymes called (5), which cut up DNA. Some of these leave fragments with two straight edges, called (6) ends. Others leave ends with uneven edges, called (7) ends. If the sequence of bases on one of these uneven ends is GAATTC, then the sequence on the other end, if read in the same direction, will be (8).

16.2 *In vivo* gene cloning – the use of vectors

Learning objectives:

- What is the importance of 'sticky ends'?
- How can a DNA fragment be inserted into a vector?
- How is the DNA of the vector introduced into host cells?
- What are gene markers and how do they work?

Specification reference: 3.5.8

Once the fragments of DNA have been obtained, the next stage is to **clone** them so that there is a sufficient quantity for medical or commercial use. This can be achieved in two ways:

- *in vivo*, by transferring the fragments to a host cell using a vector
- *in vitro*, using the polymerase chain reaction (see Topic 16.3).

Before we consider how genes can be cloned within living organisms (*in vivo* cloning), let us look at the importance of the 'sticky ends' left when DNA is cut by **restriction endonucleases**.

Importance of 'sticky ends'

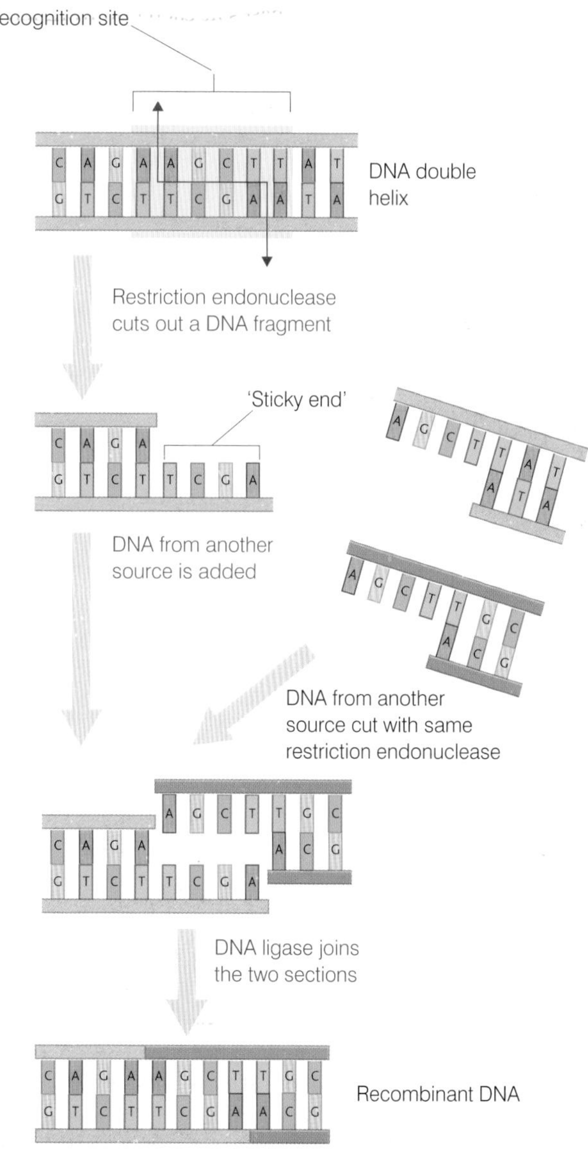

Figure 1 *The use of 'sticky ends' to combine DNA from different sources*

The sequences of DNA that are cut by restriction endonucleases are called recognition sites. If the recognition site is cut in a staggered fashion, the cut ends of the DNA double strand are left with a single strand which is a few **nucleotide** bases long. The nucleotides on the single strand at one side of the cut are obviously complementary to those at the other side because they were previously paired together.

If the same restriction endonuclease is used to cut DNA, then all the fragments produced will have ends that are complementary to one another. This means that the single-stranded end of any one fragment can be joined (stuck) to the single-stranded end of any other fragment. In other words, their ends are 'sticky'. Once the complementary bases of two 'sticky ends' have paired up, an enzyme called **DNA ligase** is used to join the phosphate-sugar framework of the two sections of DNA and so unite them as one.

'Sticky ends' have considerable importance because, provided the same restriction endonuclease is used, we can combine the DNA of one organism with that of any other organism (see Figure 1).

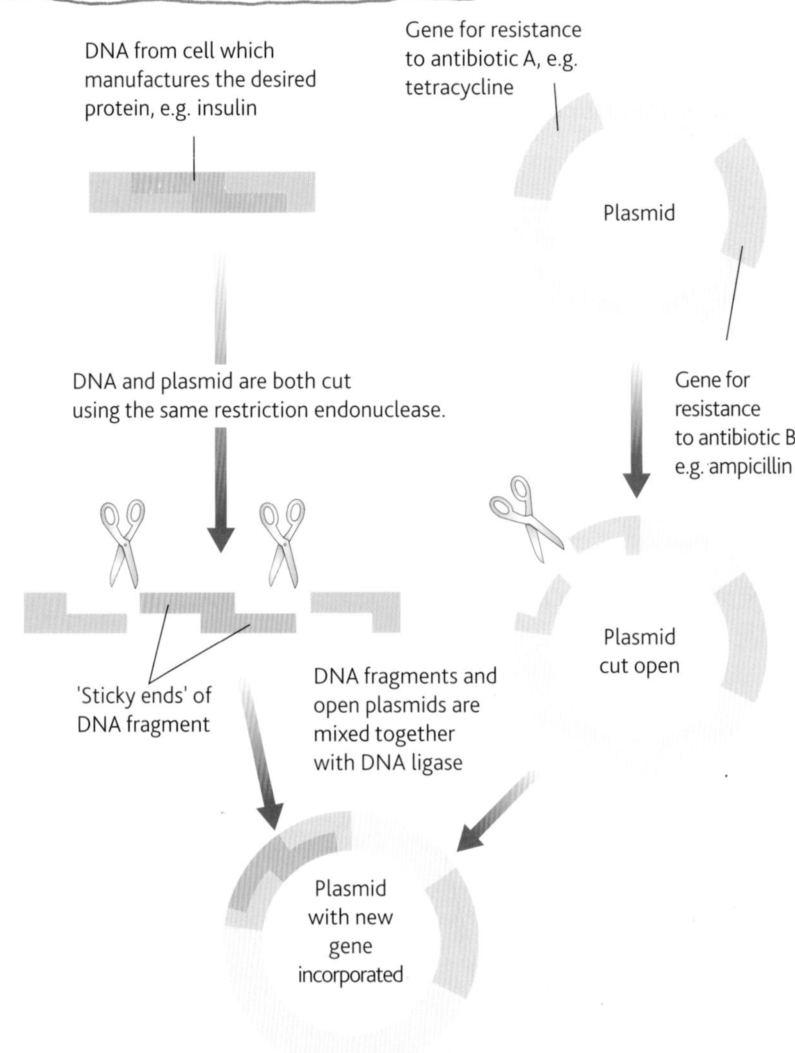

Figure 2 *Inserting a gene into a plasmid vector*

Insertion of DNA fragment into a vector

Once an appropriate fragment of DNA has been cut from the rest of the DNA, the next task is to join it into a carrying unit, known as a **vector**. This vector is used to transport the DNA into the host cell. There are different types of vector but the most commonly used is the **plasmid**. Plasmids are circular lengths of DNA, found in bacteria, which are separate from the main bacterial DNA. Plasmids almost always contain genes for antibiotic resistance, and restriction endonucleases are used at one of these antibiotic-resistance genes to break the plasmid loop.

The restriction endonuclease used is the same as the one that cut out the DNA fragment. This ensures that the 'sticky ends' of the opened-up plasmid are complementary to the 'sticky ends' of the DNA fragment. When the DNA fragments are mixed with the opened-up plasmids, they may become incorporated into them. Where they are incorporated, the join is made permanent using the enzyme DNA ligase. These plasmids now have recombinant DNA. These events are summarised in Figure 2.

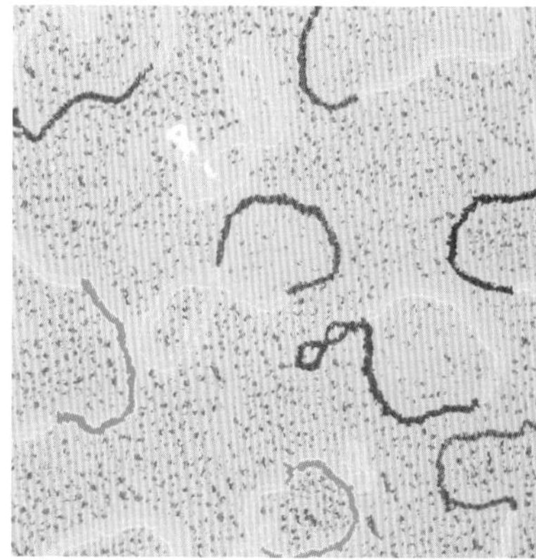

Figure 3 *Coloured TEM of genetically engineered DNA plasmids from the bacterium* Escherichia coli. *The plasmids (yellow) have had different gene sequences (various colours) inserted into them*

Introduction of DNA into host cells

Once the DNA has been incorporated into at least some of the plasmids, they must then be reintroduced into bacterial cells. This process is called **transformation** and involves the plasmids and bacterial cells being mixed together in a medium containing calcium ions. The calcium ions, and changes in temperature, make the bacteria permeable, allowing the plasmids to pass through the cell membrane into the cytoplasm. However, not all the bacterial cells will possess the DNA fragments. There are two main reasons for this:

- Only a few bacterial cells (as few as 1 per cent) take up the plasmids when the two are mixed together.
- Some plasmids will have closed up again without incorporating the DNA fragment.

The first task is to identify which bacterial cells have taken up the plasmid. To do so we use the fact that over the years, bacteria have evolved mechanisms for resisting the effects of antibiotics, typically by producing an enzyme that breaks down the antibiotic before it can destroy the bacterium. The genes for the production of these enzymes are found in the plasmids.

Some plasmids carry genes for resistance to more than one antibiotic. One example is the R-plasmid, which carries genes for resistance to two antibiotics: ampicillin and tetracycline.

The task of finding out which bacterial cells have taken up the plasmids entails using the gene for antibiotic resistance, which is unaffected by the introduction of the new gene. In Figure 2, this is the gene for resistance to ampicillin. The process works as follows:

- All the bacterial cells are grown on a medium that contains the antibiotic ampicillin.
- Bacterial cells that have taken up the plasmids will have acquired the gene for ampicillin resistance.
- These bacterial cells are able to break down the ampicillin and therefore survive.
- The bacterial cells that have not taken up the plasmids will not be resistant to ampicillin and therefore die.

Link

Antibiotics and antibiotic resistance are covered in Topics 16.2 and 16.3 of the *AS Biology* book.

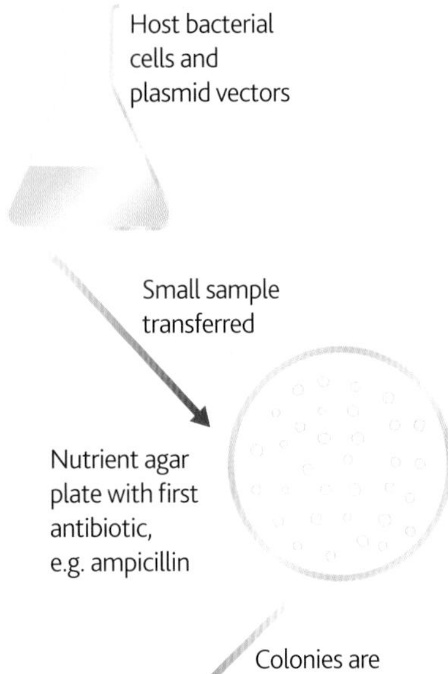

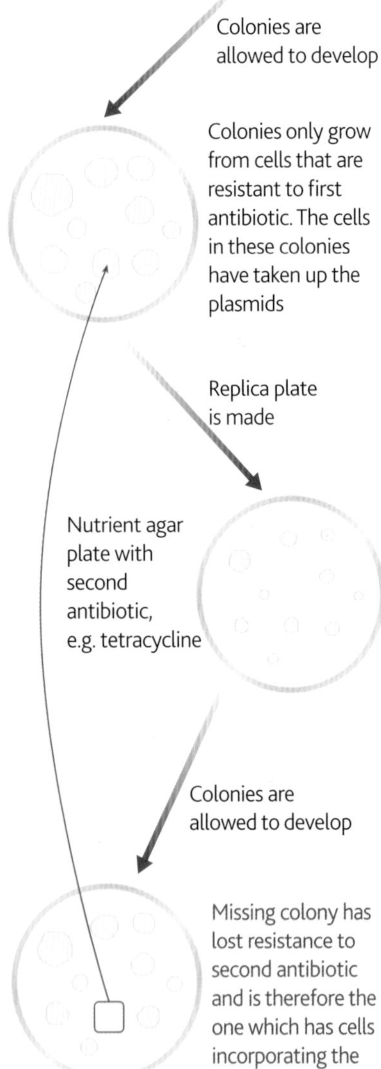

Figure 4 *Detection of transformed bacteria*

This is an effective method of showing which of the bacterial cells have taken up the plasmids. However, some cells will have taken up the plasmids and then closed up without incorporating the new gene, and these will also have survived. The next task is to identify these cells and eliminate them. This is achieved using gene markers.

Gene markers

There are a number of different ways of using gene markers to identify whether a gene has been taken up by bacterial cells. They all involve using a second, separate gene on the plasmid. This second gene is easily identifiable for one reason or another. For example:

- It may be resistant to an antibiotic.
- It may make a fluorescent protein that is easily seen.
- It may produce an enzyme whose action can be identified.

Antibiotic-resistance markers

The use of **antibiotic-resistance** genes as markers is a rather old technology and has been superseded by other methods. However, it is an interesting example of How science works, particularly of the way in which scientists use knowledge and understanding to solve new problems, use appropriate methodology and carry out relevant experiments (HSW: B, C & D).

To identify those cells with plasmids that have taken up the new gene we use a technique called **replica plating**. This process uses the other antibiotic-resistance gene in the plasmid: the gene that was cut in order to incorporate the required gene. In Figure 2 this is the gene for resistance to tetracycline. As this gene has been cut, it will no longer produce the enzyme that breaks down tetracycline. In other words, the bacteria that have taken up the required gene will no longer be resistant to tetracycline. We can therefore identify these bacteria by growing them on a culture that contains tetracycline.

The problem is that treatment with tetracycline will destroy the very cells that contain the required gene. This is where the technique of replica plating comes in. This works as follows.

- The bacterial cells that survived treatment with the first antibiotic (ampicillin) are known to have taken up the plasmid.
- These cells are cultured by spreading them very thinly on nutrient agar plates.
- Each separate cell on the plate will grow into a genetically identical colony.
- A tiny sample of each colony is transferred onto a second (replica) plate in exactly the same position as the colonies on the original plate.
- This replica plate contains the second antibiotic (tetracycline), against which the antibiotic-resistance gene will have been made useless if the new gene has been taken up.
- The colonies killed by the antibiotic must be the ones that have taken up the required gene.
- The colonies in exactly the same position on the original plate are the ones that possess the required gene. These colonies are therefore made up of bacteria that have been genetically modified, that is, have been transformed.

The process of detecting transformed bacteria is summarised in Figure 4.

Fluorescent markers

A more recent and more rapid method is the transference of a gene from a jellyfish into the plasmid. The gene in question produces a green fluorescent protein (GFP). The gene to be cloned is transplanted into the centre of the GFP gene. Any bacterial cell that has taken up the plasmid with the gene that is to be cloned will not be able to produce GFP. Unlike the cells that have taken up the gene, these cells will not fluoresce. As the bacterial cells with the desired gene are not killed, there is no need for replica plating. Results can be obtained by simply viewing the cells under a microscope and retaining those that do not fluoresce. This makes the process more rapid.

Enzyme markers

Another gene marker is the gene that produces the enzyme lactase. Lactase will turn a particular colourless substrate blue. Again, the required gene is transplanted into the gene that makes lactase. If a plasmid with the required gene is present in a bacterial cell, the colonies grown from it will not produce lactase. Therefore, when these bacterial cells are grown on the colourless substrate they will be unable to change its colour. Where the gene has not transformed the bacteria, the colonies will turn the substrate blue.

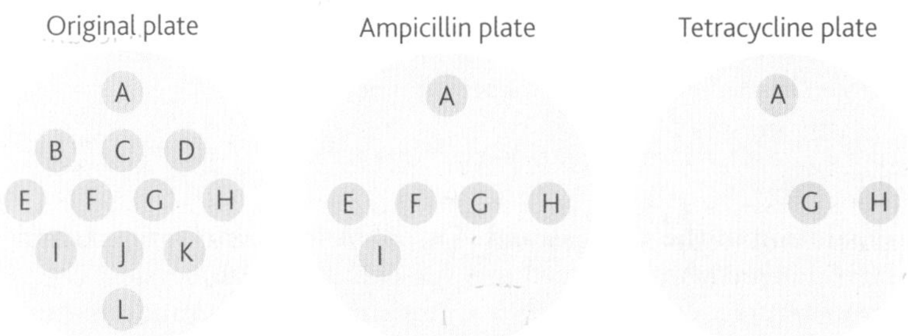

Figure 6

Summary questions

1 What is the role of a vector during *in vivo* gene cloning?

2 Why are gene markers necessary during *in vivo* gene cloning?

3 Give **one** advantage of using fluorescent gene markers rather than antibiotic gene markers. Explain your answer.

4 Figure 6 shows the results of an experiment using antibiotic-resistance gene markers to find which bacterial cells have taken up a gene X. The circles within each plate represent a colony of growing bacteria. Which colonies on the original plate:

a did not take up any plasmids with gene X

b contained plasmids possessing gene X?

Give reasons for your answers.

Isolation

Production of DNA fragments that have the required gene using reverse transcriptase or restriction endonucleases

↓

Insertion

Insertion of DNA fragment into a vector, e.g. plasmid, using DNA ligase

↓

Transformation

Introduction of DNA fragment into suitable host cell

↓

Identification

Identification of host cells that have taken up the DNA using gene markers

↓

Growth/cloning

Culturing of host cells containing the DNA to produce the protein on a large scale

Figure 5 *Outline summary of gene transfer and cloning*

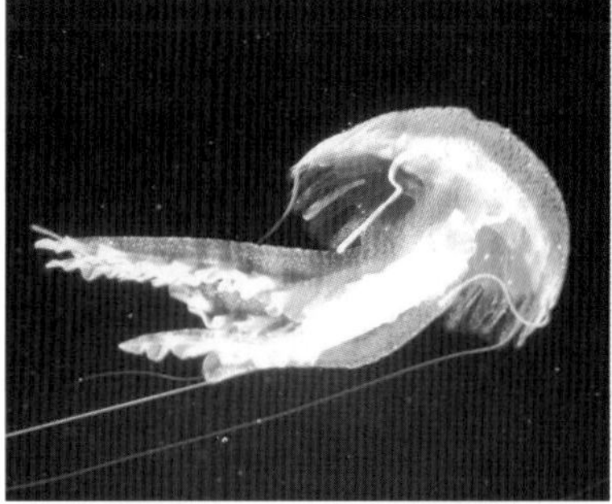

Figure 7 *The gene in this jellyfish that produces a green fluorescent protein can be transplanted into other organisms and used as a fluorescent marker*

Hint

Interestingly, the gene for the fluorescent green protein has itself been genetically modified by the same techniques it is used to support. As a result, varieties have been engineered that fluoresce more brightly and in a number of different colours.

16.3 *In vitro* gene cloning – the polymerase chain reaction

Learning objectives:

- What is the polymerase chain reaction?
- How does the process work?

Specification reference: 3.5.8

AQA Examiner's tip

Candidates frequently lose marks because they think DNA polymerase causes complementary base pairing. It does not – it causes nucleotides to join together as a strand.

Hint

The polymerase chain reaction is **not** the same as semi-conservative replication of DNA in cells.

AQA Examiner's tip

Candidates are very poor at describing the polymerase chain reaction. In particular they fail to appreciate the importance of the temperature changes.

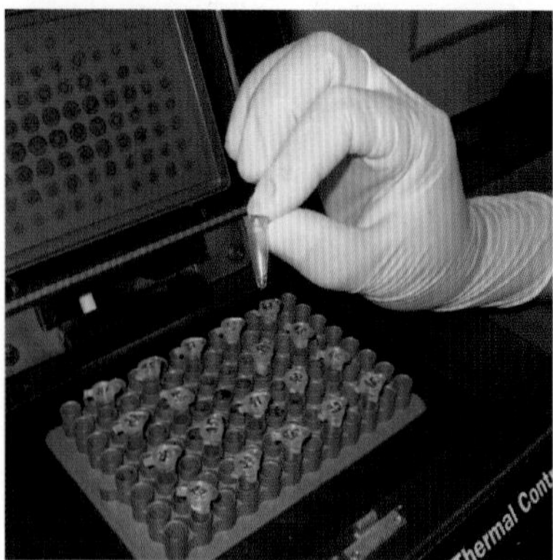

Figure 1 *This is a thermocycler, a machine that carries out the polymerase chain reaction (PCR)*

After looking at *in vivo* cloning in Topic 16.2, let us now consider *in vitro* cloning using the polymerase chain reaction.

Polymerase chain reaction

The polymerase chain reaction (PCR) is a method of copying fragments of DNA. The process is automated, making it both rapid and efficient. The process requires the following:

- **the DNA fragment** to be copied
- **DNA polymerase** – an enzyme capable of joining together tens of thousands of **nucleotides** in a matter of minutes. It is obtained from bacteria in hot springs and is therefore tolerant to heat (thermostable) and does not denature during the high temperatures of the process
- **primers** – short sequences of nucleotides that have a set of bases complementary to those at one end of each of the two DNA fragments
- **nucleotides** – which contain each of the four bases found in DNA
- **thermocycler** – a computer-controlled machine that varies temperatures precisely over a period of time (Figure 1).

The polymerase chain reaction is illustrated in Figure 2 and is carried out in three stages:

1. **separation of the DNA strand.** The DNA fragments, primers and DNA polymerase are placed in a vessel in the thermocycler. The temperature is increased to 95 °C, causing the two strands of the DNA fragments to separate.
2. **addition (annealing) of the primers**. The mixture is cooled to 55 °C, causing the primers to join (anneal) to their complementary bases at the end of the DNA fragment. The primers provide the starting sequences for DNA polymerase to begin DNA copying because DNA polymerase can only attach nucleotides to the end of an existing chain. Primers also prevent the two separate strands from simply rejoining.
3. **synthesis of DNA**. The temperature is increased to 72 °C. This is the optimum temperature for the DNA polymerase to add complementary nucleotides along each of the separated DNA strands. It begins at the primer on both strands and adds the nucleotides in sequence until it reaches the end of the chain.

Because both separated strands are copied simultaneously there are now two copies of the original fragment. Once the two DNA strands are completed, the process is repeated by subjecting them to the temperature cycle again, resulting in four strands. The whole temperature cycle takes around 2 minutes. Over a million copies of the DNA can be made in only 25 temperature cycles and 100 billion copies can be manufactured in just a few hours. The polymerase chain reaction has revolutionised many aspects of science and medicine. Even the tiniest sample of DNA from a single hair or a speck of blood can now be multiplied to allow forensic examination and accurate cross-matching. You will learn more about this in Topic 16.8.

Advantages of *in vitro* and *in vivo* gene cloning

The advantages of *in vitro* gene cloning are:

- **It is extremely rapid**. Within a matter of hours a 100 billion copies of a gene can be made. This is particularly valuable where only a minute amount of DNA is available, for example, at the scene of a crime. This can quickly be increased using the polymerase chain reaction and so there is no loss of valuable time before forensic analysis and matching can take place. *In vivo* cloning would take many days or weeks to produce the same quantity of DNA.
- **It does not require living cells**. All that is required is a base sequence of DNA that needs amplification. No complex culturing techniques, requiring time and effort, are needed.

The advantages of *in vivo* gene cloning are:

- **It is particularly useful where we wish to introduce a gene into another organism**. As it involves the use of **vectors**, once we have introduced the gene into a **plasmid**, this plasmid can be used to deliver the gene into another organism, such as a human being (i.e. it can transform other organisms). This is done by using a technique called gene therapy (see Topic 16.5).
- **It involves almost no risk of contamination**. This is because a gene that has been cut by the same **restriction endonuclease** can match the 'sticky ends' of the opened-up plasmid. Contaminant DNA will therefore not be taken up by the plasmid. *In vitro* cloning requires a very pure sample because any contaminant DNA will also be multiplied and could lead to a false result.
- **It is very accurate**. The DNA copied has few, if any, errors. At one time, about 20 per cent of the DNA cloned *in vitro* by the PCR was copied inaccurately, but modern techniques have improved the accuracy of the process considerably. However, any errors in copying DNA or any contaminants in the sample will also be copied in subsequent cycles. This problem hardly ever arises with *in vivo* cloning because, although mutations can arise, these are very rare.
- **It cuts out specific genes**. It is therefore a very precise procedure as the culturing of transformed bacteria produces many copies of a specific gene and not just copies of the whole DNA sample.
- **It produces transformed bacteria that can be used to produce large quantities of gene products**. The transformed bacteria can produce proteins for commercial or medical use (e.g. hormones such as insulin).

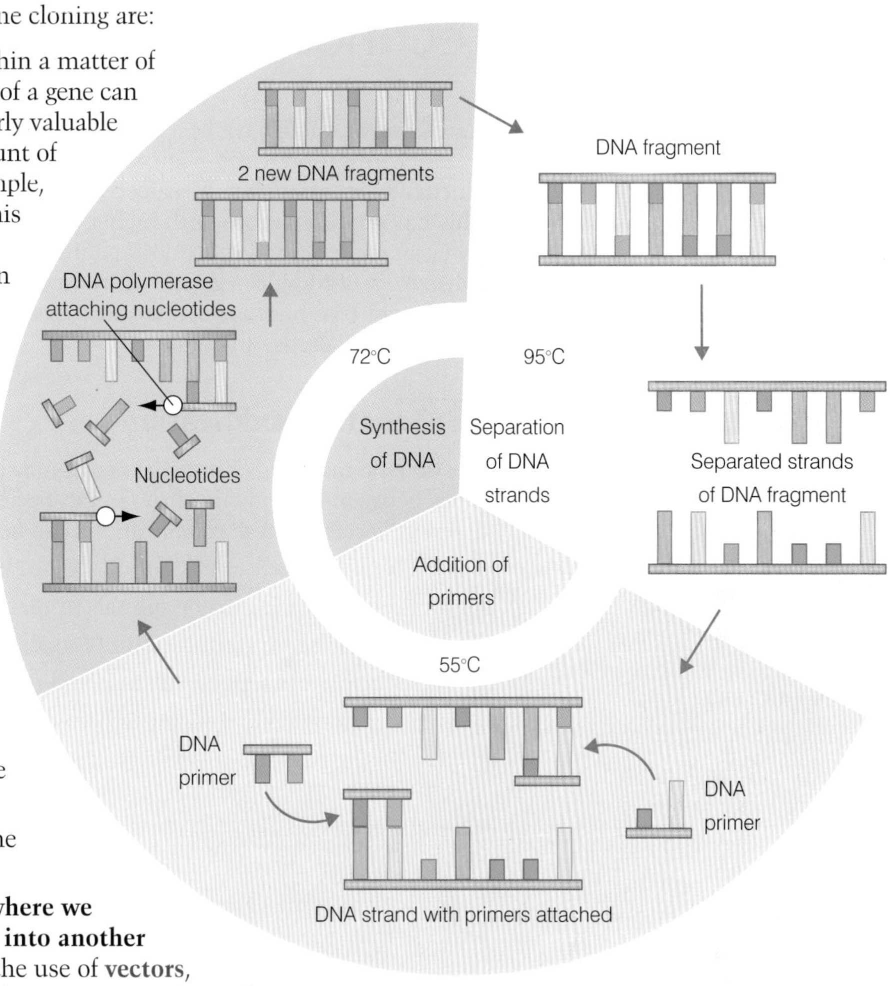

Figure 2 *The polymerase chain reaction showing a single cycle*

Summary questions

1. In the polymerase chain reaction (PCR), what are 'primers'?
2. What is the role of these primers?
3. Why are two different primers required?
4. When DNA strands are separated in the PCR, what type of bond is broken?
5. It is important in the PCR that the fragments of DNA used are not contaminated with any other biological material. Suggest a reason why.

16.4 Use of recombinant DNA technology

Learning objectives:

- How has genetic modification of organisms benefited humans?
- What roles have genetically modified microorganisms, plants and animals played in the beneficial use of recombinant DNA technology?

Specification reference: 3.5.8

Link

Selective breeding is covered in Topic 9.1 of the *AS Biology* book.

Hint

There is always a small risk that substances extracted from animals for human use may carry a disease. The use of the same substance produced by genetically modified microorganisms reduces this risk.

Plants and animals have been used for many products since the earliest time of civilisation. Crop plants and domesticated animals have been genetically manipulated, by **selective breeding**, for thousands of years. This has produced crops with higher yields, meat and milk with a lower fat content, and more docile animals. It has only been with the rediscovery of Mendel's genetic work in 1900, and with modern DNA technology, that humans can now achieve, in weeks, genetic changes that once took hundreds of years.

Genetic modification

The genetic make-up of organisms can now be altered by transferring genes between individuals of the same species or between organisms of different species. These modifications can benefit humans in many ways including:

- increasing the yield from animals or plant crops
- improving the nutrient content of foods
- introducing resistance to disease and pests
- making crop plants tolerant to herbicides
- developing tolerance to environmental conditions, e.g. extreme temperatures, drought
- making vaccines
- producing medicines for treating disease.

Let us look at some genetically modified microorganisms, plants and animals that benefit humans.

Examples of genetically modified microorganisms

There are three main groups of substances produced using genetically modified bacteria.

- **Antibiotics** are produced naturally by bacteria. Although genetic engineering has not substantially improved the quality of antibiotics, it has produced bacteria that increase the quantity of the antibiotics produced and the rate at which they are made.
- **Hormones** – **insulin** is needed daily by more than 2 million diabetics, in order for them to lead normal lives. Previously, insulin extracted from cows or pigs was used, but this could produce side-effects due to its rejection by the diabetic patient's immune system. With genetic engineering, bacterial cells have the human insulin gene incorporated into them and so the insulin produced is identical to human insulin and has no adverse effects on the patient. This method also avoids killing animals and the need to modify insulin before it is injected into humans. Other hormones produced in this way include human growth hormone, cortisone and the sex hormones – testosterone and oestrogen.
- **Enzymes** – many enzymes used in the food industry are manufactured by genetically modified bacteria. These include amylases used to break down starch during beer production, lipases used to improve the flavour of cheeses and proteases used to tenderise meat.

Examples of genetically modified plants

- **Genetically modified tomatoes** have been developed using the insertion of a gene. This gene has a base sequence that is complementary to that of the gene producing the enzyme that causes the tomatoes to soften. The mRNA transcribed from this inserted gene is therefore complementary to the mRNA of the original gene. The two therefore combine to form a double-strand. This prevents the mRNA of the original gene from being translated. The softening enzyme is therefore not produced. This allows the tomatoes to develop flavour without the problems associated with harvesting, transporting and storing soft fruit.
- **Herbicide-resistant crops** have a gene introduced that makes them resistant to a specific herbicide. When the herbicide is sprayed on the crops, the weeds that are competing with the crop plants for water, light and minerals, are killed. The crop plants are resistant to the herbicide and so are unaffected.
- **Disease-resistant crops** have genes introduced that give resistance to specific diseases. Genetically modified rice, for example, can withstand infection by a particular virus.
- **Pest-resistant crops**, e.g. maize, can have a gene added that allows the plant to make a toxin. This toxin kills insects that eat the maize, but is harmless to other animals including humans.
- **Plants that produce plastics** are a possibility currently being explored. It is hoped that we can genetically engineer plants that have the metabolic pathways necessary to make the raw material for plastic production.

Figure 1 *Genetically modified maize resistant to a herbicide. This is a field trial in Lincolnshire where the first application of herbicide has killed the competing weeds and increased the crop's productivity*

Examples of genetically modified animals

An example of genetic modification in animals is the transfer of genes from an animal that has natural resistance to a disease into a totally different animal. This second animal is then made resistant to that disease. In this way domesticated animals can be more economic to rear and hence help to reduce the price of food production.

Further examples are fast-growing food animals such as sheep and fish that have a growth hormone gene added so that, in the case of salmon, they can grow 30 times larger than normal and at 10 times the usual rate.

Another example is in the production of rare and expensive proteins for use in human medicine. Domesticated milk-producing animals such as goats can be used. The gene for the required protein is inserted alongside the gene that codes for proteins in goats' milk. In this way, the required protein is produced in the milk of the goat. The gene can be inserted into the fertilised egg of a goat, so that all the female offspring of that individual will be capable of producing the protein in their milk. One example of a protein made in this way is a protein that prevents blood from clotting (anticoagulant) called anti-thrombin.

Some individuals have an inherited disorder that affects one of the **alleles** that codes for the protein anti-thrombin. As a result, those affected are unable to produce sufficient quantities of anti-thrombin. These individuals are therefore at risk of blood clots. They are currently treated with drugs that thin the blood or are given anti-thrombin that has been extracted from donated blood.

While small amounts of anti-thrombin can be extracted from human blood, far more can be produced in the milk of genetically transformed goats. The process, summarised in Figure 2, is as follows:

- Mature eggs are removed from female goats and fertilised by sperm.
- The normal gene for anti-thrombin production from a human is added to the fertilised eggs alongside the gene that codes for proteins in goats' milk.
- These genetically transformed eggs are implanted into female goats.
- Those resulting goats with the anti-thrombin gene are crossbred, to give a herd in which goats produce milk rich in the protein anti-thrombin.
- The anti-thrombin is extracted from the milk, purified and given to humans unable to manufacture their own anti-thrombin.

In its lifetime a single genetically modified goat can produce as much anti-thrombin in its milk as can be extracted from 90 000 blood donations. Anti-thrombin is sold under its commercial name Atryn®. It is the world's first drug from a genetically modified animal to be registered for general use.

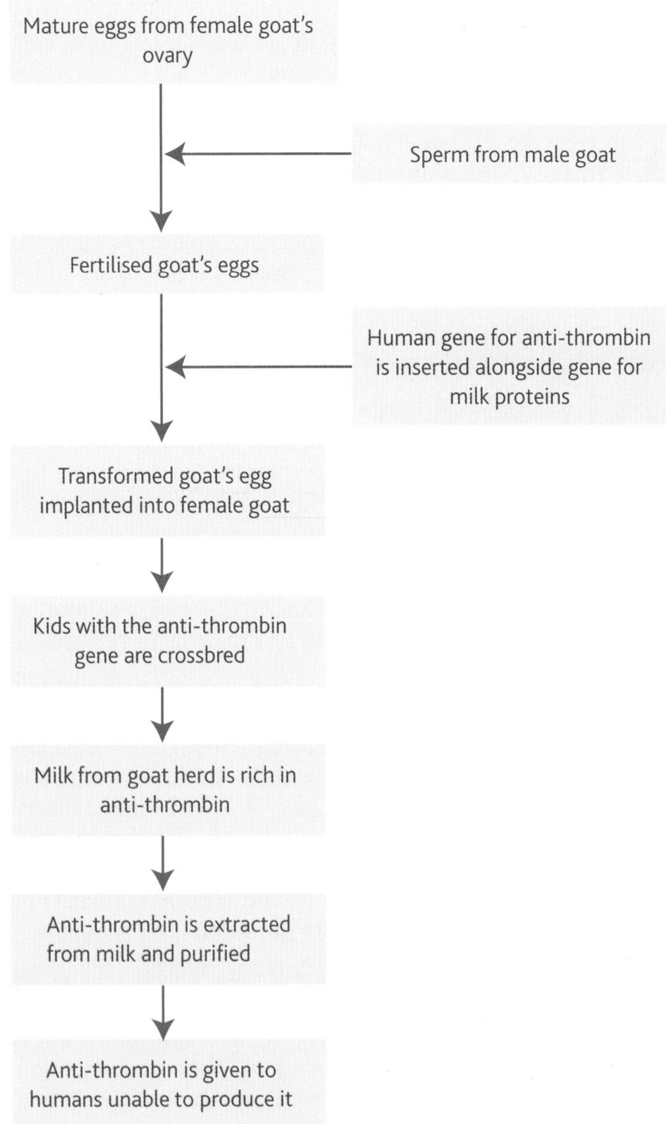

Figure 2 *Summary of genetic modification of goats in order to produce the protein anti-thrombin*

Summary questions

1. State **one** advantage to humans of genetically modified tomatoes.
2. Suggest **one** benefit and **one** possible disadvantage of using genetically modified herbicide-resistant crop plants together with the relevant herbicide. Explain your answer.
3. Why is insulin produced by recombinant DNA technology better than insulin extracted from animals?

Another means of producing drugs from genetically modified animals is being pioneered by the Edinburgh institute that produced Dolly the sheep. Domesticated chickens have had human genes for medicinal proteins added to their DNA. The eggs laid by these transgenic chickens contain the proteins in the white portion from which they can be easily extracted. The human genes are passed on from generation to generation. Large flocks can therefore be formed that offer a potentially unlimited source of cheap medicinal proteins. Drugs that have so far been produced in this way include a form of interferon used to treat multiple sclerosis and an **antibody** with the potential to treat skin cancer and arthritis. It could be five years before patient trials are carried out on these drugs and ten years before the medicines are fully developed.

Application and How science works

Evaluation of DNA technology

Genetic engineering undoubtedly brings many benefits to mankind, but it not without its risks. It is therefore important to evaluate the ethical, moral and social issues associated with its use. This is an example of How science works: I and J.

The benefits of recombinant DNA technology

- Microorganisms can be modified to produce a range of substances, e.g. antibiotics, hormones and enzymes, that are used to treat diseases and disorders.
- Microorganisms can be used to control pollution, e.g. to break up and digest oil slicks or destroy harmful gases released from factories. Care needs to be taken to ensure that such bacteria do not destroy oil in places where it is required, e.g. car engines. To do this, a 'suicide gene' can be incorporated that causes the bacteria to destroy themselves once the oil slick has been digested.
- Genetically modified plants can be transformed to produce a specific substance in a particular organ of the plant. These organs can then be harvested and the desired substance extracted. If a drug is involved, the process is called plant pharming. One promising application of this technique is in combating disease. This involves the production of plants that manufacture antibodies to pathogens and the toxins they produce. Alternatively the plants can be modified to manufacture **antigens** which, when injected into humans, induce natural **antibody** production.
- Genetically modified crops can be engineered to have economic and environmental advantages. These include making plants more tolerant to environmental extremes, e.g. able to survive drought, cold, heat, salt or polluted soils, etc. This permits crops to be grown commercially in places where they are not at present. Globally, each year, an area of land equal to half the United Kingdom becomes unfit for normal crops because of increases in soil salt concentrations. Growing of genetically modified plants, such as salt-tolerant tomatoes, could bring this land back into productivity. Other examples have been described earlier in this spread. In a world where millions lack a basic nutritious diet, and with a predicted 90 million more mouths to feed by 2025, can we ethically oppose the use of such plant crops?

Hint

MORALS are individual or group views about what is right or wrong. Such views refer to almost any subject, such as it is wrong to hunt foxes, work on a Sunday, to swear or to tell lies. Morals vary from country to country and individual to individual, and change over time. Some of the accepted moral values of a hundred years ago in Britain, most people would now disagree with.

Hint

ETHICS is a narrower concept than morals. Ethics are a set of standards that are followed by a particular group of individuals and are designed to regulate their behaviour. They determine what is acceptable and legitimate in pursuing the aims of the group.

Hint

SOCIAL ISSUES relate to human society and its organisation. They concern the mutual relationships of human beings, their interdependence and their cooperation for the benefit of all.

Hint

'Evaluate' always involves looking at the positives and negatives, i.e. the benefits and risks, of a particular issue.

- Genetically modified crops can help prevent certain diseases. Rice can have a gene for vitamin A production added. Can we morally justify not developing more vitamin A-enriched crops when 250 million children worldwide are at risk from vitamin A deficiency leading to 500 000 cases of irreversible blindness each year?
- Genetically modified animals are able to produce expensive drugs, antibiotics, hormones and enzymes relatively cheaply. Details were given earlier in this topic.
- **Gene therapy** might be used to cure certain genetic disorders, such as cystic fibrosis. Details are given in Topic 16.5.
- Genetic fingerprinting can be used in forensic science. Details are given in Topic 16.8.

The risks of recombinant DNA technology

Against the benefits of genetic engineering, must be weighed the risks – both real and potential.

- It is impossible to predict with complete accuracy what the ecological consequences will be of releasing genetically engineered organisms into the environment. The delicate balance that exists in any habitat may be irreversibly damaged by the introduction of organisms with engineered genes. There is no going back once an organism is released.
- A recombinant gene may pass from the organism it was placed in, to a completely different one. We know, for example, that viruses can transfer genes from one organism to another. What if a virus were to transfer genes for herbicide resistance and vigorous growth from a crop plant to a weed that competed with the crop plant? What if the same gene were transferred in pollen to other plants? How would we then be able to control this weed?
- Any manipulation of the DNA of a cell will have consequences for the metabolic pathways within that cell. We cannot be sure until after the event what unforeseen by-products of the change might be produced. Could these lead to metabolic malfunctions, cause cancer, or create a new form of disease?
- Genetically modified bacteria often have antibiotic resistance marker genes that have been added. These bacteria can spread antibiotic resistance to harmful bacteria.
- All genes mutate. What then, might be the consequences of our engineered gene mutating? Could it turn the organism into a **pathogen** which we have no means of controlling?
- What will be the long-term consequences of introducing new gene combinations? We cannot be certain of the effects on the future evolution of organisms. Will the artificial selection of 'desired' genes reduce the genetic variety that is so essential to evolution?
- What might be the economic consequences of developing plants and animals to grow in new regions? Developing bananas which grow in Britain could have disastrous consequences for the Caribbean economies that rely heavily on this crop for their income.
- How far can we take gene therapy (Topic 16.5)? It may be acceptable to replace a defective gene to cure cystic fibrosis, but is it equally acceptable to introduce genes for intelligence, more muscular bodies, cosmetic improvements, or different facial features?

- Will knowledge of, and ability to change, human genes lead to eugenics, whereby selection of genes leads to a means of selecting one race rather than another?
- What will be the consequences of the ability to manipulate genes getting into the wrong hands? Will unscrupulous individuals, groups or governments use this power to achieve political goals, control opposition or gain ultimate power?
- Is the cost of genetic engineering justified, or would the money be better used fighting hunger and poverty, that are the cause of much human misery. Will sophisticated treatments, with their more high-profile images, be put before the everyday treatment of rheumatoid arthritis or haemorrhoids? Will such treatments only be within the financial reach of the better-off?
- Genetic fingerprinting (Topic 16.8), with its ability to identify an individual's DNA accurately, is a highly reliable forensic tool. How easy would it be for someone to exchange a DNA sample maliciously, leading to wrongful conviction?
- Is it immoral to tamper with genes at all? Should we let nature take its own course in its own time?
- How do we deal with the issues surrounding the **human genome project**? Is it right that an individual or company can patent, and therefore effectively own, a gene?

It is inevitable that we remain inquisitive about the world in which we live, and that we will seek to try to improve the conditions around us. Genetic research is bound to continue, but the challenge will be to develop the safeguards and ethical guidelines that will allow genetic engineering to be used in a safe and effective manner.

1. Take any **three** aspects of recombinant DNA technology that are beneficial to humans (as listed above) and present a reasoned argument in each case for the continued use of that technology.
2. Using the same three aspects, present a reasoned argument that an environmentalist or anti-globalisation activist might make against the continued use of that technology.

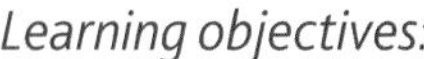

16.5 Gene therapy

Learning objectives:

- What is cystic fibrosis?
- What is the cause of cystic fibrosis?
- How can gene therapy be used in the treatment of cystic fibrosis?
- What is the difference between germ-line and somatic-cell gene therapy?

Specification reference: 3.5.8

Link

Details of osmosis and carrier proteins are given in Topics 3.7 and 3.8 of the *AS Biology* book.

Up to 2 per cent of the human population are affected by one of the 4000 diseases caused by a missing gene or a gene that does not express itself properly. One example is cystic fibrosis. A potential cure is to replace defective genes with genes cloned from healthy individuals in a technique called **gene therapy**. Gene therapy has many potential uses, including the treatment of cancer. In 2008, a young man with a type of genetic blindness had his sight partially restored using this technique.

Cystic fibrosis

Cystic fibrosis is the most common genetic disorder among the white population of Europe and North America, with around 1 in every 20 000 people having the disease. It is caused by a mutant **recessive allele** in which three DNA bases, adenine-adenine-adenine, are missing. It is therefore an example of a deletion **mutation** (see Topic 14.4). The normal gene, called the cystic fibrosis trans-membrane-conductance regulator (CFTR) gene, normally produces a protein of some 1480 amino acids. The deletion results in a single amino acid being left out of the protein. This, however, is enough to make the protein unable to perform its role of transporting chloride ions across epithelial membranes. CFTR is a chloride-ion channel protein that transports chloride ions out of epithelial cells, and water naturally follows, by the process of **osmosis**. In this way, epithelial membranes are kept moist.

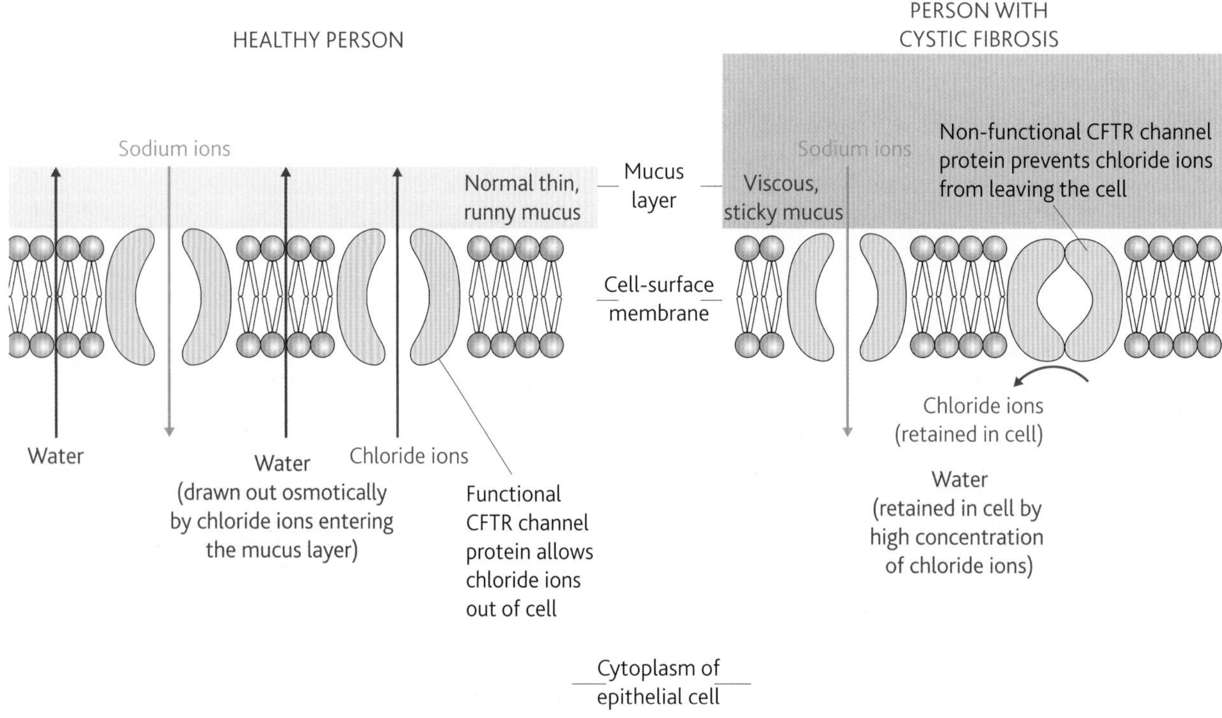

Figure 1 *Transport of ions by CFTR channel protein in a healthy person and a person with cystic fibrosis*

In a patient with cystic fibrosis, the defective gene means that the protein is either not made or does not function normally. The epithelial membranes are therefore dry and the mucus they produce remains viscous and sticky. The symptoms this causes include:

- mucus congestion in the lungs, leading to a much higher risk of infection because the mucus, which traps disease-causing organisms, cannot be removed
- breathing difficulties and less efficient gaseous exchange
- accumulation of thick mucus in the pancreatic ducts, preventing pancreatic enzymes from reaching the duodenum and leading to the formation of fibrous cysts (hence the name of the disease)
- accumulation of thick mucus in the sperm ducts in males, possibly leading to infertility.

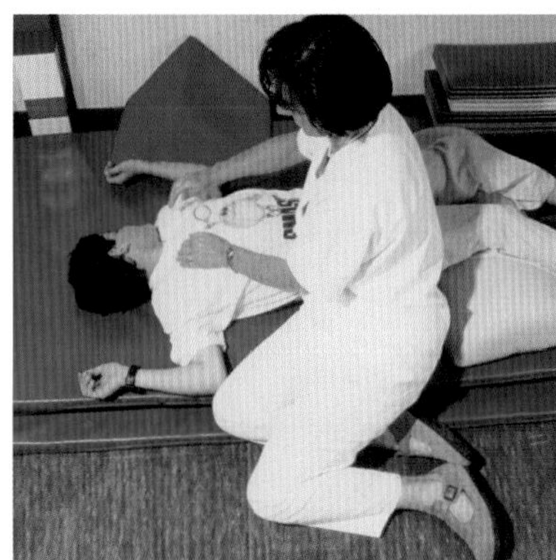

Figure 2 *Physiotherapist treating a young cystic fibrosis patient*

As cystic fibrosis is caused by a recessive allele, it is possible that two outwardly normal parents could have a child who suffers from the disease. Where there is a history of the disease in both families, the parents may choose to be genetically screened to see whether they carry the allele. More details on genetic screening are given in Topic 16.7.

Treatment of cystic fibrosis using gene therapy

There are two ways in which gene therapy may be used to treat cystic fibrosis:

- **gene replacement**, in which the defective gene is replaced with a healthy gene
- **gene supplementation**, in which one or more copies of the healthy gene are added alongside the defective gene. As the added genes have dominant alleles, the effects of the recessive alleles of the defective gene are masked.

> **Hint**
>
> Gene supplementation **does not involve replacing** a gene, it simply introduces a dominant allele alongside the existing recessive ones. The dominant allele therefore masks the effect of the recessive alleles.

In addition, there are two different techniques of gene therapy that may be adopted, according to which type of cell is being treated:

- **Germ-line gene therapy** involves replacing or supplementing the defective gene in the fertilised egg. This ensures that all cells of the organism will develop normally, as will all the cells of their offspring. This is therefore a much more permanent solution, affecting future generations. However, the moral and ethical issues of manipulating such a long-term genetic change mean that the process is currently prohibited.
- **Somatic-cell gene therapy** targets just the affected tissues, such as the lungs, and the additional gene is therefore not present in the sperm or egg cells, and so is not passed on to future generations. As the cells of the lungs are continually dying and being replaced, the treatment needs to be repeated periodically – as often as every few days. At present, the treatment has had limited success. The long-term aim is therefore to target undifferentiated **stem cells** that give rise to mature tissues. The treatment would then be effective for the lifespan of the individual.

> **Hint**
>
> Inserting a functional gene does not remove the defective genes – it just means that cells produce both functional and non-functional proteins at the same time.

Delivering cloned CFTR genes

The aim of somatic-cell gene therapy is to introduce cloned normal genes into the epithelial cells of the lungs. This can be carried out in two ways.

Using a harmless virus

Viruses, called adenoviruses, cause colds and other respiratory diseases by injecting their DNA into the epithelial cells of the lungs. They therefore

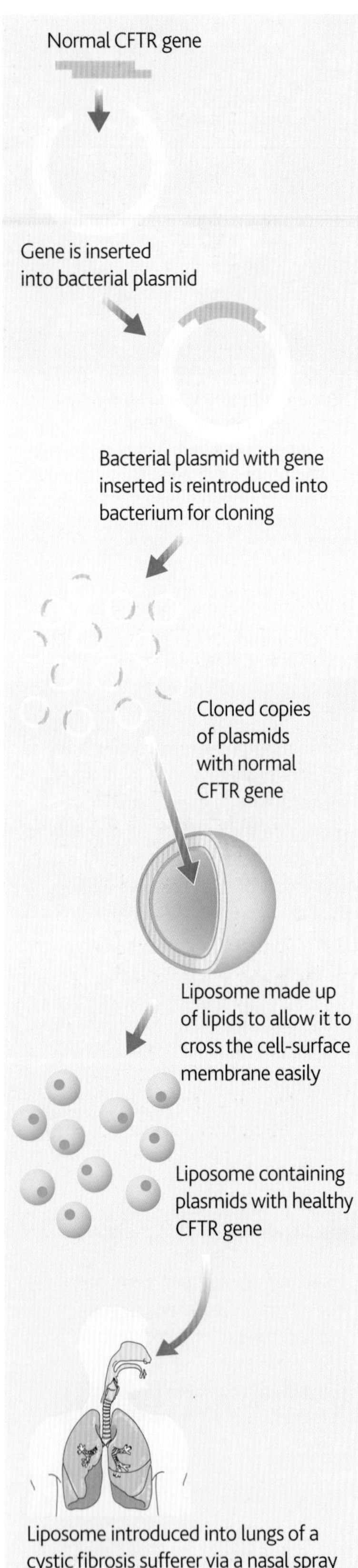

Figure 3 *Summary of treatment of cystic fibrosis by gene therapy*

make useful vectors for the transfer of the normal CFTR gene into the host cells. The process works as follows:

- The adenoviruses are made harmless by interfering with a gene involved in their replication.
- These adenoviruses are then grown in epithelial cells in the laboratory, along with plasmids that have had the normal CFTR gene inserted.
- The CFTR gene becomes incorporated into the DNA of the adenoviruses.
- These adenoviruses are isolated from the epithelial cells and purified.
- The adenoviruses with the CFTR gene are introduced into the nostrils of the patients.
- The adenoviruses inject their DNA, which includes the normal CFTR gene, into the epithelial cells of the lungs.

Wrapping the gene in lipid molecules

Genes are wrapped in lipid molecules because lipid molecules can relatively easily pass through the phospholipid portion of the cell-surface membranes. The process of delivering the CFTR genes to their target cells (i.e. the epithelial cells of the lungs) is as follows:

- CFTR genes are isolated from healthy human tissue and inserted into bacterial plasmid vectors (see Topic 16.2).
- The plasmid vectors are reintroduced into their bacterial host cells and gene markers are used to detect which bacteria have successfully taken up the plasmids with the CFTR gene (see Topic 16.2).
- These bacteria are cloned to produce multiple copies of the plasmids with the CFTR gene.
- The plasmids are extracted from the bacteria and wrapped in lipid molecules to form a liposome.
- The liposomes containing the CFTR gene are sprayed into the nostrils of the patient as an aerosol and are drawn down into the lungs during inhalation.
- The liposomes pass across the phospholipid portion of the cell-surface membrane of the lung epithelial cells.

These events are summarised in Figure 3.

These forms of delivery are not always effective because:

- adenoviruses may cause infections
- patients may develop immunity to adenoviruses
- the liposome aerosol may not be fine enough to pass through the tiny bronchioles in the lungs
- even where the CFTR gene is successfully delivered to the epithelial cells, very few are actually expressed.

Treatment of severe combined immunodeficiency using gene therapy

Severe combined immunodeficiency (SCID) is a rare inherited disorder. People who suffer from this condition do not show a cell-mediated immune response nor are they able to produce antibodies. Until recently, SCID was found only in young children because individuals with the disorder soon died of an infection. The disorder arises when individuals inherit a defect in the gene that codes for the enzyme ADA (adenosine deaminase). This enzyme destroys toxins that would otherwise kill white blood cells.

Survival has depended upon patients being raised in the strictly sterile environment of an isolation tent, or 'bubble', and giving them bone marrow transplants and/or injections of ADA. There have been recent attempts to treat the disorder using gene therapy as follows:

- The normal ADA gene is isolated from healthy human tissue using restriction endonucleases (see Topic 16.1).
- The ADA gene is inserted into a retrovirus.
- The retroviruses are grown with host cells in the laboratory to increase their number and hence the number of copies of the ADA gene.
- The retroviruses are mixed with the patient's T cells (a type of white blood cell).
- The retroviruses inject a copy of the normal ADA gene, into the T cells.
- The T cells are reintroduced into the patient's blood to provide the genetic code needed to make ADA.

The effectiveness of this treatment is limited because T cells live for only 6–12 months and so the treatment has to be repeated at intervals. More recent treatment involves transforming bone marrow stem cells rather than T cells. Because bone marrow stem cells divide to produce T cells, there is a constant supply of the ADA gene and hence the enzyme ADA. Although not totally effective because there is an increased risk of leukaemia, the results have been promising.

Application

Effectiveness of gene therapy

Gene therapy is very much in its infancy. It has great potential but, as yet, much of it is unrealised. It is in its experimental stage and so far there has been only limited success in the somatic-cell gene therapy trials that have been conducted. There are a number of reasons for this:

- **The effect is short-lived**. Because the somatic cells, which have a cloned gene added, are not passed on to the daughter cells repeat treatments are necessary for the therapy to have any effect.
- **It can induce an immune response**. Both the gene that is being introduced and the structure used to deliver it (vector or liposome) can induce an immune response in the recipient. This means that it is often rejected. This is made worse by the fact that the immune system typically responds to foreign material by making antibodies, some of which remain to initiate an even greater response to a future infection (the secondary response).

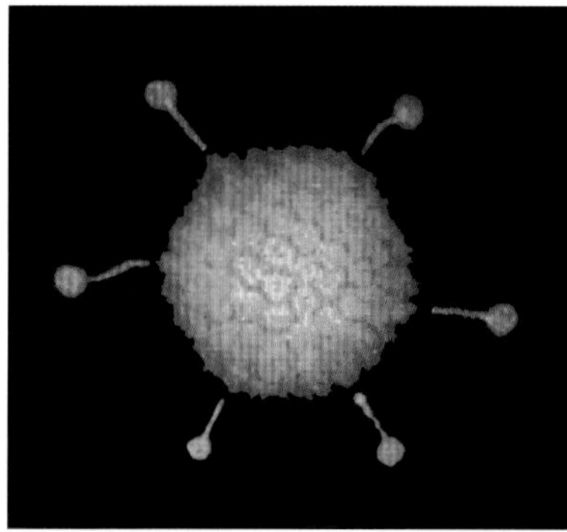

Figure 4 *TEM of an adenovirus. Adenoviruses can infect the respiratory tract, causing cold-like symptoms. This makes them useful as vectors to transfer healthy genes into host cells*

Link

Cell-mediated immunity and antibodies are covered in Topics 6.3 and 6.5 of the *AS Biology* book.

Summary questions

1. Explain how two parents, neither of whom suffers from cystic fibrosis, might have a child with the disease.
2. Why does somatic-cell gene therapy fail to provide a permanent cure for cystic fibrosis?
3. In which **two** ways can a normal CFTR gene be delivered to the lungs of a patient with cystic fibrosis?

Link

The secondary immune response is explained in Topic 6.4 of the *AS Biology* book.

- **Using viral vectors to deliver the gene presents problems.** Viruses are the usual means of getting genes to their target cells. Unfortunately these viruses can lead to toxic, inflammatory and immune responses in the recipients. There is also the fear that the 'disabled' virus might recover the ability to cause disease once inside the patient.
- **The genes are not always expressed.** Even if successfully delivered to their target cells, only a small proportion of the introduced genes are usually expressed.
- **It is not effective in treating conditions that arise in more than one gene.** Gene therapy works best in disorders that are the result of a single mutation. However, many commonly occurring disorders, such as arthritis, heart disease, diabetes and Alzheimer's disease, are the result of variations in a number of genes.

Most of the problems that relate to somatic-cell gene therapy do not apply to germ-line gene therapy. However, although germ-line therapy research is allowed in other organisms, it is not permitted in humans.

1 Why does the secondary immune response present a particular difficulty for patients undergoing gene therapy?

2 Explain why the first three problems of somatic-cell gene therapy listed above do not apply to germ-line gene therapy.

Application and How science works

Risks and benefits of gene therapy

As with many scientific applications, the use of gene therapy brings risks and benefits and raises ethical issues. This is an example of How science works (HSW: I & J). These issues include:

- Who decides what is normal and what is a disability? While serious diseases are clearly disabling, there are many other conditions that some individuals consider disabling, for example, being very tall or very short, having a particular hair colour, skin colour, body shape or birthmark. Are these candidates for gene therapy?
- Do disabilities need to be cured or prevented or are they just part of the genetic variety that makes up all species?
- Gene therapy research and treatment is very expensive. Could the money be better spent on more proven treatments, where success is guaranteed?
- Germ-line therapy would be very effective but what might be the long-term consequences of introducing inheritable genes into the population? Could it lead to the selection of one race in favour of another (eugenics)?

16.6 Locating and sequencing genes

Learning objectives:

- How are DNA probes and DNA hybridisation used to locate specific genes?
- How can the exact order of nucleotides on a strand of DNA be determined?
- What is restriction mapping and how does it help to determine the sequence of nucleotides in a gene?

Specification reference: 3.5.8

Many human diseases have a genetic origin. These are often the result of a **gene mutation**. Sometimes, as in the case of sickle-cell anaemia, the mutation may be an advantage in one situation but a disadvantage in another (see the Application in Topic 16.7).

Recombinant DNA technology has enabled us to diagnose and treat many of these genetic disorders. In doing so, it is often necessary to know exactly where a particular DNA sequence (gene) is located. To achieve this we use labelled DNA probes and DNA hybridisation.

DNA probes

A DNA probe is a short, single-stranded section of DNA that has some sort of label attached that makes it easily identifiable. The two most commonly used probes are:

- **radioactively labelled probes**, which are made up of **nucleotides** with the **isotope** ^{32}P. The probe is identified using a photographic plate that is exposed by radioactivity.
- **fluorescently labelled probes**, which emit light (fluoresce) under certain conditions.

DNA probes are used to identify particular genes in the following way:

- A DNA probe is made that has bases that are complementary to the portion of the DNA sequence that makes up part of the gene whose position we want to find.
- The DNA that is being tested is treated to separate its two strands.
- The separated DNA strands are mixed with the probe, which binds to the complementary bases on one of the strands. This is known as **DNA hybridisation**.
- The site at which the probe binds can be identified by the radioactivity or fluorescence that the probe emits.

Before we can make a specific probe we need to know the sequence of nucleotides in the particular gene that we are trying to locate.

DNA sequencing

A number of different methods are used to sequence the exact order of nucleotides in a section of DNA, one of which is the Sanger method. This method uses modified nucleotides that cannot attach to the next base in the sequence when they are being joined together. They therefore act as terminators, ending the synthesis of a DNA strand. Four different terminator nucleotides are used, each with one of the four bases adenine, thymine, guanine or cytosine. The sequencing process is described below.

The first stage is to set up four test tubes, each containing:

- many single-stranded fragments of the DNA to be sequenced. This acts as a template for the synthesis of its complementary strand
- a mixture of nucleotides with the bases adenine, thymine, guanine and cytosine

Hint

A primer is essential to start DNA synthesis because it makes a double strand of DNA, and DNA polymerase only works on double-stranded DNA.

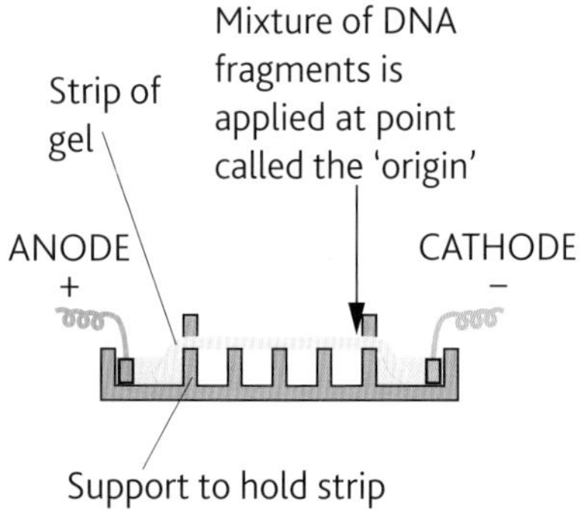

Figure 1 *Apparatus for carrying out electrophoresis*

- a small quantity of one of the four terminator nucleotides (test tube 1 – adenine terminator nucleotide, test tube 2 – thymine terminator nucleotide, etc.)
- a primer to start the process of DNA synthesis. This primer is radioactively labelled or labelled with a fluorescent dye
- DNA polymerase to catalyse DNA synthesis.

As the binding of nucleotides to the template is a random process, the addition of a normal nucleotide or a terminator nucleotide, is equally likely. Depending upon exactly where the terminator nucleotide binds to the DNA template, DNA synthesis may be terminated after only a few nucleotides or after a long fragment of DNA has been synthesised. As a result, the DNA fragments in each test tube will be of varying lengths. One thing that they have in common is that all the fragments of new DNA in any of the test tubes will end with a nucleotide that has the same base: adenine in tube 1, thymine in tube 2, etc. These fragments can be identified because the primer attached to the other end of the DNA section is labelled radioactively or with a fluorescent dye.

The next stage is to separate out these different length fragments of DNA. One way of doing this is to use a technique called gel electrophoresis.

Gel electrophoresis

The DNA fragments are placed on to an agar gel and a voltage is applied across it (Figure 1). The resistance of the gel means that the larger the fragments, the more slowly they move. Therefore, over a fixed period, the smaller fragments move further than the larger ones. In this way DNA fragments of different lengths are separated. A sheet of photographic film is then placed over the agar gel for several hours. The radioactivity from each DNA fragment exposes the film and shows where it is situated on the gel.

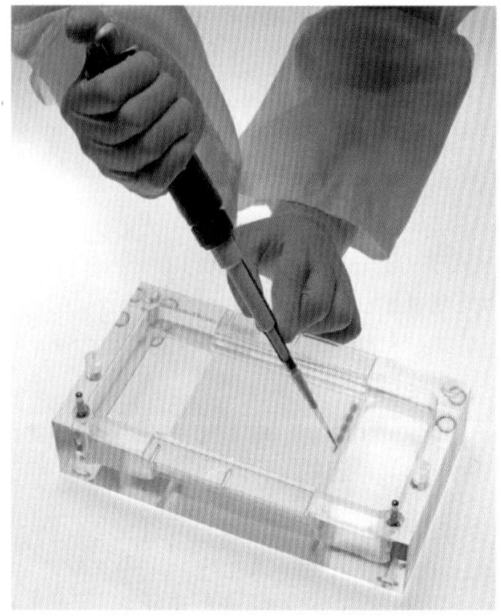

Figure 2 *An electrophoretic gel being loaded with DNA samples*

The results of one such experiment in DNA sequencing is shown in Figure 3. This used a DNA fragment that was just eight nucleotides long. The results are read from the bottom up because the shortest fragments move the furthest distance. The smallest fragment (labelled 1) is just one nucleotide long and is therefore nearest the bottom. This fragment has a terminator nucleotide with the base adenine. The second fragment (labelled 2) is two bases long and has a terminator nucleotide with the base guanine, and so on. In this way the whole sequence of bases on the terminator nucleotides was found to be AGCTTGAC and this is the sequence on one of the strands of the newly formed DNA.

Hint

Remember that, in gel electrophoresis, the smallest DNA fragments travel the furthest. The DNA sequence must always start with the terminator nucleotide base on the smallest fragment.

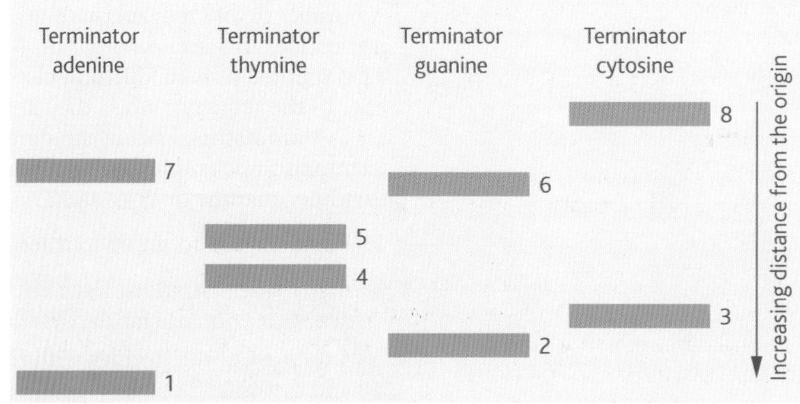

Figure 3 *Results of a DNA sequencing experiment*

Only DNA fragments up to around 500 bases can be sequenced in this way. Larger genes and whole genomes must therefore be cut into smaller fragments by **restriction endonucleases** and each fragment sequenced. The problem then is to piece these sequenced fragments back together to make up the original gene or genome. This is done using restriction mapping.

Restriction mapping

We saw in Topic 16.1 how restriction endonucleases can be used to cut DNA at specific recognition sites. Restriction mapping involves cutting DNA with a series of different restriction endonucleases. The fragments produced are then separated by gel electrophoresis. The distance between the **recognition sites** can be determined by the patterns of fragments that are produced.

To illustrate the principles involved, let us consider an example of a plasmid that has 100 000 bases (100 kilobases, or 100 kb). Suppose each of the three restriction endonucleases cuts the plasmid as shown in Figure 4. As the plasmid is circular, if we use only one of the restriction endonucleases at a time then, we will always get a single piece of DNA that is 100 kb long, regardless of which restriction endonuclease we use. But what happens if we use the restriction endonucleases in pairs? Two cuts mean that we will always be left with two fragments of DNA from a circular plasmid. The length of each of the two fragments depends on which two restriction endonucleases are used to make the cuts. So what will be the length of the DNA fragments produced if the following pairs of restriction endonucleases are used together?

- ***Hind*III and *Bam*HI**. From Figure 4, we can see that as, we move clockwise from the *Hind*III cut, the distance to the *Bam*HI cut is 10 kb. This means that one fragment is 10 kb in length. Continuing in a clockwise direction, we see that the distance from the *Bam*HI cut to the *Hind*III cut is (30 + 60) = 90 kb. (We can ignore the cut made by *Not*I as this restriction endonuclease is not used in this example.) The second fragment is therefore 90 kb long. Alternatively, as we know that the total length of the plasmid is 100 kb and that the length of one fragment is 10 kb, it follows that the second fragment must be (100 − 10) = 90 kb long.
- ***Hind*III and *Not*I**. Using the same method as above, we can show that the lengths of the two fragments will be 40 kb and 60 kb.
- ***Bam*HI and *Not*I**. The lengths of the two fragments will be 30 kb and 70 kb.

What results would be obtained from separating out these various DNA fragments using gel electrophoresis? Figure 5 provides the answer.

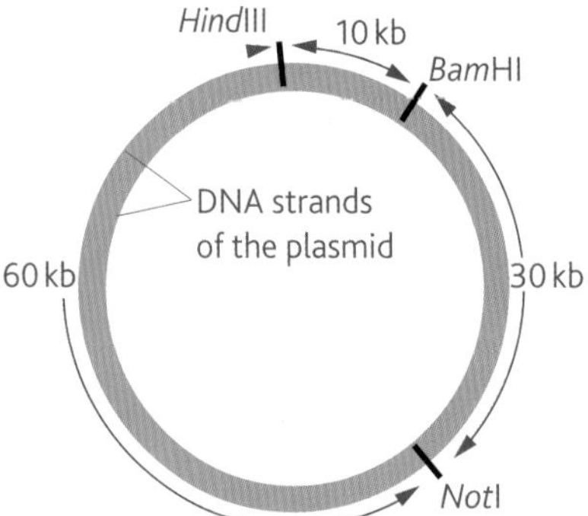

Figure 4 *Three different restriction endonucleases showing the points at which they cut a plasmid*

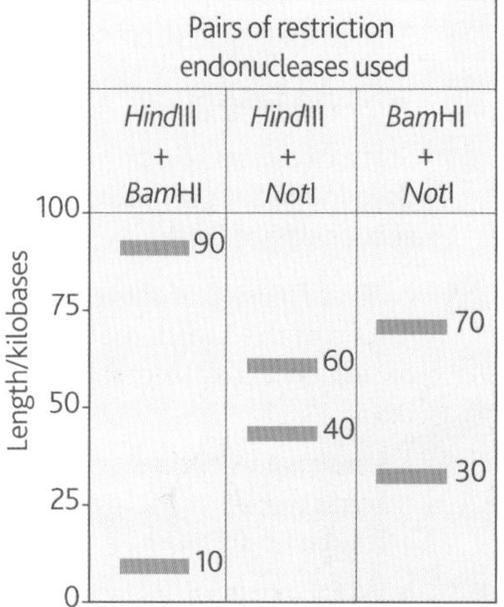

Figure 5 *Results of gel electrophoresis of DNA fragments produced when pairs of restriction endonucleases cut a plasmid*

Hint

The restriction endonucleases in Figure 4 are examples to illustrate how restriction mapping works. Their names and where they cut do **not** have to be learned.

Automation of DNA sequencing and restriction mapping

Both DNA sequencing and restriction mapping are now routinely carried out by automatic machines and computers analyse the data that they produce.

In these computerised systems, instead of radioactively labelling the DNA primer, the four types of terminators are labelled with fluorescent dye. Each type takes up a different colour: adenine (green), thymine (red), cytosine (blue) and guanine (yellow) (Figure 6). The DNA synthesis takes place in a single test tube and is speeded up by using PCR cycles (see Topic 16.3).

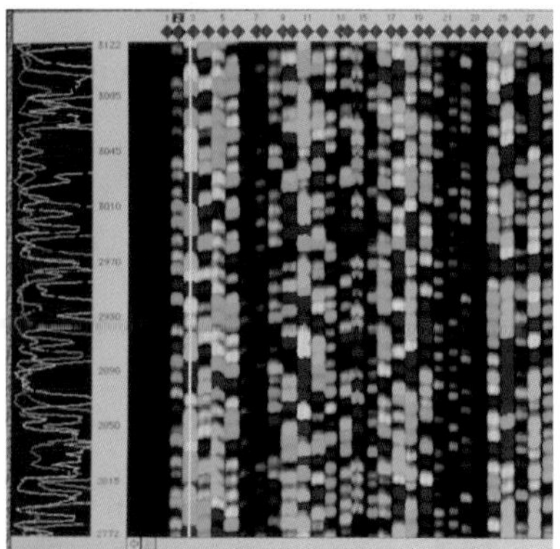

Figure 6 *Computer screen display of a DNA sequence. Each coloured band represents one of the four bases*

Summary questions

1 What is a DNA probe?

2 State **two** roles of a primer used in the Sanger method of sequencing of DNA.

3 Look at Figure 8. It shows the results of the Sanger method of sequencing a fragment of DNA.

a How many adenine bases were present in the fragment of DNA?

b Which nucleotide starts the shortest fragment?

c What is the nucleotide sequence of the longest DNA fragment that has been produced?

Electrophoresis is carried out in a single narrow capillary gel. The results are scanned by lasers and interpreted by computer software, giving the DNA sequence in a fraction of the time taken by conventional methods. Further automation includes the use of a polymerase chain reaction (PCR) machine to produce the DNA fragments required in these techniques.

These methods are continually being updated as further innovations in the fields of DNA technology and computer software are developed.

Application

Compiling a restriction map

Plasmids were cut using three restriction endonucleases: A, B and C. The restriction endonucleases were used in pairs. The DNA fragments produced were separated using gel electrophoresis. Figure 8 shows the results of this separation.

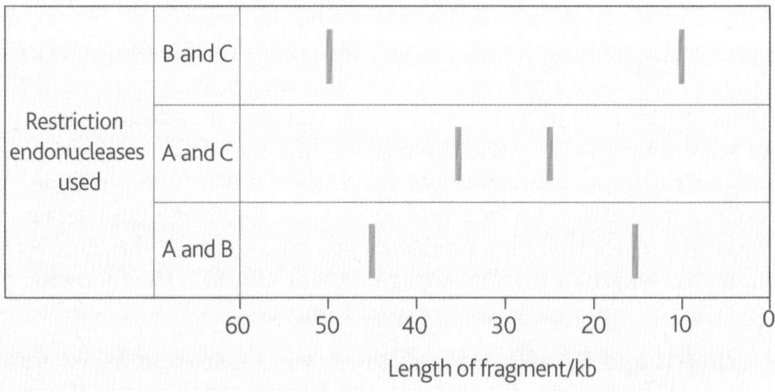

Figure 7 *Results of separation of DNA fragments using gel electrophoresis*

1 Using this information, draw a restriction map, like that in Figure 4 showing where on the plasmid the three restriction endonucleases cut the DNA relative to each other.

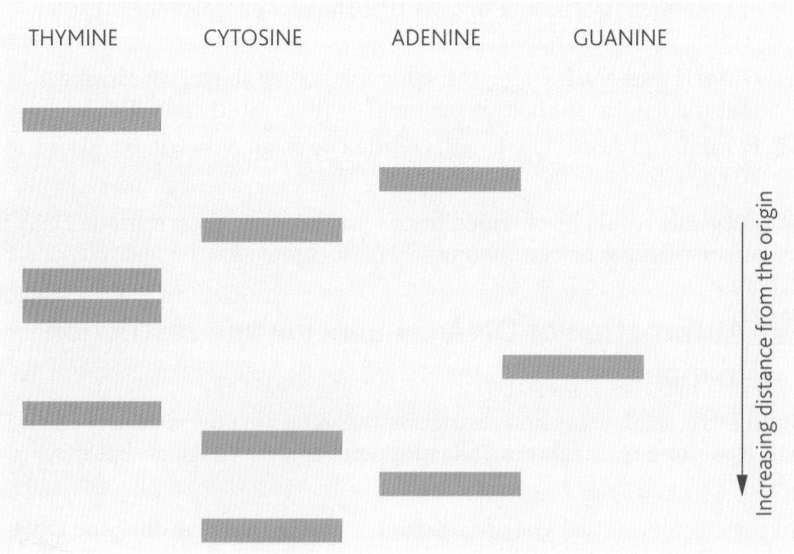

Figure 8

16.7 Screening for clinically important genes

Learning objectives:

- How can DNA probes be used to screen patients for gene mutations?
- What role does genetic counselling play in the process?

Specification reference: 3.5.8

Genetic techniques can determine whether an unborn child might be affected by a genetic disorder. This process is called **genetic screening**.

Genetic screening

Many genetic disorders, such as sickle-cell anaemia, are the result of **gene mutations**. In Topic 14.4, we saw that gene mutations may arise if one or more **nucleotides** in DNA are either deleted or substituted by other nucleotides. If the mutation results in a **dominant allele**, all individuals will have the genetic disorder. If the allele is **recessive**, it will only be apparent in those individuals that have two recessive alleles, that is, who are **homozygous** recessive. Individuals that are **heterozygous** will not display symptoms of the disease but will carry one copy of the mutant allele. They have the capacity to pass the disease to their offspring if the other parent is also heterozygous or homozygous recessive.

It is important to screen individuals who may carry a mutant allele. Such individuals often have a family history of a disease. Screening can determine the probabilities of a couple having offspring with a genetic disorder. As a result, potential parents who are at risk can obtain advice from a genetic counsellor (see page 273) about the implications of having children, based on their family history and the results of genetic screening.

The screening of individuals to find out whether they have a mutated gene, such as the one causing sickle-cell anaemia, is illustrated in Figure 1 overleaf.

It is possible to fix hundreds of different DNA probes in an array (pattern) on a glass slide. By adding a sample of DNA to the array, any complementary DNA sequences in the donor DNA will bind to one or more probes. In this way it is possible to test simultaneously for many different genetic disorders.

Another area where genetic screening can be valuable is in the detection of oncogenes, which are responsible for cancer. Cancers may develop as a result of mutations that prevent the **tumour suppressor genes** inhibiting cell division. Mutations of both alleles must be present to inactivate the tumour suppressor genes and to initiate the development of a tumour. Some people inherit one mutated tumour suppressor gene. These individuals are at greater risk of developing cancer.

If a mutated gene is detected by genetic screening, individuals who are at greater risk of cancer can then make informed decisions about their lifestyle and future treatment. They can choose to give up smoking, lose weight, eat more healthily and avoid **mutagens** as far as possible. They can also check themselves more regularly for early signs of cancer, which can lead to an early diagnosis and a better chance of successful treatment. They may choose to undergo gene therapy (see Topic 16.5). This treatment is in the developmental stage but trials are taking place on gene therapy for bladder, brain, liver, ovarian and prostate cancers.

Genetic screening goes hand in hand with genetic counselling. The expert advice provided by a counsellor helps individuals to understand the results and implications of the screening and so make appropriate decisions.

Hint

The donor's DNA must be single-stranded to allow the use of DNA probes and hybridisation because both depend on the formation of complementary base pairs.

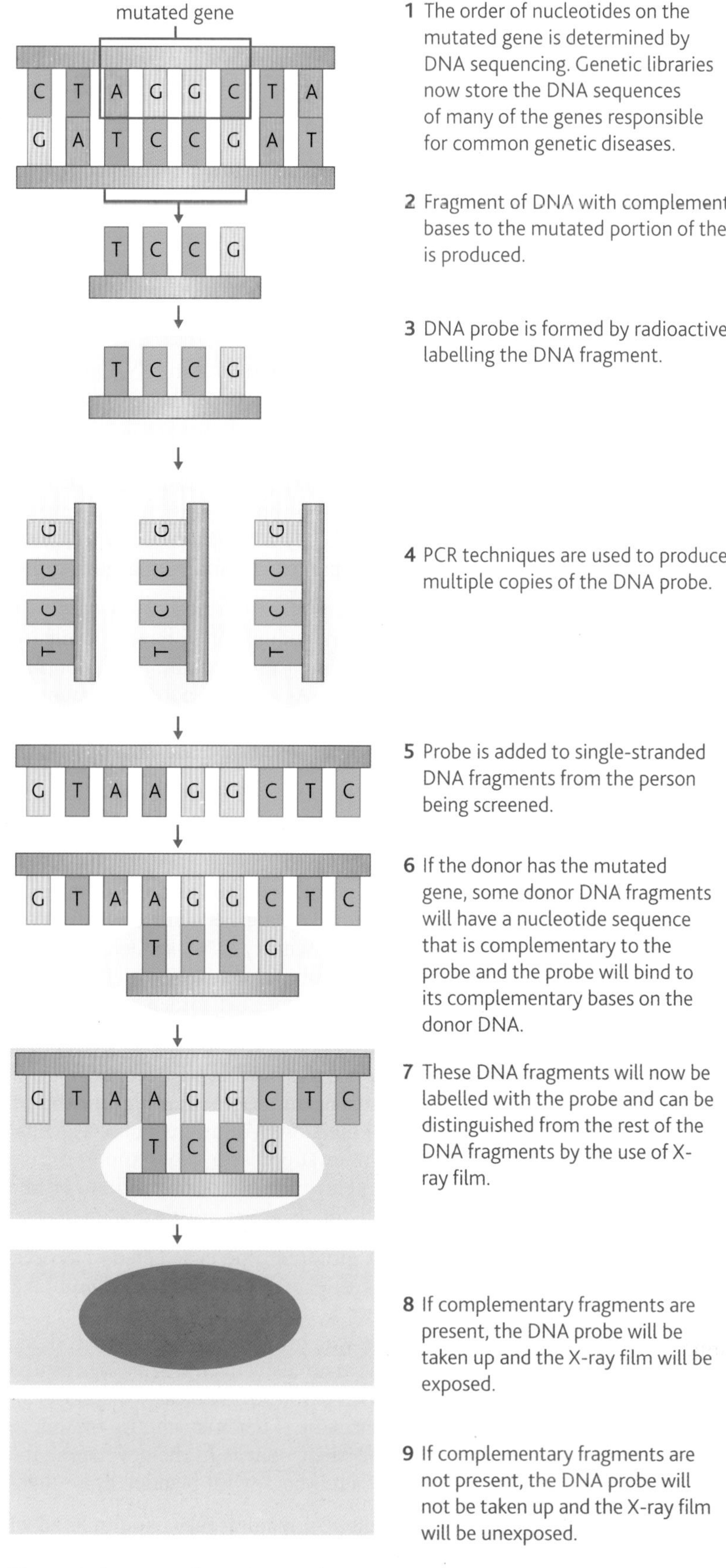

Figure 1 *Summary of genetic screening*

Genetic counselling

Genetic counselling is like a special form of social work, where advice and information are given that enable people to make personal decisions about themselves or their offspring. One important aspect of genetic counselling is to research the family history of an inherited disease and to advise parents on the likelihood of it arising in their children.

Consider a mother who has a family history of sickle-cell anaemia. If the mother herself is unaffected but carries the gene for sickle-cell anaemia, she must be heterozygous for the condition (see Application below). Suppose that she wishes to have children with a man who has no history of sickle-cell anaemia. In this case, it can be assumed that the man does not carry the allele for the disease, and therefore none of the children will develop the disease, although they may be carriers. On the other hand, if the man has a family history of the disease, it is possible that he too carries the allele. In this case, the genetic counsellor can make the couple aware that there is a one in four chance of their children being affected and a two in four chance that their children will be carriers.

A counsellor can also inform the couple of the effects of sickle-cell anaemia and its emotional, psychological, medical, social and economic consequences. On the basis of this advice the couple can then choose whether or not to have children. Counselling can also make them aware of any further medical tests that might give a more accurate prediction of whether their children will have the condition.

Genetic counselling is closely linked to genetic screening and the screening results provide the genetic counsellor with a basis for informed discussion. For example, in cases of cancer, screening can help to detect:

- oncogene mutations, which can determine the type of cancer that the patient has and hence the most effective drug or radiotherapy to use
- gene changes that predict which patients are more likely to benefit from certain treatments and have the best chance of survival. For example, the drug herceptin is most effective at treating certain types of breast cancer
- a single cancer cell among millions of normal cells, thus identifying patients at risk of relapse from certain forms of leukaemia.

This information can help a counsellor to discuss with the patient the best course of treatment and their prospects of survival.

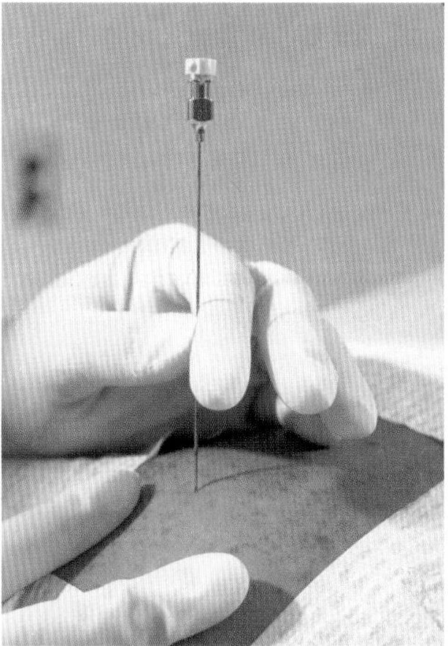

Figure 2 *Amniotic fluid is being taken from this pregnant woman. The fluid can be used to screen the unborn baby for genetic disorders*

Link

To remind yourself about how oncogenes and tumour suppressor genes work, re-read Topic 14.4 in this book.

Application

Sickle-cell anaemia

The following is a synoptic exercise.

Sickle-cell anaemia was the first human disease to be successfully understood at the molecular level, its cause being suggested by Linus Pauling in 1949. Sickle-cell anaemia illustrates how the smallest of mutations can significantly influence the phenotype. It is the result of a gene mutation in the gene producing haemoglobin.

In the DNA molecule that produces one of the amino acid chains in haemoglobin, the single nucleotide base adenine is substituted by the nucleotide base, thymine. The normal DNA triplet on the template strand is hence changed from CTC to CAC. As a result, the mRNA produced has a different code.

Figure 3 *Genetic counsellor talking with potential parents. Genetic counselling is used to advise on the risk of passing on genetic disorders and is given to couples in certain risk groups who are intending to have children*

Summary questions

1 Outline the process of genetic screening for a disease such as sickle-cell anaemia.

2 Genetic screening shows that a person has one mutant allele of the tumour suppressor gene.

a What is the role of the tumour suppressor gene?

b How might the person use the information revealed by genetic screening?

1 What was the original mRNA code and what has it been changed to following the mutation?

The changed mRNA codes for the amino acid valine rather than for glutamic acid. This minor difference produces a molecule of haemoglobin (called haemoglobin S) that has a 'sticky patch'. When the haemoglobin molecules are not carrying oxygen (i.e. at low oxygen concentrations) they tend to adhere to one another by their sticky patches and become insoluble, forming long fibres within the red blood cells. These fibres distort the red blood cells, making them inflexible and sickle (crescent) shaped. These sickle cells are unable to carry oxygen and may block small capillaries because their diameter is greater than that of the capillaries.

2 Explain how a change in a single amino acid might lead to the change in protein structure described.

3 Sufferers of sickle-cell anaemia become tired easily. Explain why.

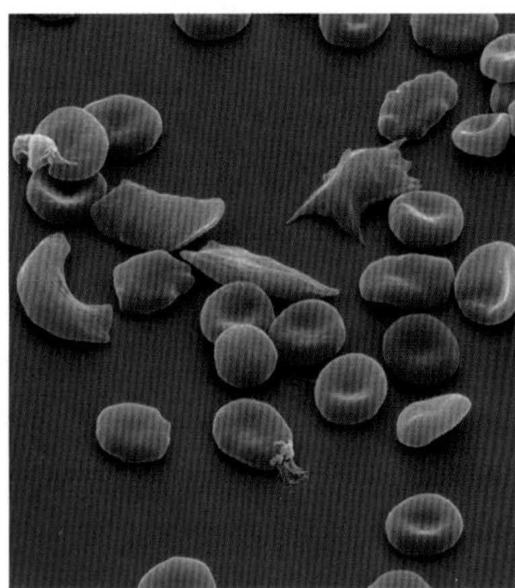

Figure 4 *SEM of red blood cells from a person with sickle cell anaemia. Normal red blood cells (rounded) contrast with the distorted sickle-shaped cells (pink)*

Because sickle-cell anaemia disables and kills individuals, we might expect it to have been eliminated by the process of natural selection. However, the condition is relatively common in some parts of Africa and among black populations of African origin. This is because sickle-cell anaemia is the result of a gene that has two **co-dominant** alleles, Hb^A (normal) and Hb^S (sickled). The malarial parasite, *Plasmodium*, is unable to exist in sickled red blood cells. The three possible **genotype** combinations of these two alleles, their corresponding **phenotypes** and the selection pressures on each type differ as follows:

- **homozygous for haemoglobin S (Hb^SHb^S).** Individuals suffer from sickle-cell anaemia and are considerably disadvantaged if they do not receive medical attention. They rarely live long enough to pass their genes on to the next generation. Their anaemia is so severe it outweighs the advantage of being resistant to one form of malaria and so individuals are always selected against.
- **homozygous for haemoglobin A (Hb^AHb^A).** Individuals lead normal healthy lives, but are susceptible to malaria in areas of the world where the disease is endemic and they are therefore selected against only in these regions.
- **heterozygous for haemoglobin (Hb^AHb^S).** Individuals are said to have sickle-cell trait, but are not badly affected, except when the oxygen concentration of their blood is low, for example, in exercising muscles. Sufferers may therefore become tired more easily but, in general, the condition is symptomless. They do, however, have resistance to malaria and this advantage outweighs the disadvantage of tiredness in areas of the world where malaria occurs. Heterozygous individuals are therefore selected *against* in areas without malaria, but selected *for* in areas where malaria is common.

4 What is meant by the term 'co-dominant'?

5 When two heterozygous individuals produce offspring, what is the chance of any one of them having sickle-cell anaemia?

6 In a population of 175 individuals the frequency of the Hb^A allele is 0.6 and the frequency of the Hb^S allele is 0.4.

a If the individuals mated randomly, what frequencies of the Hb^A and Hb^S alleles would be expected in the next generation?

b Using the Hardy–Weinberg equation, calculate the number of individuals with each phenotype. Show your working.

7 In parts of the world where malaria is prevalent, the heterozygous state (Hb^AHb^S) is selected for at the expense of both homozygous states. What form of selection is this? Explain your answer.

8 Genetic screening for sickle-cell anaemia can be carried out. Explain why the advice given by a genetic counsellor to individuals with the same genotypes might differ depending on where they live.

Application and How science works

Implications of genetic screening

This is an example of How science works: (HSW: I, J & L).

Techniques such as genetic screening provide us with more information. With information comes power and opportunity: the power to make decisions and the opportunity to change what we do. The challenge is how to use this power and opportunity responsibly.

Many people might be in favour of screening for inherited disorders that are apparent soon after birth and result in a short life of considerable suffering. But what about screening for Alzheimer's disease, which often arises much later in life? Should we use a technology simply because it exists? Other social, ethical and legal implications include:

- **Who decides who should be screened?** Screening is expensive and decisions have to be made about who has priority when financial resources are scarce.
- **Who has access to the test results?** Do employers or insurers have a right to the information? If they knew a person had a genetic disorder would they refuse employment or insurance cover?
- **What are the responsibilities of someone who carries a gene for an inherited disease?** Is all life sacrosanct or does it make sense to abort a fetus whose life will be only 4 years of constant pain, which is the case with Tay-Sachs disease?
- **Does mankind have a responsibility to maintain genetic diversity?** Should we preserve mutant genes? After all evolution depends on genetic diversity. Can we be sure that a cure will not soon be found to enable those with genetic disorders to lead normal lives? Will screening lead to the elimination of certain genes and is this a form of eugenics?
- **Who decides what is a defect or a disease?** Should screening be available to allow parents to produce 'designer' babies? Is it legitimate to use screening to let someone decide the sex of his or her child?

1 Give **two** reasons in favour of carrying out genetic screening and **two** reasons against doing so.

16.8 Genetic fingerprinting

Learning objectives:

- What is genetic fingerprinting?
- How is genetic fingerprinting carried out?
- How are the results interpreted?
- For what purposes is it used?

Specification reference: 3.5.8

Link

Not all DNA sequences carry obvious genetic information. You will have already come across this idea with the processing of DNA described in Topic 14.2 of this book.

Hint

In theory, the inheritance of non-coding DNA does not have any influence on the phenotype of an organism.

Father | Child | Mother

Figure 1 *DNA fingerprints of a child and each parent. Note that each band on the child's fingerprint corresponds to a band on one or other parent's fingerprint*

Genetic fingerprinting is a diagnostic tool used widely in forensic science. It is based on the fact that the DNA of every individual, except identical twins, is unique.

Genetic fingerprinting

This technique relies upon the fact that the genome of any organism contains many repetitive, non-coding bases of DNA. Indeed, 95 per cent of human DNA does not code for any characteristic. These non-coding DNA bases are known as **introns** and they contain repetitive sequences of DNA called core sequences. For every individual the number and length of core sequences has a unique pattern. They are different in all individuals except identical twins, and the probability of two individuals having identical sequences of these repetitive non-coding bases is extremely small. However, the more closely related two individuals are, the more similar the core sequences will be.

The making of a genetic fingerprint consists of five main stages: extraction, digestion, separation, hybridisation and development. The complete process of genetic fingerprinting is summarised in Figure 2, opposite.

Extraction

Even the tiniest sample of animal tissue, such as a drop of blood or a hair root, is enough to give a genetic fingerprint. Whatever the sample, the first stage is to extract the DNA by separating it from the rest of the cell. As the amount of DNA is usually small, its quantity can be increased by using the polymerase chain reaction (see Topic 16.3).

Digestion

The DNA is then cut into fragments, using **restriction endonucleases** (see Topic 16.1). The endonucleases are chosen for their ability to cut close to, but not within, groups of core sequences.

Separation

The fragments of DNA are next separated according to size by **gel electrophoresis** under the influence of an electrical voltage (see Topic 16.6). The gel is then immersed in alkali in order to separate the double strands into single strands. The single strands are then transferred on to a nylon membrane by a technique called Southern blotting, which involves a series of stages:

- A thin nylon membrane is laid over the gel.
- The membrane is covered with several sheets of absorbent paper, which draws up the liquid containing the DNA by capillary action.
- This transfers the DNA fragments to the nylon membrane in precisely the same relative positions that they occupied on the gel.
- The DNA fragments are then fixed to the membrane using ultraviolet light.

Hybridisation

Radioactive (or fluorescent) DNA probes are now used to bind with the core sequences. The probes have base sequences which are complementary to the core sequences, and bind to them under specific conditions, such as temperature and pH. The process is carried out with different probes, each of which binds with a different core sequence.

Development

Finally, an X-ray film is put over the nylon membrane. The film is exposed by the radiation from the radioactive probes. (If using fluorescent probes, the positions are located visually.) Because these points correspond to the position of the DNA fragments as separated during electrophoresis, a series of bars is revealed. The pattern of the bands (see Figure 1) is unique to every individual except identical twins.

Interpreting the results

DNA fingerprints from two samples, for example, from blood found at the scene of a crime and from a suspect, are visually checked. If there appears to be a match, the pattern of bars of each fingerprint is passed through an automated scanning machine, which calculates the length of the DNA fragments from the bands. It does this using data obtained by measuring the distances travelled during electrophoresis by known lengths of DNA. Finally, the odds are calculated of someone else having an identical fingerprint. The closer the match between the two patterns, the greater the probability that the two sets of DNA have come from the same person.

Uses of DNA fingerprinting

DNA fingerprinting is used extensively in forensic science to indicate whether or not an individual is connected with a crime, for example, from blood or semen samples found at the scene. It can also be used to help resolve questions of paternity. Individuals inherit half their genetic material from their mother and half from their father. Therefore each band on a DNA fingerprint of an individual should have a corresponding band in the parents' DNA fingerprint (Figure 1). This can be used to establish whether someone is the genetic father of a child.

Genetic fingerprinting is also useful in determining genetic variability within a population. The more closely two individuals are related the closer the resemblance of their genetic fingerprints. A population whose members have very similar genetic fingerprints has little genetic diversity. A population whose members have a greater variety of genetic fingerprints has greater genetic diversity.

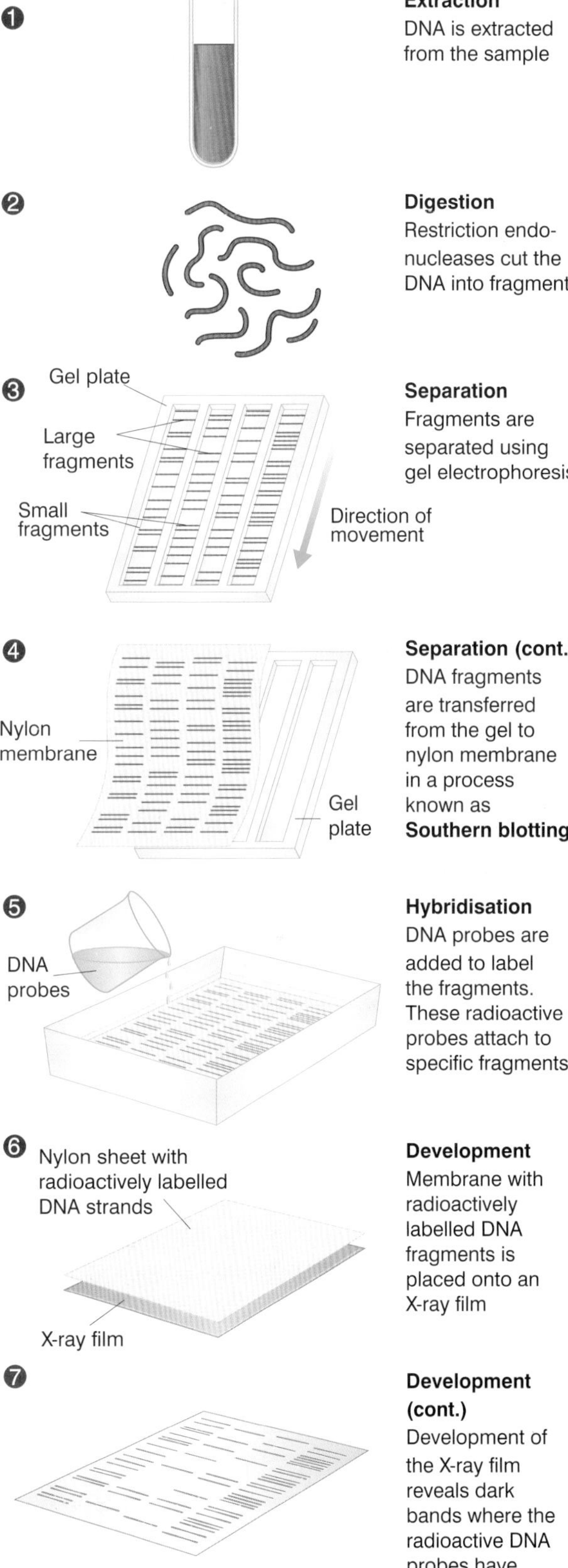

Figure 2 *Summary of genetic fingerprinting technique*

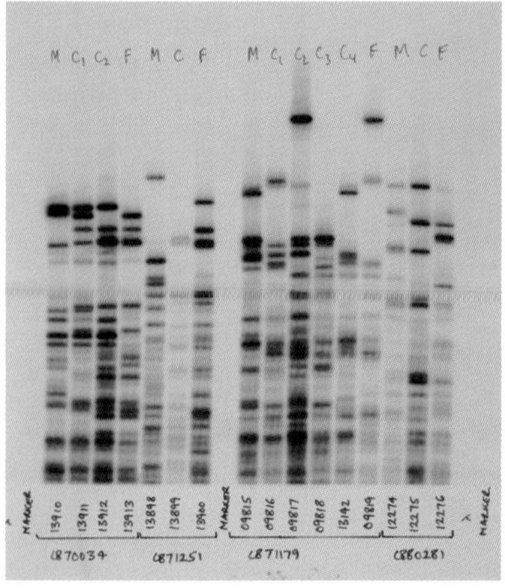

Figure 3 *The bands in these DNA fingerprints are marked M for mother, C for child and F for father*

Figure 4

Application

Other uses of genetic fingerprinting

In addition to determining genetic relationships and genetic variability, genetic fingerprinting has a number of other applications.

Forensic science

DNA is often left at the scene of a crime, e.g. blood at the scene of a violent crime, semen at the scene of a rape and hair at the scene of a robbery. Genetic fingerprinting can establish whether a person is likely to have been present at the crime scene, although this does not prove they actually carried out the crime. Even if there is a close match between a suspect's DNA and the DNA found at the crime scene, it does not follow that the suspect carried out the crime. Other possible explanations need to be investigated. For example:

- The DNA may have been left on some other, innocent occasion.
- The DNA may belong to a very close relative.
- The DNA sample may have been contaminated after the crime, either by the suspect's DNA or by chemicals that affected the action of the restriction endonucleases used in preparing the fingerprint.

1 Explain how chemicals that affect the action of restriction endonucleases can alter the genetic fingerprint of a DNA sample.

Finally, the probability that someone else's DNA might match that of the suspect has to be calculated. This calculation is based on the assumption that the DNA which produces the banding patterns is randomly distributed in the community. This may not always be the case, e.g. it may not apply where religious or ethnic groups tend to have partners from within their own small community.

Medical diagnosis

Genetic fingerprints can help in diagnosing diseases such as Huntington's disease. This is a genetic disorder of the nervous system. It results from a three-nucleotide segment (AGC) at one end of a gene on chromosome 4 being repeated over and over again – a sort of genetic stutter. People with fewer than 30 repeats are unlikely to get the disease, while those with more than 38 repeats

Summary questions

1 Explain why it is often necessary to use the polymerase chain reaction when producing a genetic fingerprint.

2 Figure 4 shows the genetic fingerprints of four DNA samples collected following a crime.

a Which of the two suspects do you think was present at the scene of the crime? Give a reason for your answer.

b Suggest why a genetic fingerprint of a DNA sample from the victim was made.

are almost certain to do so. If they have over 50 repeats, the onset of the disease will occur earlier than average.

A sample of DNA from a person with the allele for Huntington's disease can be cut with restriction endonucleases and a DNA fingerprint prepared. This can then be matched with fingerprints of people with various forms of the disease and those without the disease. In this way, the probability of developing the symptoms, and when, can be determined.

2 Suggest how the genetic fingerprint of someone with the allele for Huntington's disease might differ from that of someone who does not have the allele.

Genetic fingerprints are also used to identify the nature of a microbial infection by comparing the fingerprint of the microbe found in patients with that of known pathogens.

Plant and animal breeding

Genetic fingerprinting can be used to prevent undesirable inbreeding during breeding programmes on farms or in zoos. It can also identify plants or animals that have a particular allele of a desirable gene. Individuals with this allele can be selected for breeding in order to increase the probability of their offspring having the characteristic that it produces. Another application is the determination of paternity in animals and thus establishing the pedigree (family tree) of an individual.

3 Explain how genetic fingerprinting can be used to ensure that inbreeding is avoided.

Application

Locating DNA fragments

A section of DNA was cut into fragments and these fragments were separated by electrophoresis. Table 1 shows the number of base pairs in each fragment. The position of the fragments after gel electrophoresis is shown in Figure 5.

1 Name an enzyme that could have been used to cut the DNA.

2 Using the letters (A–F) in the boxes in Table 1, indicate which of the fragments (1–6) in Figure 5 are located in each box. Explain your answer.

3 The enzyme used to cut the DNA does so at a particular sequence of nucleotide bases. How many times does this base sequence occur in the original section of DNA?

Table 1

Fragment	Number of base pairs (kilobases)
A	8.02
B	5.43
C	4.78
D	11.31
E	2.46
F	6.12

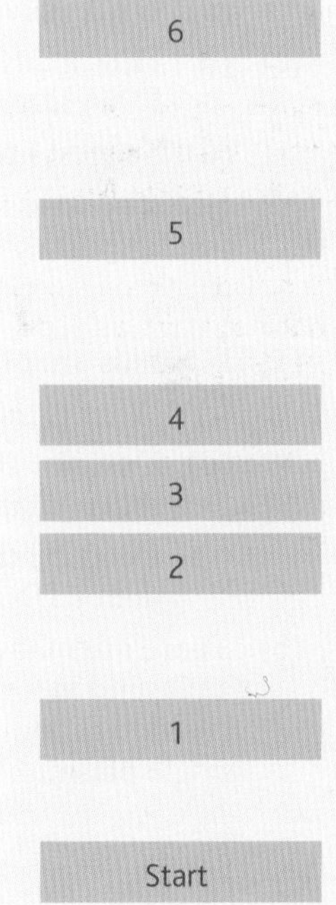

Figure 5

AQA Examination-style questions

1 The polymerase chain reaction is a process which can be carried out in a laboratory to replicate DNA. **Figure 1** shows the main stages involved in the polymerase chain reaction.

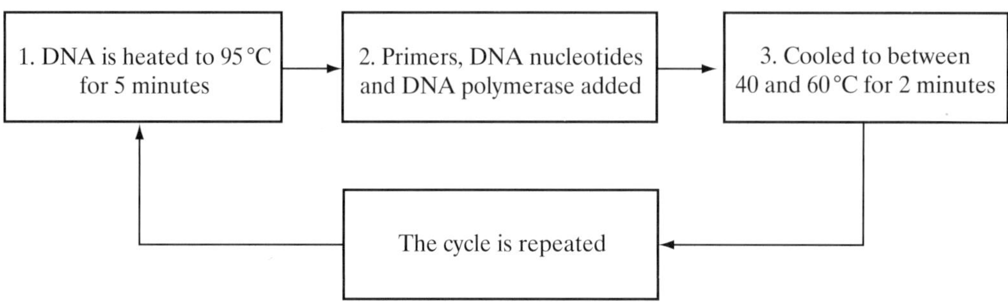

Figure 1

(a) Explain why DNA is heated to 95 °C. *(1 mark)*

(b) What is the role of:

(i) a primer in this process;

(ii) DNA polymerase? *(2 marks)*

(c) (i) How many DNA molecules will have been produced from one molecule of DNA after 6 complete cycles?

(ii) Suggest **one** use of the polymerase chain reaction. *(2 marks)*

(d) Give **two** ways in which the polymerase chain reaction differs from the process of transcription. *(2 marks)*

AQA, 2005

2 (a) Plasmids are often used as vectors in genetic engineering.

(i) What is the role of a vector?

(ii) Describe the role of restriction endonucleases in the formation of plasmids that contain donor DNA.

(iii) Describe the role of DNA ligase in the production of plasmids containing donor DNA. *(4 marks)*

(b) There are many different restriction endonucleases. Each type cuts the DNA of a plasmid at a specific base sequence called a restriction site. **Figure 2** shows the position of four restriction sites, **J**, **K**, **L** and **M**, for four different enzymes on a single plasmid. The distances between these sites is measured in kilobases of DNA.

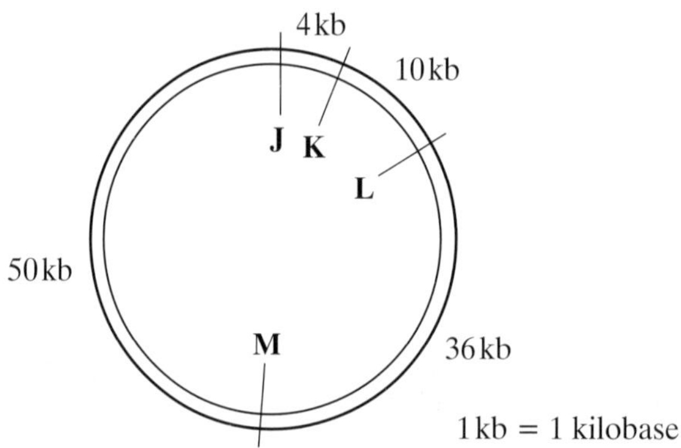

Figure 2

The plasmid was cut using only two restriction endonucleases. The resulting fragments were separated by gel electrophoresis. The positions of the fragments are shown in **Figure 3**.

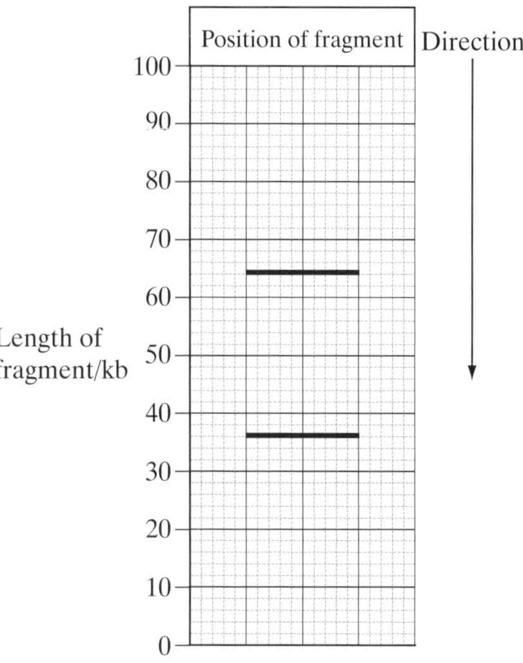

Figure 3

(i) Which of the restriction sites were cut?

(ii) Explain your answer. *(2 marks)*

AQA, 2006

3. Gene therapy is used to treat the genetic disorder, ADA deficiency. Affected individuals are unable to produce the enzyme adenosine deaminase (ADA). Without this enzyme, T lymphocytes, a type of white blood cell, cannot provide immunity to infection. **Figure 4** shows the processes involved in the treatment of ADA deficiency by gene therapy.

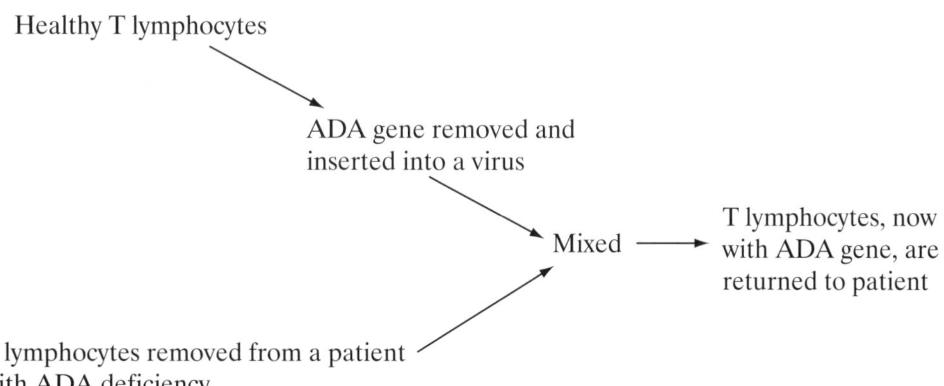

Figure 4

(a) What is meant by gene therapy? *(1 mark)*

(b) Individuals who have been treated by this method of gene therapy do not pass on the ADA gene to their children. Explain why. *(1 mark)*

(c) T lymphocytes are produced in the bone marrow. A bone marrow transplant from a genetically matched donor can provide a permanent cure for ADA deficiency.

(i) Suggest why bone marrow for a transplant is obtained from a genetically matched donor.

(ii) Explain why treatment of ADA deficiency by gene therapy must be repeated at regular intervals, whereas a single bone marrow transplant can provide a permanent cure. *(3 marks)*

AQA, 2005

4 One technique used to determine the sequence of nucleotides in a sample of DNA is the Sanger procedure. This requires four sequencing reactions to be carried out at the same time. The sequencing reactions occur in four separate tubes. Each tube contains;

- a large quantity of the sample DNA
- a large quantity of the four nucleotides containing thymine, cytosine, guanine and adenine
- DNA polymerase
- radioactive primers.

A modified nucleotide is also added to each tube, as shown in **Figure 5**.

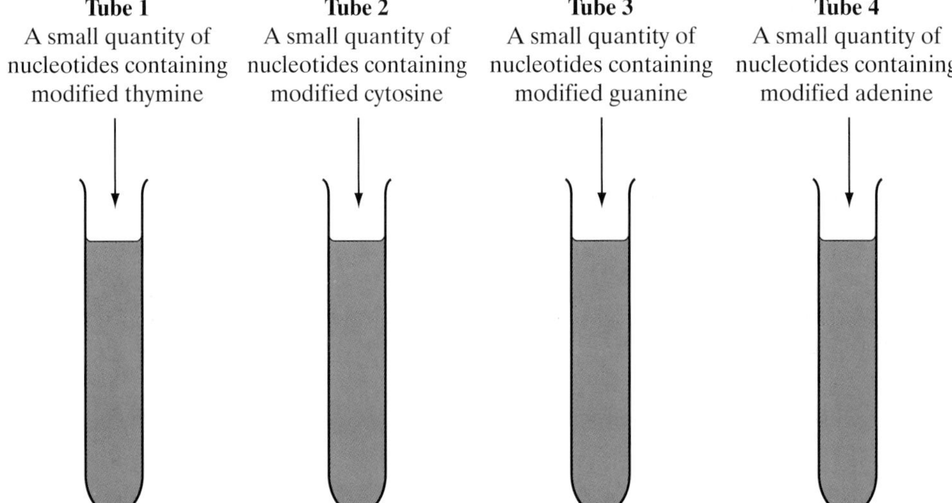

Figure 5

(a) A large quantity of the DNA sample is required for this procedure. Name the reaction used to amplify small amounts of DNA into quantities large enough for this procedure. *(1 mark)*

(b) Explain the reason for adding each of the following to the tubes.

(i) DNA polymerase

(ii) Primers *(2 marks)*

(c) (i) When a modified nucleotide is used to form a complementary DNA strand, the sequencing reaction is terminated. Suggest how this sequencing reaction is terminated.

(ii) A sample of DNA analysed by this technique had the following nucleotide base sequence.

T G G T C A C G A

Give the base sequence of the shortest DNA fragment which would be produced in **Tube 2**. *(2 marks)*

(d) A different sample of DNA was then analysed. The DNA fragments from the four tubes were separated in a gel by electrophoresis and analysed by autoradiography. **Figure 6** shows the banding pattern produced.

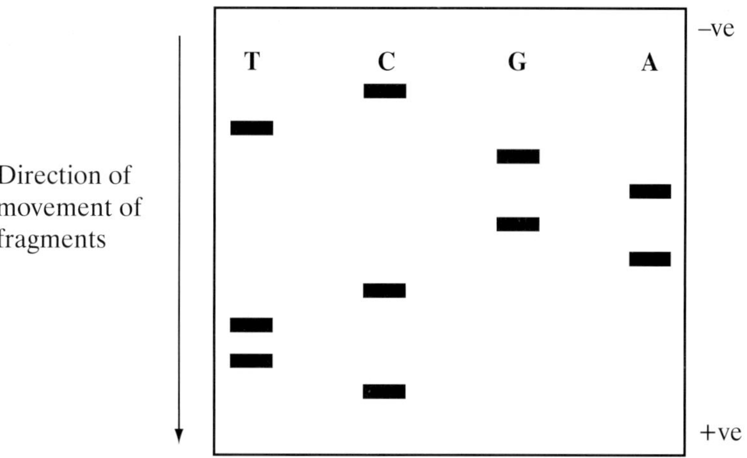

Figure 6

(i) Explain why the DNA fragments move different distances in the gel.

(ii) What makes the DNA fragments visible on the autoradiograph?

(iii) Use **Figure 6** to determine the sequence of nucleotides in this sample of DNA. *(3 marks)*

AQA, 2005

5 Read the following passage:

The giant panda is one of the rarest animals in the world and is considered to be on the brink of extinction in the wild. Giant pandas have been kept and bred in zoos with the hope that they could be released in to the wild. One worry is that small populations, like those in zoos, reduce the genetic variation needed to allow species to adapt to changing situations.

Unfortunately, pandas find it difficult to reproduce in captivity. Fertilization of the females is guaranteed only by insemination with semen from several males. With so many potential fathers, the true paternity of the cubs is not clear. It is important to identify the fathers to maintain genetic variation.

Panda faeces can be collected in the wild. The faeces contain DNA from the panda, from the bamboo on which they feed and bacteria. The DNA is subjected to the polymerase chain reaction (PCR). The primers used only attach to the panda DNA. The resulting DNA is subjected to genetic fingerprinting. This can help us count the number of individuals in the wild because it allows us to identify individual pandas.

Use information in the passage and your own biological knowledge to answer the questions.

(a) Describe how genetic fingerprinting may be carried out on a sample of panda DNA. *(6 marks)*

(b) (i) Explain how genetic fingerprinting allows scientists to identify the father of a particular panda club.

(ii) When pandas are bred in zoos, it is important to ensure only unrelated pandas breed. Suggest how genetic fingerprints might be used to do this. *(3 marks)*

(c) (i) Suggest why panda DNA is found in faeces. (line 10)

(ii) Explain why the PCR is carried out on the DNA from the faeces. (line 12)

(iii) Explain why the primers used in the PCR will bind to panda DNA, but not to DNA from bacteria or bamboo. (line 12) *(4 marks)*

(d) DNA from wild pandas could also be obtained from blood samples. Suggest **two** advantages of using faeces, rather than blood samples, to obtain DNA from pandas. *(2 marks)*

AQA, 2006

AQA Examination-style questions

Unit 5 questions – Control in cells and in organisms

1 **Figure 1** shows part of the retina in a human eye.

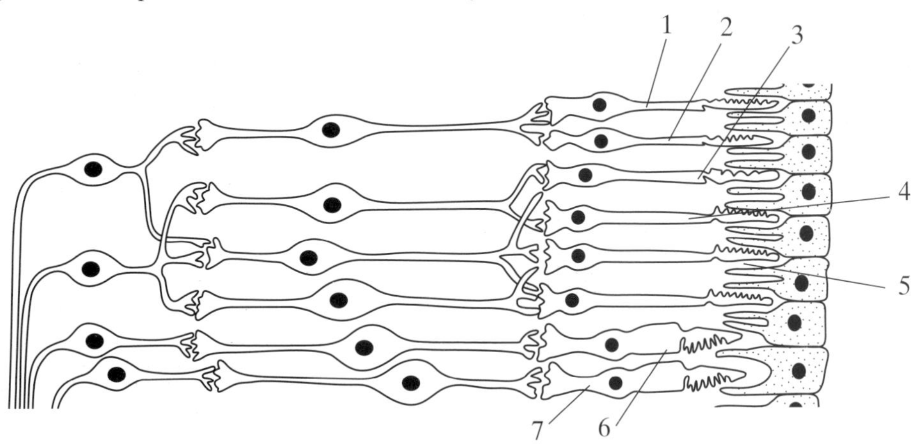

Figure 1

(a) Explain each of the following observations.

(i) When light falls on cells **1** and **2**, only one spot of light is seen. But, when light falls on cells **2** and **3**, two spots of light are seen.

(ii) When one unit of light energy falls on cell **3**, no light is seen. But, when one unit of light energy falls on cell **3**, one unit falls on cell **4** and one unit falls on cell 5, light is seen. *(4 marks)*

(b) Cells of the same type as cells **6** and **7** are found in large numbers at the fovea. This results in high visual acuity. Explain what causes high visual acuity at the fovea. *(1 mark)*

AQA, 2005

2 **Figure 2** shows a neuromuscular junction.

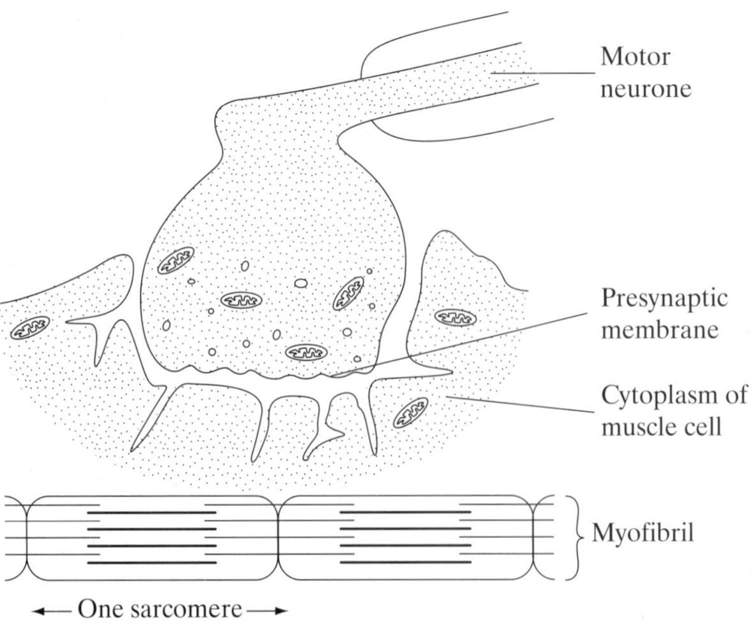

Figure 2

(a) (i) On a copy of **Figure 2**, label the myelin sheath.

(ii) The myelin sheath is not formed in new-born babies. Explain how this leads to slower reflexes in babies. *(3 marks)*

(b) Nerve impulses arriving at the presynaptic membrane at the neuromuscular junction result in shortening of myofibrils. Describe how. *(7 marks)*

(c) Myofibrils are made up of many sarcomeres. The sarcomeres in a myofibril are always the same length as each other. However this length changes with the state of contraction of the muscle. The table shows the force produced by a myofibril in relation to the length of its sarcomeres.

Sarcomere length/μm	Force produced as percentage of maximum	Degree of overlap between filaments
1.3	0	
2.3	100	
3.7	0	

One sarcomere

(i) Give the sarcomere length at which the H-zone will be at its minimum length. Explain your answer.

(ii) Explain why no force is generated when the sarcomere length is 3.7 μm.

(iii) Explain why the maximum force is produced when the sarcomere length is 2.3 μm. *(5 marks)*

AQA, 2007

3 (a) One effect of getting into a cold shower is a reduction in the amount of blood flowing through the capillaries near the surface of the skin. Explain how the cold water causes this response. *(4 marks)*

(b) (i) When exercising at 30 °C, the body is more likely to overheat in humid conditions than in dry conditions. Explain why.

(ii) Strenuous exercise leads to exhaustion more quickly in hot conditions than in cool conditions. One reason for this is a reduced blood supply to the muscles, which means that they receive less oxygen.

Suggest an explanation for the reduced blood supply to the muscles. *(4 marks)*

AQA, 2004

4 Lysozyme is an enzyme consisting of a single polypeptide chain of 129 amino acids.

(a) What is the minimum number of nucleotide bases needed to code for this enzyme? *(1 mark)*

(b) The diagram shows the sequence of bases in a section of the mRNA strand used to synthesise this enzyme.

G G U C U U U C U U A U G G U A G A U A U

(i) Give the DNA sequence which would be complementary to the first four bases in this section of mRNA.

(ii) How many different types of tRNA molecule would attach to the section of mRNA shown in the diagram? *(2 marks)*

(c) Give **two** factors which might increase the frequency at which a mutation in DNA occurs. *(2 marks)*

(d) Two single base mutations occurred in the DNA coding for this section of mRNA. These mutations caused an alteration in the sequence of amino acids in the enzyme. **Figure 3** shows the original and altered sequence of amino acids.

Original amino acid sequence	Gly	Leu	Ser	Tyr	Gly	Arg	Tyr
Original mRNA base sequence	GGU	CUU	UCU	UAU	GGU	AGA	UAU

Altered amino acid sequence	Gly	Leu	Tyr	Leu	Trp	Arg	Tyr
Altered mRNA base sequence	GGU	CUU				AGA	UAU

Figure 3

(i) Use the mRNA codons provided in the table to complete the altered mRNA base sequence in **Figure 3**.

Amino acid	**mRNA codons which can be used**
Arg	AGA
Gly	GGU
Leu	CUU or UUA
Ser	UCU
Trp	UGG
Tyr	UAU or UAC

(ii) Use the information provided to determine the precise nature of the two single base mutations in the DNA. *(4 marks)*

AQA, 2005

5 Read the following passage:

Scientists discovered a protein in milk. This protein had antibacterial properties.

They extracted the protein and determined its amino acid sequence. This information allowed them to make the gene for the protein. A vector was used to insert this gene into rice cells, together with a marker gene.

The scientists hope to use the genetically modified rice to make drinks that would combat diarrhoea. Diarrhoea is often caused by bacterial infections, and is a major killer of children worldwide. Many tests, however, would need to be carried out before the rice drinks could be sold to consumers.

The scientists have been given permission to plant a trial field of the genetically modified rice. They were told, however, that the trial field had to be a long way from any commercial rice farm.

Use information from the passage and your own knowledge to answer the following questions.

(a) Scientists know the order of amino acids in milk protein.

Explain how they can use this information to obtain a gene that codes for this protein. *(2 marks)*

(b) (i) A vector was used to insert the new genes into rice cells.

Explain what is meant by a vector.

(ii) A marker gene was used in genetically modifying the rice.

Explain why. *(3 marks)*

(c) Many tests would be needed before the rice drinks could be sold to consumers. Give **two** reasons why these tests would be needed. *(2 marks)*

(d) The genetically modified rice will be grown a long way from any commercial rice farms. Explain why this is important. *(2 marks)*

(e) Describe how the rice cells produce the milk protein from the inserted gene. *(6 marks)*

AQA, 2008

6 **Figure 6** shows part of a myofibril from a relaxed muscle fibre.

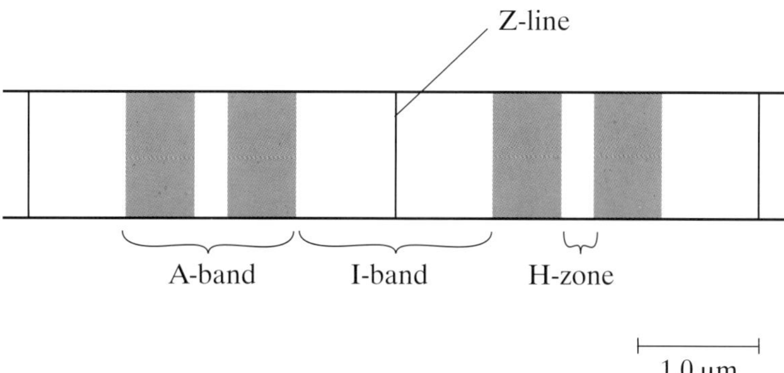

Figure 6

(a) When the muscle fibres contract, which of the **A**-band, **I**-band and **H**-zone:
  (i) remain unchanged in length;
  (ii) decrease in length? *(2 marks)*

(b) Explain what caused the decrease in length in part (a)(ii). *(2 marks)*

(c) The table gives some properties of the two different types of muscle fibre found in skeletal muscle.
  (i) Copy and complete the table by writing the words 'high' or 'low' for the remaining three properties of each type of muscle fibre.

	Type of muscle fibre	
	Fast	**Slow**
Speed of contraction	high	low
Force generated	high	low
Activity of the enzymes of glycolysis	high	low
Number of mitochondria		
Activity of Krebs cycle enzymes		
Rate of fatigue		

  (ii) The myosin-ATPase of **fast** muscle fibres has a faster rate of reaction than that in **slow** fibres. Use your knowledge of this mechanism of muscle contraction to explain how this will help **fast** muscle fibres to contract faster than **slow** fibres.
  (iii) The blood leaving an active muscle with a high percentage of **fast** muscle fibres contained a higher concentrate of lactate than that leaving a muscle with a high percentage of **slow** muscle fibres. Explain why. *(9 marks)*

AQA, 2006

7 Scientists believe that the tendency to develop cancer can be inherited. It is thought that some people possess cancer-causing genes. These genes only become functional when activated by an environmental factor. The functional genes then cause the production of abnormal cells. The abnormal cells multiply and spread, causing cancer.

(a) Explain why medical screening of people for the presence of these cancer-causing genes is recommended. *(2 marks)*

(b) Cells contain suppressor genes, which code for proteins that control cell division and growth. Describe what is meant by a mutation, and explain how a mutation in a suppressor gene might lead to the development of a malignant tumour. *(8 marks)*

AQA, 2002

Glossary

A

abiotic: an ecological factor that makes up part of the non-biological environment of an organism, e.g. temperature, pH, rainfall and humidity. See also *biotic*.

acetylcholine: one of a group of chemicals, called *neurotransmitters*, released by *neurones*. It diffuses across the gap (*synapse*) between adjacent neurones and so passes an impulse from one neurone to the next.

action potential: change that occurs in the electrical charge across the membrane of an *axon* when it is stimulated and a nerve impulse passes.

actin: filamentous protein which is involved in contraction within cells, especially muscle cells. See also *myosin*.

activation energy: energy required to bring about a chemical reaction. The activation energy is lowered by the presence of enzymes.

active transport: movement of a substance across a membrane from a region where it is in a low concentration to a region where it is in a high concentration. The process requires the use of *ATP*.

adenosine triphosphate (ATP): an activated nucleotide found in all living cells that acts as an energy carrier. The *hydrolysis* of ATP leads to the formation of adenosine diphosphate (ADP) and inorganic phosphate, with the release of energy.

adrenaline: a hormone produced by the adrenal glands in times of stress that prepares the body for an emergency.

aerobic: connected with the presence of oxygen. Aerobic respiration requires oxygen to release energy from glucose and other foods. See also *anaerobic*.

allele: one form of a *gene*. For example, the gene for the shape of pea seeds has two alleles: one for 'round' and one for 'wrinkled'.

allele frequency: the number of times an allele occurs within the *gene pool*.

allergen: a normally harmless substance that causes the immune system to produce an immune response.

anaerobic: connected with the absence of oxygen. Anaerobic respiration releases energy from glucose or other foods without the presence of oxygen. See also *aerobic*.

antibiotic resistance: the development, in populations of microorganisms, of mechanisms that prevent antibiotics from killing them.

antibody: a protein produced by lymphocytes in response to the presence of a specific *antigen*.

anticodon: a sequence of three adjacent nucleotides on a molecule of transfer RNA that is complementary to a particular codon on a messenger RNA molecule.

antigen: a molecule that triggers an immune response by causing lymphocytes to produce specific *antibodies* which bind to it.

ATP: see *adenosine triphosphate*.

autonomic nervous system: part of the nervous system, controlling the muscles and glands, that is not under voluntary control.

axon: a process extending from a *neurone* that conducts *action potentials* away from the cell body.

B

biodiversity: the range and variety of living organisms within a particular area.

biomass: the total mass of living material in a specific area at a given time. It is usually measured as dry mass because the amount of water in an organism is very variable.

biotic: an ecological factor that makes up part of the living environment of an organism. Examples include food availability, competition and predation. See also *abiotic*.

C

Calvin cycle: a biochemical pathway that forms part of the *light-independent reaction* of photosynthesis, during which carbon dioxide is reduced to form carbohydrate.

cardiac muscle: type of muscle found only in the heart. It has fewer striations than *skeletal muscle* and can contract continuously throughout life without stimulation by nerve impulses. See also *smooth muscle*.

clone: a group of genetically identical organisms formed from a single parent as a result of asexual reproduction or by artificial means.

cholinesterase: enzyme that breaks down and therefore inactivates the *neurotransmitter*, *acetylcholine*, in the *synapse*.

climax community: the organisms that make up the final stage of ecological succession.

co-dominance: condition in which both *alleles* for one *gene* in a *heterozygous* organism contribute to the *phenotype*.

codon: a sequence of three adjacent *nucleotides* in mRNA that codes for one amino acid.

community: the organisms of all species that live in the same area. See also *population*.

condensation reaction: chemical process in which two molecules combine to form a more complex one with the elimination of a simple substance, usually water. Many biological polymers are formed by *condensation reactions*. See also *hydrolysis*.

conservation: method of maintaining *ecosystems* and the living organisms that occupy them. It requires planning and organisation to make best use of resources while preserving the natural landscape and wildlife.

consumer: any organism that obtains energy by 'eating' another. Organisms feeding on plants are known as primary consumers and organisms feeding on primary consumers are known as secondary consumers. See also *producer*.

continuous variation: variation in which organisms do not fall into distinct categories, but rather there are gradations from one extreme to the other, e.g. height in humans. See also *discontinuous variation*.

cuticle: exposed non-cellular outer layer of certain animals and the leaves of plants. It is waxy and impermeable to water. It therefore helps to reduce water loss.

cystic fibrosis: inherited disease in which the body produces abnormally thick mucus that obstructs breathing passages and prevents secretion of pancreatic enzymes. It is a recessive condition that leads to production of a non-functioning membrane protein needed to transport chloride *ions*.

D

deciduous: term applied to plants that shed all their leaves together at one season.

denaturation: permanent changes to the structure, and hence the three-dimensional shape, of a protein due to factors such as changes in temperature or pH.

dendrite: a process, usually branched, extending from the cell body of a *neurone*, which conducts impulses towards the cell body.

denitrifying bacteria: bacteria that convert nitrates to nitrogen gas as part of the nitrogen cycle.

depolarisation: temporary reversal of charges on the cell-surface membrane of a *neurone* that takes place when a nerve impulse is transmitted.

diabetes: a metabolic disorder in which the body is unable to regulate the level of blood glucose. There are two forms of the disorder – Type I and Type II diabetes.

diploid (2n): a term applied to cells in which the nucleus contains two sets of chromosomes. See also *haploid*.

directional selection: selection that operates towards one extreme in a range of variation.

discontinuous variation: variation shown when the characteristics of organisms fall into distinct categories, e.g. blood groups in humans. See also *continuous variation*.

DNA helicase: enzyme that acts on a specific region of the DNA molecule to break the hydrogen bonds between the bases causing the two strands to separate and expose the *nucleotide* bases in that region.

DNA replication: the process in which the double helix of a DNA molecule unwinds and each strand acts as a template on which a new strand is constructed.

dominant allele: a term applied to an *allele* that is always expressed in the *phenotype* of an organism. See also *recessive allele*.

E

ecological niche: all conditions and resources required for an organism to survive, reproduce and maintain a viable population.

ecosystem: more or less self-contained functional unit in ecology made up of all the interacting *biotic* and *abiotic* factors in a specific area.

ectothermic: an animal that uses the environment to regulate its body temperature. See also *endotherm*.

effector: an organ that responds to stimulation by a nerve impulse resulting in a change or response.

electron: negatively charged sub-atomic particle that orbits the positively charged nucleus of all atoms. See also *proton*.

electron carrier molecules: a chain of carrier molecules along which *electrons* pass, releasing energy in the form of *ATP* as they do so.

endotherm: an animal maintaining its body temperature by physiological mechanisms. See also *ectotherm*.

eukaryotic cell: a cell with a membrane-bound nucleus that contains chromosomes. The cells also possess a variety of other membranous organelles, e.g. mitochondria and endoplasmic reticulum. See also *prokaryotic cell*.

eutrophication: consequence of an increase in nutrients, especially nitrates and phosphates, in fresh-water lakes and rivers, that often leads to a decrease in *biodiversity*.

G

gamete: reproductive (sex) cell that fuses with another gamete during fertilisation.

gel electrophoresis: a technique used to separate DNA fragments of different lengths by placing them on agar gel and passing a voltage across them.

gene: length of DNA on a chromosome normally coding for a particular polypeptide.

gene pool: total number of *alleles* in a particular *population* at a specific time.

gene marker: a section of DNA that is used to indicate the location of a *gene* or other section of DNA.

gene therapy: a mechanism by which genetic diseases, e.g. *cystic fibrosis*, may be cured by masking the effect of the defective *gene* by inserting a functional gene.

generator potential: *depolarisation* of the membrane of a receptor cell as a result of a stimulus.

genetic engineering: see *recombinant DNA technology*.

genetically modified organism (GMO): organism that has had its DNA altered as a result of *recombinant DNA technology*.

genotype: the genetic composition of an organism. See also *phenotype*.

glucagon: a hormone produced by α cells of the *islets of Langerhans* in the pancreas that increases blood glucose levels by initiating the breakdown of glycogen to glucose.

gluconeogenesis: the conversion of non-carbohydrate molecules to glucose.

glycogenesis: the conversion of glucose to glycogen.

glycogenolysis: the conversion of glycogen to glucose.

glycolysis: first part of cellular respiration in which glucose is broken down anaerobically in the cytoplasm to two molecules of pyruvate.

glycoprotein: substance made up of a carbohydrate molecule and a protein molecule. Parts of cell-surface membrane and certain hormones are glycoproteins.

granum (plural grana): a stack of *thylakoids* in a chloroplast that resembles a pile of coins. This is the site of the *light-dependent reaction* of photosynthesis. See also *stroma*.
greenhouse gases: gases such as methane and carbon dioxide which in the atmosphere cause more heat energy to be trapped, so raising the temperature at the Earth's surface.
guard cell: one of a pair of cells that surround a *stoma* in plant leaves and control its opening and closing.

H

habitat: the place where an organism normally lives, which is characterised by physical conditions and the species of other organisms present.
haploid (n): term referring to cells that contain only a single copy of each chromosome, e.g. the sex cells or gametes. See also *diploid*.
heterozygous: condition in which the *alleles* of a particular gene are different.
hibernation: an inactive, dormant state, accompanied by a very low body temperature, that certain animals go into during periods of prolonged cold.
homeostasis: the maintenance of a more or less constant internal environment.
homologous chromosomes: a pair of chromosomes that have the same gene loci and therefore determine the same features. They are not necessarily identical, however, as individual *alleles* of the same genes may vary, e.g. one chromosome may carry the allele for blue eyes, the other the allele for brown eyes. Homologous chromosomes are capable of pairing during meiosis.
homozygous: condition in which the *alleles* of a particular *gene* are identical.
human genome project: international scientific project to map the entire sequence of all the base pairs of the genes in a single human cell.
hydrogen bonds: chemical bonds formed between the positive charge on a hydrogen atom and the negative charge on another atom of an adjacent molecule, e.g. between the hydrogen atom of one water molecule and the oxygen atom of an adjacent water molecule.
hydrolysis: the breaking down of large molecules into smaller ones by the addition of water molecules. See also *condensation reaction*.
hyperthermia: a condition that results from the core body temperature rising above normal.
hypothalamus: region of the brain adjoining the pituitary gland that acts as the control centre for the *autonomic nervous system* and regulates body temperature and fluid balance.
hypothermia: a condition that results from the core body temperature falling below normal.

I

insulin: a hormone, produced by the β cells of the *islets of Langerhans* in the pancreas, which decreases blood glucose levels by, amongst other things, increasing the rate of conversion of glucose to glycogen.
interspecific competition: competition between organisms of different species.
intraspecific competition: competition between organisms of the same species.
intrinsic proteins: proteins of the cell-surface membrane that completely span the *phospholipid* bilayer from one side to the other.
introns: portions of DNA within a gene that do not code for a polypeptide. The introns are removed from pre-messenger RNA after *transcription*.
ion: an atom or group of atoms that has lost or gained one or more *electrons*. Ions therefore have either a positive or negative charge.
islets of Langerhans: groups of cells in the pancreas comprising large α cells, which produce the hormone *glucagon*, and small β cells, which produce the hormone *insulin*.
isotope: variations of a chemical element which have the same number of *protons* and *electrons* but different numbers of neutrons. While their chemical properties are similar, they differ in mass. One example is carbon, which has a relative atomic mass of 12 and an isotope with a relative atomic mass of 14.
***in vitro*:** refers to experiments carried out outside the living body, e.g. in test tubes.
***in vivo*:** refers to experiments that are carried out within living bodies.

K

kinetic energy: energy that an object possesses due to its motion.
Krebs cycle: series of aerobic biochemical reactions in the matrix of the mitochondria of most *eukaryotic cells* by which energy is obtained through the oxidation of acetylcoenzyme A produced from the breakdown of glucose.

L

ligament: a tough, fibrous connective tissue, rich in collagen, that joins bone to bone. See also *tendon*.
light-dependent reaction: stage of photosynthesis in which light energy is required to produce *ATP* and reduced *NADP*. See also *light-independent reaction*.
light-independent reaction: stage of photosynthesis which does not require light energy directly but does need the products of the *light-dependent reaction* to reduce carbon dioxide and so form carbohydrate.
limiting factor: a variable that limits the rate of a chemical reaction.
link reaction: the process linking *glycolysis* with the *Krebs cycle* in which hydrogen and carbon dioxide are removed from pyruvate to form acetylcoenzyme A in the matrix of the mitochondria.

M

mesophyll: tissue found between the two layers of epidermis in a plant leaf, comprising an upper layer of palisade cells and a lower layer of spongy cells.
migration: the temporary movement of a population of organisms from one locality to another.
monoculture: term used to describe a large area of land in which only one type of crop is grown.

motor neurone: *neurone* that transmits *action potentials* from the central nervous system to an effector, e.g. a muscle or gland.
multiple alleles: term used to describe a *gene* that has more than two possible *alleles*.
mutagen: any agent that induces a *mutation*.
mutation: a change in the sequence of bases in DNA.
myelin: a fatty substance that surrounds *axons* and *dendrites* in certain *neurones*.
myosin: the thick filamentous protein found in *skeletal muscle*.

N

NAD (nicotinamide adenine dinucleotide): a molecule that carries *electrons* and hydrogen ions during *aerobic* respiration.
NADP (nicotinamide adenine dinucleotide phosphate): a molecule that carries *electrons* produced in the *light-dependent reaction* of photosynthesis.
negative feedback: a series of changes, important in homeostasis, that result in a substance being restored to its normal level. See also *positive feedback*.
neurone: a nerve cell, comprising a cell body, *axon* and *dendrites*, which is adapted to conduct *action potentials*.
neuromuscular junction: a *synapse* that occurs between a neurone and a muscle.
neurotransmitter: one of a number of chemicals that are involved in communication between adjacent neurones or between nerve cells and muscles. Two important examples are *acetylcholine* and *noradrenaline*.
niche: see *ecological niche*.
nitrifying bacteria: microorganisms that convert ammonium compounds to nitrites and nitrates.
nitrogen fixation: incorporation of atmospheric nitrogen gas into organic nitrogen-containing compounds. It can be brought about by lightning, industrial processes and by both free-living and mutualistic bacteria.
node of Ranvier: a gap in the *myelin* sheath that surrounds the axon of a neurone.
normal distribution: a bell-shaped curve produced when a certain distribution is plotted on a graph.
nucleotides: complex chemicals made up of an organic base, a sugar and a phosphate. They are the basic units of which the nucleic acids DNA and RNA are made.

O

oestrus: the period in the oestrous cycle immediately after ovulation when the female is most fertile.
oncogenes: mutated versions of proto-oncogenes that result in increased cell division leading to the growth of a tumour.
osmosis: the net passage of water from a region of higher water potential to a region of lower water potential through a partially permeable membrane, as a result of the random motion of the water molecules.
oxidation: chemical reaction involving the loss of *electrons*.
oxidation-reduction: a chemical reaction in which electrons are transferred from one substance to another substance. The substance losing electrons is oxidised and the substance gaining the electrons is reduced.
oxidative phosphorylation: the formation of *ATP* in the electron transport system of *aerobic* respiration.

P

parasite: an organism that lives on or in a host organism. The parasite gains a nutritional advantage and the host is harmed in some way.
pathogen: any microorganism that causes disease.
pentose sugar: a sugar that possesses five carbon atoms. Two examples are ribose and deoxyribose.
peptide bond: the chemical bond formed between two amino acids during a *condensation reaction*.
phagocytosis: mechanism by which cells transport large particles across the cell-surface membrane.
phenotype: the characteristics of an organism, often visible, resulting from both its *genotype* and the effects of the environment.
phloem: plant tissue that transports the products of photosynthesis from leaves to the rest of the plant. See also *xylem*.
phospholipids: lipid molecules in which one of the three fatty acid molecules is replaced by a phosphate molecule. They are important in the structure and functioning of all plasma membranes.
photolysis: splitting of a water molecule by light such as occurs during the *light-dependent reaction* of photosynthesis.
pioneer species: a species that can colonise bare rock or ground.
plasmid: small circular piece of DNA found in bacterial cells and used as a *vector* in *recombinant DNA technology*.
polygenes: group of genes that are responsible for controlling a characteristic.
polymer: large molecule made up of repeating smaller molecules.
polymerase chain reaction (PCR): process of making many copies of a specific sequence of DNA or part of a gene. It is used extensively in *gene technology* and genetic fingerprinting.
population: a group of individuals of the same species that occupy the same *habitat* at the same time.
positive feedback: process which results in a substance that departs from its normal level becoming further from its norm. See also *negative feedback*.
primary succession: the progressive colonisation of bare rock or other barren terrain by living organisms.
producer: an organism that synthesises organic molecules from simple inorganic ones such as carbon dioxide and water. Most producers are photosynthetic and form the first trophic level in a food chain. See also *consumer*.
prokaryotic cell: a cell, belonging to the kingdom Prokaryotae, which is characterised by being less than 5 μm in diameter and which lacks a nucleus and membrane-bound organelles. See also *eukaryotic cell*.
proton: positively charged sub-atomic particle found in the nucleus of an atom. See also *electron*.

R

receptor: a cell adapted to detect changes in the environment.

recessive allele: the condition in which the effect of an allele is apparent in the *phenotype* of a *diploid* organism **only** in the presence of another identical allele. See also *dominant allele.*

recognition site: a nucleotide sequence, usually of 4, 6 or 8 *nucleotides*, that is recognised by a *restriction endonuclease* and to which it attaches.

recombinant DNA technology: general term that covers the processes by which genes are manipulated, altered or transferred from organism to organism. Also known as *genetic engineering*.

reflex arc: the nerve pathway in the body taken by an action potential that leads to a rapid, involuntary response to a stimulus.

refractory period: period during which the membrane of the axon of a neurone cannot be *depolarised* and no new *action potential* can be initiated.

repolarisation: return to the *resting potential* in the axon of a neurone after an *action potential*.

resting potential: the difference in electrical charge maintained across the membrane of the axon of a neurone when not stimulated. See also *action potential*.

restriction endonucleases: a group of enzymes that cut DNA molecules at a specific sequence of bases called a recognition sequence.

RNA polymerase: enzyme that joins together *nucleotides* to form messenger RNA during *transcription*.

S

saltatory conduction: propagation of a nerve impulse along a *myelinated dendron* or *axon* in which the *action potential* jumps from one *node of Ranvier* to another.

saprobiotic microorganism: also known as a saprophyte, this is an organism that obtains its food from the dead or decaying remains of other organisms.

sarcomere: a section of myofibril between two Z-lines that forms the basic structural unit of *skeletal muscle*.

Schwann cell: cell around a neurone whose cell-surface membrane wraps around the *dendron* or *axon* to form the *myelin* sheath.

secondary succession: the recolonisation of an area after an early community has been removed or destroyed.

selection: process that results in the best-adapted individuals in a *population* surviving to breed and so pass their favourable *alleles* to the next generation.

selection pressure: the environmental force altering the frequency of alleles in a *population*.

selective breeding: breeding of organisms by human selection of parents/*gametes* in order to perpetuate certain characteristics and/or eliminate others.

sensory neurone: a *neurone* that transmits an *action potential* from a sensory receptor to the central nervous system.

skeletal muscle: the muscle that makes up the bulk of the body and which works under conscious control. Also known as voluntary muscle. See also *smooth muscle*.

sickle-cell anaemia: inherited blood disorder in which abnormal haemoglobin leads to red cells becoming sickle-shaped and less able to carry oxygen.

sinoatrial node (SAN): an area of heart muscle in the right atrium that controls and coordinates the contraction of the heart. Also known as the pacemaker.

smooth muscle: also known as involuntary or unstriated muscle, smooth muscle is found in the alimentary canal and the walls of blood vessels. Its contraction is not under conscious control. See also *skeletal muscle*.

sodium–potassium pump: protein channels across cell-surface membranes that use ATP to move sodium *ions* out of the cell in exchange for potassium ions that move in.

speciation: the evolution of two or more species from existing species.

species: a group of similar organisms that can breed together to produce fertile offspring.

species diversity: the number of different species and the number of individuals of each species within any one *community*.

stabilising selection: selection that tends to eliminate the extremes of the *phenotype* range within a *population*. It arises when environmental conditions are constant.

stem cells: undifferentiated dividing cells that occur in embryos and in adult animal tissues that require constant replacement, e.g. bone marrow.

stimulus: a detectable alteration in the internal or external environment of an organism that produces some change in that organism.

stoma (plural stomata)**:** a pore, surrounded by two *guard cells*, mostly in the lower epidermis of a leaf, through which gases diffuse in and out of the leaf.

stroma: matrix of a chloroplast where the *light-independent reaction* of photosynthesis takes place.

substrate-level phosphorylation: the formation of ATP by the direct transfer of a phosphate group from a reactive intermediate to ADP.

synapse: a junction between *neurones* in which they do not touch but have a narrow gap, the synaptic cleft, across which a *neurotransmitter* can pass.

T

tendon: tough, flexible, but inelastic, connective tissue that joins muscle to bone. See also *ligament*.

threshold level/value: the minimum intensity that a stimulus must reach in order to trigger an *action potential* in a *neurone*.

thylakoid: series of flattened membranous sacs in a chloroplast that contain chlorophyll and the associated molecules needed for the *light-dependent reaction* of photosynthesis.

tissue fluid: fluid that surrounds the cells of the body. Its composition is similar to that of blood plasma except that it lacks some of the larger proteins, in particular those that cause the blood to clot. It supplies nutrients to the cells and removes waste products.

transcription: formation of messenger RNA molecules from the DNA that makes up a particular *gene*. It is the first stage of protein synthesis.

transducer cells: cells that convert a non-electrical signal, such as light or sound, into an electrical (nervous) signal and vice versa.

transduction: the process by which one form of energy is converted into another. In microbiology, the natural process by which genetic material is transferred between one host cell and another by a virus.

transpiration: evaporation of water from a plant.

trophic level: the position of an organism in a food chain.

tumour suppressor gene: a gene that maintains normal rates of cell division and so prevents the development of tumours.

vasoconstriction: narrowing of the internal diameter of blood vessels. See also *vasodilation*.

vasodilation: widening of the internal diameter of blood vessels. See also *vasoconstriction*.

vector: a carrier. The term may refer to something such as a *plasmid*, which carries DNA into a cell, or to an organism that carries a *parasite* to its host.

voltage-gated channel: protein channel across a cell-surface membrane that opens and closes according to changes in the electrical potential across the membrane.

water potential: measure of the extent to which a solution gains or loses water. The greater the number of water molecules present, the higher (less negative) the water potential. Pure water has a water potential of zero.

xerophyte: a plant adapted to living in dry conditions.

xylem vessels: dead, hollow, elongated tubes with lignified side walls and no end walls, that transport water in most plants.

Answers

1.1

1 ecology

2 biosphere

3 biotic

4 abiotic

5 community

6 population

7 habitat

1.2

1 $\frac{100 \times 80}{5} = 1600$

2 a Population over-estimated (appears larger) as there will be proportionally fewer marked individuals in the second sample.

b Population over-estimated / appears larger as there will be proportionally fewer marked individuals in the second sample because all the 'new' individuals will be unmarked.

c No difference because the proportion of marked and unmarked individuals killed should be the same.

Ethics and fieldwork

1 They can be eaten by other organisms and so provide energy and nutrients to the ecosystem.

2 It allows the habitat to recover from any disturbance / removal of organisms. The results of a further study carried out too soon after may result in data that are not typical of the habitat under 'normal' conditions.

3 The organisms live beneath stones so they remain moist when not covered by the tide. If the stone is left upside down the organism may become desiccated and die.

4 A selection from each of the following:

For	Against
• Practical experience aids learning / is better than theoretical study	• Students are inexperienced and therefore more likely to damage habitats
• Experienced ecologists have to start somewhere	• Information could be provided by theoretical means / use of data obtained by experienced ecologists / videos
• Students may become ecologists and so aid conservation in the long term	• Large number of A-level students puts pressure on / increases damage to popular sites
• Students can be taught conservation through ecology which makes people more aware of the environment and so more likely to support conservation	• Many students will not continue studies in biology / ecology and so may never use the information again

1.3

1 Certain factors limit growth, e.g. availability of food, accumulation of waste, disease.

2 Biotic factors involve the activities of living organisms.

Abiotic factors involve the non-living part of the environment.

3 a low light intensity

b lack of water

c low temperature

The influence of abiotic factors on plant populations

1 a 1

b 3

c 2

d 3

2 The pH is too high for species X and the temperature is too low for species Y.

1.4

1 Intraspecific competition occurs when individuals of the **same** species compete with one another for resources.

Interspecific competition occurs when individuals of **different** species compete for resources.

2 Any 2 from: food / water / breeding sites (or any other relevant factor, e.g. light, minerals).

The effects of interspecific competition on population size

1 After 1985 the rise in the grey squirrel population is mirrored by a fall in the red squirrel population.

2 Lack of food / adverse weather, e.g. cold winters / increase in number of squirrel predators / new disease.

3 Grey squirrels have more chance of finding fruits / nuts / seeds that have fallen to the ground as well as those that are still on the trees / bushes.

4 The sea presents a barrier to the grey squirrel reaching islands. The red squirrels already present on the islands have little or no competition from grey squirrels and so flourish.

Competing to the death

1 Population increases slowly at first and then at an accelerating / exponential / logarithmic rate to around 8 days. The growth rate then slows, reaching a maximum at around 12 days which is maintained at a constant level up to 20 days.

2 Population growth is faster initially. Maximum size is reached earlier. Maximum size is reduced to less than half. Size is not maintained at a constant level (it falls to zero).

3 *P. caudatum* is unsuccessful in competing with *P. aurelia* for yeast / food. Most available food is taken by *P. aurelia* and *P. caudatum* starves, leading to a population crash.

4 Some of the yeast / food is taken by *P. caudatum*, leaving less for the population growth of *P. aurelia*.

5 After 20 days all *P. caudatum* have died. *P. aurelia* has no competition for food and so it reaches its previous maximum. *P. aurelia* is in effect 'alone' again.

Effects of abiotic and biotic factors on population size

1 mark-release-recapture technique

2 Increase in population is 1320 (in 1995) minus 260 (in 1993) = 1060.

Time period = 2 years.

Mean annual growth in population is therefore: $\frac{1060}{2} = 530$.

3 The more acorns produced in the autumn, the larger the deer mice population the following spring. The fewer acorns produced, the smaller the deer mice population.

4 1992

5 The population of deer mice would fall as more oak leaves are eaten by gypsy moth caterpillars so there will be less food (acorns) to support the deer mice population / some deer mice will starve.

6 a warm spring → more acorn seed is set → more acorns produced in autumn → more food is available over winter → more deer mice survive and breed → deer mice population increases → more predation of gypsy moth pupae by deer mice → smaller gypsy moth population

b more owls → more predation on deer mice → smaller deer mice population → fewer gypsy moth pupae are eaten → larger gypsy moth population → more oak leaves eaten → less energy available from photosynthesis for the production of acorns → fewer acorns

1.5

1 The range and variety of laboratory habitats is much smaller than in natural ones. This means that in nature there is a greater range of hiding places and so the prey has more space and places to escape the predator and survive.

2 With fewer predators, fewer prey are taken as food. The death rate of prey is reduced. Assuming the birth rate remains unchanged the population size increases.

3 Graph showing population fluctuations (peaks and troughs) of A. Species B mirrors these changes after a time lag. The population size of B is, for the most part, smaller than A.

B eats A → population of A falls → fewer A for B to eat → population of B falls → fewer B means fewer A are eaten → population of A rises → more A means more food for B → population of B rises.

The Canadian lynx and the snowshoe hare

1 The assumption is made that the relative numbers of each type of fur traded represents the relative size of each animal's population at the time.

2 The population size of the snowshoe hare fluctuated in a series of peaks and troughs. Each peak and trough was repeated about every 10 years. The population size of the Canadian lynx also fluctuated in a 10-year cycle of peaks and troughs. The relative pattern of peaks and troughs is similar for the lynx and the snowshoe hare. The rise in the population size of the lynx often (but not always) followed that of the snowshoe hare.

3 The snowshoe hare population increases due to the low numbers of Canadian lynx that feed on them → more hares mean more food for the lynx, whose population therefore increases as fewer starve / more are able to raise young → more lynx means there is more predation of hares, whose population therefore decreases → fewer hares means less food for the lynx, many of which starve and so their population decreases.

4 4 times

5 Addition of food – because the population increased more in every year that data were collected.

6 Both food supply and predation influence hare population size. Food supply has a greater influence than predation but a combination of both factors has an even greater influence than either of the other two separately.

1.6

1 development of agriculture and the industrial revolution

2 a Number of births = $\frac{25 \times 1\,000\,000}{1000} = 25\,000$

b Number of deaths = $\frac{20 \times 1\,000\,000}{1000} = 20\,000$

c $\frac{\text{population change during 2007}}{\text{population at start of 2007}} \times 1000$

$= \frac{(25\,000 - 20\,000)}{1\,000\,000} \times 100$

$= \frac{5000}{1\,000\,000} \times 100 = 0.5\%$.

3 Economically less developed country = approx. 45 years.

Economically more developed country = approx. 80 years.

Demographic transition of the human population

1 a Stage 3

b Stage 1

c Stage 2

d Stage 1

e Stage 4

2 Pyramid A represents stage 4 because there is a low birth rate (narrow base to the pyramid) and a low death rate (sides are fairly vertical and many people live beyond 65 years).

Pyramid B represents stage 2 because there is a high birth rate (wide base to the pyramid) and a falling death rate (sides slope upwards and some, but not many, people live beyond 65 years).

2.1

1 ATP releases its energy very rapidly. This energy is released in a single step and is transferred directly to the reaction requiring it.

2 ATP provides a phosphate that can attach to another molecule, making it more reactive and so lowering its activation energy. As enzymes work by lowering activation energy they have less 'work' to do and so function more readily.

3 Any 3 from: building up macromolecules (or named example of macromolecule) / active transport / secretions (formation of lysosomes) / activation of molecules.

3.1

1 carbon dioxide and water

2 glucose and oxygen

3 a grana / thylakoids

b stroma

4 a reduced NADP, ATP and oxygen

b sugars and other organic molecules

3.2

1 on the thylakoid membranes (of the grana in the chloroplast)

2 Water molecules are split to form electrons, protons and oxygen, as a result of light exciting electrons / raising the energy levels of electrons in chlorophyll molecules.

3 a reduction

b reduction

c oxidation

Chloroplasts and the light-dependent reaction

1 A = (double) membrane of chloroplast / chloroplast envelope

C = granum

D = stroma

2 C

3 starch

4 Any 2 from: the light-dependent reaction does not produce sufficient ATP for the plants' needs / photosynthesis does not take place in the dark / cells without chlorophyll cannot produce ATP in this way and ATP cannot be transported around the plant.

5 Length X–Y on Figure 4 = 24 mm (= 24 000 μm)

Actual length X–Y = 2 μm

Magnification = $\frac{24\,000}{2}$ = 12 000 times

3.3

1 It accepts / combines with a molecule of CO_2 (to produce 2 molecules of glycerate-3-phosphate).

2 It is used to reduce (donate hydrogen) glycerate-3-phosphate to triose phosphate.

3 ATP

4 Stroma of the chloroplasts.

5 The Calvin cycle requires ATP and reduced NADP in order to operate. Both are the products of the light-dependent reaction, which needs light. No light means no ATP or reduced NADP are produced and so the Calvin cycle cannot continue once any ATP or reduced NADP already produced have been used up.

Using a lollipop to work out the light-independent reaction

1 To allow the substances into which it becomes incorporated to be identified / to allow the sequence of substances produced to be identified.

2 The radioactive carbon is initially found in glycerate-3-phosphate (5 seconds) and is next found in triose phosphate (10 seconds).

3 The high temperature and / or the methanol denature the enzymes that catalyse reactions.

4 The quantity of GP begins to decrease almost immediately. The rate of decrease becomes less until, after about 4.5 minutes, the quantity of GP becomes constant, but at around a quarter of its original level. The quantity of RuBP rises almost immediately. The rate of increase is steady at first, but then slows, peaking at 3.5 minutes. The quantity of RuBP then falls until it becomes constant at around 4.5 minutes, but at around double its original level. The quantities of GP and RuBP are the same after 2.5 minutes.

5 RuBP combines with CO_2 to form GP during the light-independent reaction / Calvin cycle of photosynthesis. GP is ultimately used to regenerate RuBP. When the CO_2 level is decreased, there is less to combine with RuBP and so less GP is formed but it is still being used up and so its level falls. There is still some CO_2 and so some GP is made, but much less than originally. With less CO_2 to combine with, the RuBP accumulates because it cannot be converted to GP. Its quantity rises to a new higher level due to the lower level of CO_2.

3.4

1 Volume of oxygen produced / CO_2 absorbed.

2 Light intensity – because an increase in light intensity produces an increase in photosynthesis over this region of the graph.

3 Raising the CO_2 level to 0.1% – because this increases the rate of photosynthesis more than increasing the temperature to 35 °C.

4 Because light is limiting photosynthesis and so an increase in temperature will not increase the rate of photosynthesis.

5 More CO_2 is available to combine with RuBP to form more GP, then more triose phosphate and ultimately more glucose.

Measuring photosynthesis

1 Because any air escaping from or entering the apparatus will respectively decrease or increase the volume of gas measured, which will give an unreliable result.

2 So that any changes in the rate of photosynthesis can be said to be the result of changes in light intensity and not changes in temperature.

3 To ensure there is sufficient CO_2 and so it does not limit the rate of photosynthesis.

4 To prevent other light falling on the plant as this may fluctuate and will affect the light intensity and hence the rate of photosynthesis, leading to an unreliable result.

5 To prevent photosynthesis and to allow any oxygen produced before the experiment begins, to disperse.

6 Because the volume of oxygen produced will be less than that produced by photosynthesis as some of the oxygen will be used up in cellular respiration / dissolved oxygen (and other gases) may be released from, or absorbed by the water.

4.1

1 cytoplasm

2 glucose

3 phosphate

4 ATP

5 phosphorylated glucose

6 triose phosphate

7 hydrogen

8 NAD

9 pyruvate

10 ATP

4.2

1. 3
2. acetylcoenzyme A
3. matrix of mitochondria
4. 1 = True 2 = True 3 = True
5. 4 = True 5 = False 6 = False
6. 7 = True 8 = True 9 = False
7. 10 = False 11 = False 12 = True
8. 13 = False 14 = True 15 = False
9. 16 = False 17 = False 18 = False

Coenzymes in respiration

1. To show that the yeast suspension was responsible for any changes that occurred and the glucose did not change methylene blue nor did methylene blue change by itself.
2. a Yeast uses glucose as a respiratory substrate producing hydrogen atoms that are taken up by methylene blue causing it to become reduced and changing from blue to colourless.

   b As in (2a), except that the yeast uses stored carbohydrate as a respiratory substrate that has to be converted to glucose and so the production of hydrogen atoms is slower / reduced.
3. Contents of tube might have remained blue because the enzymes involved in respiration are denatured at 60 °C and so respiration, and hence the reduction of methylene blue, ceases / the enzymes involved in hydrogen transport have been denatured and so the indicator is not reduced by hydrogen.
4. Air contains oxygen, which would re-oxidise methylene blue, turning it blue.
5. This is a single experiment. The same results would need to be obtained on many occasions to increase reliability.

4.3

1. The movement of electrons along the chain is due to oxidation. The energy from the electrons combines inorganic phosphate and ADP to form ATP = phosphorylation.
2. It provides a large surface area for the attachment of the coenzymes (NAD / FAD) and electron carriers that transfer the electrons along the chain.
3. Oxygen is the final acceptor of the electrons and hydrogen ions (protons) in the electron transport chain. Without it the electrons would accumulate along the chain and respiration would cease.
4. water molecule

Sequencing the chain

1. Sequence – C, A, D, B

   Explanation – Electron carriers become reduced by electrons from glycolysis and the Krebs cycle. Enzymes catalyse the transfer of these electrons to the next carrier. If an enzyme is inhibited all molecules prior to that enzyme will not be able to pass on their electrons and so will be reduced and those after it will be oxidised. The first molecule in the chain will be reduced with all inhibitors, the second with 2 out of 3 inhibitors, the third with 1 out of 3 and the last in the chain with none (i.e. it is always oxidised).

4.4

1. a D

   b A, C, D

   c A, D

   d A, B

   e A, D

   f B, C, D

   g A

Investigating where certain respiratory pathways take place in cells

1. Homogenate is spun at slow speed. Heavier particles (e.g. nuclei) form a sediment. Supernatant is removed, transferred to another tube and spun at a greater speed. Next heaviest particle is removed. Process is repeated.
2. Nuclei and ribosomes – because neither CO_2 nor lactate (products of respiration) are formed in any of the samples.
3. a mitochondria

   b Krebs cycle produces CO_2 and results show that CO_2 is produced when mitochondria only are incubated with pyruvate.
4. (remaining) cytoplasm (Note: The complete homogenate is not a 'portion' of the homogenate.)
5. Cyanide prevents electrons passing down the transport chain. Reduced NAD therefore accumulates and blocks Krebs cycle where CO_2 is produced. Glycolysis can still occur because the reduced NAD it produces is used to make lactate. Glucose can therefore be converted to lactate, but not into CO_2, in the presence of cyanide.
6. The conversion of glucose to CO_2 involves glycolysis (occurs in cytoplasm) and Krebs cycle (occurs in mitochondria). Only the complete homogenate contains both cytoplasm and mitochondria.
7. ethanol and CO_2
8. liver cell, epithelial cell and muscle cell

5.1

1. dragonfly nymphs
2. unicellular and filamentous algae
3. sticklebacks
4. the direction of energy flow
5. decomposers / saprobiotic (micro)organisms

5.2

1. Any 3 from: some of the organism is not eaten; some parts are not digested and so are lost as faeces; some energy is lost as excretory materials; some energy is lost as heat
2. The proportion of energy transferred at each trophic level is small (less than 20%). After four trophic levels there is insufficient energy to support a large enough breeding population.
3. a $\frac{1250 \times 100}{6300} = 19.84\%$

   b $\frac{50 \times 100}{42\,000} = 0.12\ (0.119)\%$

Adding up the totals

1. decomposers / saprobiotic (micro)organisms
2. insect-eating birds

3 $\frac{42\,500 \times 100}{1.7 \times 10^6} = 2.5\%$

4 Any 3 from: most (90%+) solar energy is reflected by clouds, dust or absorbed by the atmosphere / not all light wavelengths are used in photosynthesis / much of the light does not fall on the chloroplast/chlorophyll molecule / factors may limit the rate of photosynthesis or photosynthesis is inefficient / respiration by producers means energy is lost (as heat).

5 4120 – (1010 + 810) = 2300 kJ m^{-2} year^{-1}

5.3

1 In a pyramid of numbers: no account is taken of size / the number of individuals of one species may be so great that it is impossible to represent them on the same scale as other species in the food chain.

2 At certain times of year (e.g. spring) zooplankton consume phytoplankton so rapidly that their biomass temporarily exceeds that of phytoplankton.

3 grams per square metre (g m^{-2}) / grams per cubic metre (g m^{-3})

A woodland food chain

1 moth caterpillars

2 $\frac{92 \times 100}{806} = 11.41\%$

3 kJ m^{-2} year^{-1}

4 Each beech tree has many caterpillar moths living on it, so the beech tree block is smaller in the number pyramid. Each beech tree is very large and so has a greater biomass and more energy than all the moth caterpillars together, so the beech tree block is larger in the biomass and energy pyramids.

5 Any 2 from: lost in respiration / lost as heat / lost to decomposers (in excretion, faeces, death and decay) / part of robin is not eaten or not digested.

6 The parasitic flea block must be at the top of each pyramid. It is larger than the sparrowhawk block in the number pyramid but smaller in the biomass and energy pyramids.

5.4

1 gross productivity minus respiratory losses

2 kJ m^{-2} year^{-1}

3 In an agricultural ecosystem additional energy is put in to remove other species, add fertilisers and pesticides. These reduce competition for light, water, CO_2, etc., provide mineral ions, destroy pests and reduce disease. All these increase photosynthesis and hence productivity.

4 Natural ecosystems use only solar energy, agricultural ecosystems use solar energy and additional energy from food (labour) and fossil fuels (machinery and transport).

Increasing productivity

1 On a cloudy day, light intensity will be the limiting factor and only an increase in this will increase the rate of photosynthesis. Increases in CO_2 concentration will have no effect and would be economically wasteful.

2 With the greenhouse open to the air any additional CO_2 would simply disperse outside. It therefore makes economic sense to set the level to around that found in the air, namely 400 ppm.

3 Long periods of CO_2 levels at 5000 ppm would cause stomata to close. Diffusion of gases, and therefore photosynthesis, would be considerably reduced. Keeping levels this high would be counterproductive. At 1000 ppm the higher level of CO_2 increases photosynthesis without causing stomata to close. For short durations there may be some benefit in bright sun of having a level of CO_2 as high as 5000 ppm as the plant takes some time to close its stomata. During this period 5000 ppm will increase photosynthesis as CO_2 is likely to be the limiting factor.

5.5

1 If the pesticide kills most of the pests then the population of organisms (predators) feeding on it will fall. With no predators controlling it, the pest population will increase again, possibly to a level higher than before. The crop will be even more affected by the pest, leading to lower productivity.

2 Advantages – any 2 from: highly specific, targeted only on pest / once introduced it reproduces itself and does not need to be re-applied / pests do not become resistant.

Disadvantages – any 2 from: effect is slow as there is a time lag between application and results / may itself become a pest or may disrupt the ecological balance.

3 Weeds compete for light, water, mineral ions, CO_2, space, etc. If these are in limited supply there will be less available to the crop plants. One or more may limit the rate of photosynthesis and hence productivity.

To weed or not to weed?

1 As the number of weeds increases, the productivity of wheat decreases. The reduction in productivity is initially large, between 0 and 40 weeds m^{-2}, but lessens as the number of weeds increases, from 40 to 50 m^{-2}, the curve then flattens out.

2 Soya bean because it has an increase in productivity of 50% (1000 to 1500 kg ha^{-1}) while wheat only increases by 33% (4500 to 6000 kg ha^{-1}).

3 No. Cost of herbicide per hectare = £100. Reducing weeds from 40–20 m^{-1} increases wheat productivity from 4500 to 5000 kg ha^{-1} – an increase of 500 kg or half a tonne. Wheat is sold at £150 per tonne, so increased income is £75 per hectare – £25 per hectare less than the cost of treating with herbicide.

A mighty problem

1 Description – In both experiments the spider mite populations rise slowly during the first 15 days and then very rapidly up to around 50 days.

In experiment 1 the spider mite population remains high up to 150 days but fluctuates (between 400 and 900). (Note: The scale is the square root of the numbers and so the figures on the *y*-axis need to be squared to give actual numbers.)

In experiment 2 the spider mite population falls over the period 50–150 days until it reaches the starting level.

Explanation – In experiment 1 the population of the spider mite increases until some factor (e.g. food supply) limits its size. It remains fairly constant as an equilibrium is reached with the limiting factor, fluctuating slightly as the factor fluctuates.

In experiment 2 the population of the spider mite increases up to 50 days, by which time the population of the predatory mite has increased considerably. The predatory mites feed on the spider mites, causing their population to drop to a very low level by 150 days.

2 Predatory mites are effective in controlling the population of spider mites as their presence reduces the spider mite population from around 400–900 when the predatory mite is absent to around 4 when the predatory mite is present.

3 The two populations will probably remain small as they remain in balance. They will fluctuate because, as the spider mite population falls, there will be less food for the predatory mite and so, a short time later, its population will also fall. The fall in the predatory mite's population means there will be less predation on the spider mite, whose population will then increase, followed in turn by an increase in the predatory mite's population.

5.6

1 Movement is restricted so less energy is expended in muscle contraction / heat loss is reduced so less energy is expended maintaining body temperature / the optimum amount and type of food for rapid growth can be provided / predators are excluded so no energy is lost to other organisms.

2 A longer dark period means more time is spent resting, less energy is expended, and more energy is converted into body mass.

Features of intensive rearing of livestock

1 The gene / allele / DNA for antibiotic resistance can pass from the animal disease-causing bacteria to the human disease-causing bacteria along a conjugation tube formed between the two (horizontal gene transmission).

2 A selection from each of the following:

For	Against
• Efficient energy conversion	• Lower quality eggs / less taste
• Economic – makes good use of resources	• Disease spreads more rapidly
• Cheaper eggs	• Antibiotic resistance – can spread to human disease-causing bacteria
• Less land required / more land for natural habitats	• Hens kept unnaturally – may be stressed or aggressive and therefore need to be de-beaked
• Safer / more easily regulated farms	• Restricted movement means more osteoporosis / joint pain
• Disease / predators more easily excluded	• More pollution / smells / harmful to environment
• Animals are warm and well fed	• Use of fossil fuels means more CO_2 / global warming
	• Hens require drugs as a consequence of the environment in which they are kept / to make them grow more rapidly

3 Any balanced discussion of issues such as: crowded, confined conditions versus being warm, fed, kept healthy, free from predators / need for drugs versus becoming diseased / de-beaking versus harming one another / keeping animals unnaturally versus our demand for cheap food / treating animals badly (osteoporosis, boredom, frustration) versus animal welfare legislation to prevent cruelty.

Economic and environmental issues concerned with intensive food production

1 Any 2 from: leaves more space to grow crops / reduces competition for light, water, etc. / hedges may harbour weeds and other pests that harm the crops or animals thereby reducing productivity / easier to manoeuvre large machinery / time and energy has to be expended on maintaining the hedge.

2 Any 1 from: they might be natural predators of crop pests and thereby provide a means of biological control / they might be pollinators of crop plants and so help fruit and seed production.

3 Most conservation techniques lead to reduced productivity. This means that food is more expensive to produce and some farmers might be unable to compete with cheaper food produced by other farmers. Those who carry out conservation might therefore go out of business without subsidies.

4 Advantage – Less pesticide use means reduced risk from pesticide residues in food.

Disadvantage – Food is more expensive. (Note: 'Organic' food is not necessarily better tasting or more nutritious than intensively produced food.)

6.1

1 Light intensity is greater / longer period of light in summer and the temperature is usually higher. Both factors increase the rate of photosynthesis (assuming there is no other limiting factor). More CO_2 is taken from the atmosphere during photosynthesis and so its concentration falls.

2 A = combustion (burning)

B = respiration

C = feeding (and digestion)

D = photosynthesis

3 d

6.2

1 Fluctuations are the result of annual changes in CO_2 concentration due to seasonal temperature changes. CO_2 levels fall in the summer as warmer temperatures and longer periods of more intense light lead to more photosynthesis and hence more CO_2 being fixed into organic molecules. In winter, colder temperatures, less light and the loss of leaves in deciduous plants means less photosynthesis and less CO_2 absorption by plants. As respiration in organisms continues to produce CO_2, its level increases in winter.

2 Palm tree plantations – less productive than tropical rain forest therefore there is less photosynthesis → less CO_2 is absorbed → CO_2 level in the atmosphere increases → greenhouse gas → more global warming.

Burning of forest – heat directly contributes to global warming, the CO_2 produced does so indirectly (as above).

Palm oil extraction – manufacture of biofuel and its transport use fossil fuels that produce CO_2 and hence increase global warming.

Burning of biofuel – also produces CO_2 but as this has only recently been absorbed from the atmosphere by the palm trees it will have little or no effect on global warming.

3 Cattle produce methane, a greenhouse gas that contributes to global warming.

Digging into the past

1 There is a correlation because the graphs have similar shapes. Rises and falls in CO_2 concentration are reflected in similar rises and falls in temperature.

2 No. To prove this we would need to establish a causal link between changes in CO_2 concentration and temperature change. It is equally possible from the data to suggest that the temperature changes cause the changes in CO_2 concentration rather than the other way around.

Global warming and crop yields

1 The yield decreases slightly as the minimum temperature increases from 22.0 °C to 22.5 °C.

The yield decreases more significantly/to a greater extent as the minimum temperature increases from 22.5 °C to 24.0 °C.

The total decrease in yield is from 9.4 tonnes ha^{-1} at 22.0 °C down to 6.5 tonnes ha^{-1} at 24.0 °C.

2 Above-ground biomass at 22.0 °C = 1800 g m^{-2}

Above-ground biomass at 24.0 °C = 1500 g m^{-2}

Decrease in biomass = 1800 – 1500 = 300 g m^{-2}

% decrease = 300/1800 × 100 = 16.7%.

3 As the data are collected at night there is no photosynthesis but respiration is still taking place. An increase in temperature increases enzyme activity and hence respiration rate. More carbohydrate is used up in respiration leaving less available to form grain/biomass.

Global warming and insect pests

1 Description – The higher the mean winter temperature, the more rice stem-borer larvae there are per rice plant. For each 0.5 °C rise in mean winter temperature, the number of larvae per plant doubles. A temperature rise from 6.0 °C to 6.5 °C doubles the number of larvae from 10 to 20 and a rise from 6.5 °C to 7.0 °C doubles the number of larvae from 20 to 40.

Explanation – The larvae over-winter in the paddy fields and so the warmer the winter temperature the more are likely to survive and infect rice plants and the earlier they emerge in spring and so the longer they have to infect rice plants and the more larvae there are per plant.

2 From graph A – global warming from 1950 to 2000 has led to an increase in mean winter temperatures.

From graph B – this means more rice stem-borer larvae per rice plant.

From graph C – this means that the crop yield of rice is reduced.

3 10% – because in 1980 the mean winter temperature was 6.5 °C (graph A). The predicted number of larvae per plant when the mean winter temperature is 6.5 °C is 20 (graph B). When there are 20 larvae per plant, the loss in crop yield is 10% (graph C).

4 Larvae bore into stems of rice plants and so may: block xylem/prevent transpiration so no water for photosynthesis reaches leaves, no photosynthesis, no sugars for respiration / products of photosynthesis/sugars cannot reach roots from leaves, no sugars for respiration and so roots, and hence plant, dies / damage to stem may cause plant to collapse/fall over.

6.3

1 nitrogen fixation

2 plants

3 nitrate ions

4 root hairs

5 proteins / amino acids / nucleic acids

6 decomposers / saprobiotic microorganisms

7 ammonia / ammonium ions

8 nitrifying

9 nitrate

10 denitrifying

6.4

1 Crops are grown repeatedly and intensively on the same area of land. Mineral ions are taken up by the crops, which are transported and consumed away from the land. The mineral ions they contain are not returned to the same area of land and so the levels in the soil are reduced, which can limit the rate of photosynthesis. Fertilisers need to be applied to replace them if photosynthesis / productivity is to be maintained.

2 100 kg ha^{-1} – although 150 kg ha^{-1} gives a slightly better yield, this is marginal and the cost of using 50% more fertiliser makes it uneconomical.

3 Some other factor is limiting photosynthesis, e.g. light, CO_2, and only the addition of this factor will increase photosynthesis and hence productivity.

4 Natural fertilisers are organic and come from living organisms in the form of dead remains, urine or faeces (manure).

Artificial fertilisers are inorganic and are mined from rocks and deposits.

Different forms of nitrogen fertilisers

1 manure, bone meal and urea

2 To act as a control to show that any changes in productivity were the result of the nitrogen fertiliser being added.

3 Nitrogen is needed for proteins / amino acids / chlorophyll and DNA and therefore for plant growth. Nitrogen shortage may limit the production of proteins and DNA and hence growth. Its addition increases productivity.

4 Some forms of fertilisers contain more actual nitrogen than others and so different masses are added to ensure that the total nitrogen added was always the same (140 kg ha^{-1}).

5 The data do not support the view. While ammonium nitrate brings about the greatest increase in productivity, ammonium sulphate produces a smaller increase than both urea and bone meal. Therefore the investigation suggests that only some ammonium salts are better.

6 The farmer should spread the manure a few months before the main growing season for the crop.

6.5

1 Eutrophication is the process by which salts build up in bodies of water.

2 The concentration of algae near the surface becomes so dense that no light penetrates to deeper levels. No light means no photosynthesis and hence no carbohydrate for respiration and so plants at lower levels die.

3 Dead plants are used as food by saprobiotic microorganisms. With an increased supply of this food, the population of saprobiotic microorganisms increases exponentially. Being aerobic they use up the oxygen in the water leading to the death of the fish, which cannot respire without it.

Troubled waters

1 It has taken 10 days for the fertiliser that has dissolved in the rainwater to leach through the soil and into the lake.

2 In normal circumstances, a low level of nitrate (or other ions) is the limiting factor to algal growth. The fertiliser leaching into the lake contains nitrate (and other ions) and removes this limit on growth. The algal population grows rapidly, increasing in density.

3 Description – As the density of the algae increases so the clarity of the water decreases, i.e. there is a negative correlation. For the first 20 days the algal density (30 cells cm^{-3}) and water clarity (Secchi = 9 m) remain constant. From day 20 to day 100 the algal density increases from 30 to 120 cells cm^{-3} while the water clarity decreases from 9 to 1 m (Secchi depth). However, there is an anomaly between day 40 and day 50 when the water clarity suddenly falls from 7 to 4 m.

Explanation – As the density of algae increases, more light is absorbed / reflected by them and so less light penetrates / water clarity is reduced. Between day 40 and day 50 some factor (e.g. water turbulence stirring up sediment) other than algal density is reducing the water clarity.

4 Days 0–10: oxygen level is constant (at 10 ppm) because there is a balance between oxygen produced in photosynthesis of plants and algae, and oxygen used up in respiration of all organisms.

Days 10–25: oxygen level rises (up to around 13 ppm) due to increased photosynthesis by the larger population of algae.

Days 25–100: oxygen level decreases (more rapidly at first and then less so down to around 3 ppm) due to higher density of algae blocking out the light to lower depths and reducing the rate of photosynthesis of plants / algae at these depths. In time, light is blocked out altogether at lower depths → no photosynthesis → plants / algae die → saprobiotic microorganisms decompose them → their population increases → they use up much oxygen in respiration → oxygen levels fall.

7.1

1 pioneer species

2 primary colonisers (pioneer species) photosynthesise and fix nitrogen → these die and form a soil with nutrients → further colonisers can survive in this soil → environment is a little less hostile → more habitats and food sources available → other species are able to survive → increased biodiversity

3 climax community

Warming to succession

1 Biomass increases very slowly and so the line curves gently upwards at first (up to 60 years) because there is little nitrogen in the soil and therefore growth and hence net production of the pioneer species (*Dryas*) is small.

Biomass increases at a greater but constant rate and so the curve becomes a straight line with an upward gradient, from 60 to 120 years as soil nitrogen levels rise. Increased levels of soil nitrogen remove this limit on growth (net productivity) therefore large species, such as alder, and later spruce, establish themselves and hence biomass increases more rapidly.

Biomass increase slows and finally stops and so the curve flattens out after 150 years because soil nitrogen levels fall as plants take it into their biomass – nitrogen again limits plant growth (net productivity).

2 a Nitrogen from the atmosphere is fixed into compounds, e.g. proteins and amino acids by the nitrogen-fixing species (lichens, *Dryas* and alder). When these die or shed their leaves this nitrogen is released when decomposers break them down into ammonium compounds (ammonification) which are then broken down by nitrifying bacteria into nitrites and nitrates.

b More nitrogen is being absorbed by the increased biomass of the plants. The nitrogen-fixing lichens, *Dryas* and alder have been replaced by spruce that does not fix nitrogen therefore less nitrogen is being added to the soil.

3 a (Pioneer) species are taking advantage of new habitats and lack of competition to rapidly colonise the empty land.

b Spruce is becoming dominant and out-competing the other species, such as lichen, *Dryas* and alder, for light, nutrients, etc. These other species are eliminated from the community.

4 Transects are better because there is a gradient of environmental factors that produce a series of changes over a long distance. Transects also ensure that every community is sampled, which may not be the case with random sampling.

7.2

1 The species within the habitat possess unique genes that at some point in the future may be useful. Conserving habitats maintains biodiversity. The greater the variety of habitats, the greater their potential to enrich our lives and provide enjoyment.

2 Cut back reeds to prevent them becoming dominant. Remove dead vegetation to prevent build-up and thus stop fens drying out. Pump water into fens to keep them waterlogged. Cut back grasses and shrubs to prevent succession.

Conflicting interests

1 96 (32% of 300)

2 It might increase the population of grouse as harriers would have alternative sources of food and therefore eat fewer grouse chicks. Alternatively it might lead to a large increase in harriers that then prey on grouse (especially once the supply of voles and meadow pipits has been exhausted). This would lead to a decrease in the grouse population.

3 The moorland would undergo secondary succession, finally reaching its climax community of deciduous (oak) woodland.

4 A selection from each of the following arguments:

For	Against
• The harrier is a very rare bird – there are only 750 pairs in the UK.	• The harrier is a major predator of grouse and so could threaten the already declining grouse population.
• Previous persecution led to its extinction on the UK mainland and this could happen again.	• If the grouse population is reduced / eliminated and / or the harrier population is not controlled, this could adversely affect the populations of alternative harrier prey, such as voles and meadow pipits.
• Harriers are part of our natural heritage and their population should not be controlled other than by natural means.	• Reduction / elimination of grouse population could make grouse shooting uneconomic and, unless money is found from elsewhere, the moorland habitat might be lost along with the species that live there, and so reduce biodiversity.

5 a A long time is needed to allow population changes in both species as each only breeds once a year.

b The conflicting interests of conservationists and grouse managers mean that agreement on issues such as the population ceiling for hen harriers is unlikely without independent arbitration. An independent body can ensure that the experiment is carried out properly and that the results are interpreted without bias from parties with a vested interest.

c This ensures that hen harrier populations rise within as short a time as possible so that results can be analysed and decisions on future policy made. If this takes too long, the harrier may already be eliminated from some regions.

d This ensures a wide range of different biotic and abiotic conditions as well as a range of different individuals. Some areas may not be typical and some individuals may not be totally cooperative and this may skew the results. A number of varied sites / individuals will dilute any such anomalies.

e They fear it might further reduce the currently dangerously low harrier population. They fear it might set a precedent for other species and other experiments.

6 Where views conflict, evidence is essential to support or discount any claims made. Scientists can produce this evidence in carefully devised, controlled and unbiased experiments. The scientific evidence helps decisions to be made that are more likely to have the desired effect.

8.1

1 genotype
2 mutation
3 phenotype
4 nucleotides/bases
5 polypeptides
6 locus
7 homozygous
8 heterozygous
9 recessive
10 co-dominant

8.2

1 Let allele for Huntington's disease = H

Let allele for normal condition = h

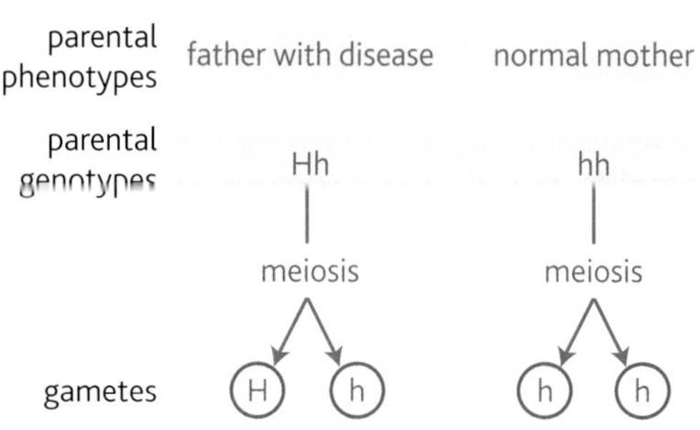

Offspring:

	Father's gametes	
Mother's gametes	**(H)**	**(h)**
(h)	Hh	hh
(h)	Hh	hh

Half (50%) of offspring will have Huntington's disease (Hh).

Half (50%) of offspring will be normal (hh).

2 a Let allele for black coat = B

Let allele for red coat = b

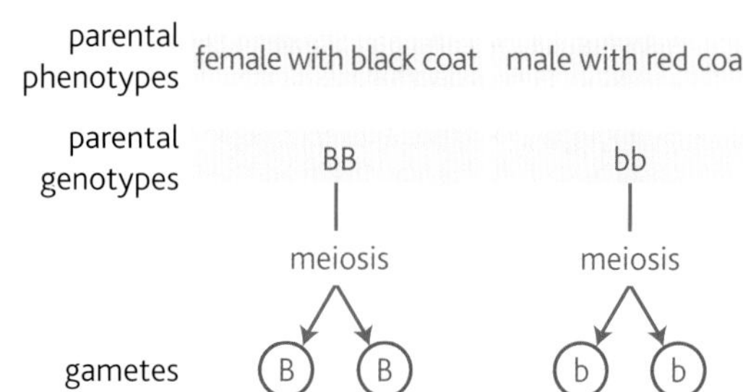

Offspring:

	Male gametes	
Female gametes	**(b)**	**(b)**
(B)	Bb	Bb
(B)	Bb	Bb

All (100%) offspring will have black coats (Bb).

b Let allele for black coat = B

Let allele for red coat = b

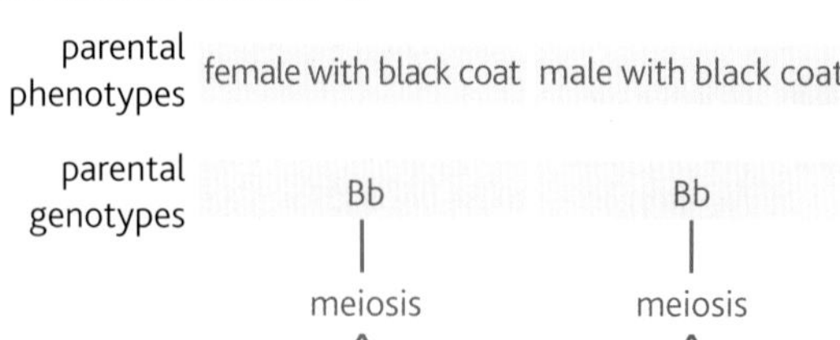

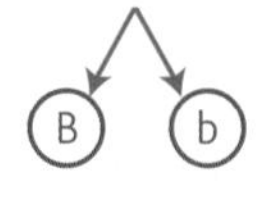

Offspring:

	Male gametes	
Female gametes	(B)	(b)
(B)	BB	Bb
(b)	Bb	bb

3 offspring (75%) with black coat (BB, Bb and Bb).

1 offspring (25%) with red coat (bb).

Probability of offspring having red coat = 1 in 4 (25%/0.25)

Determining genotypes

1 Let allele for green pods = G

Let allele for yellow pods = g

a **Where a plant is homozygous dominant:**

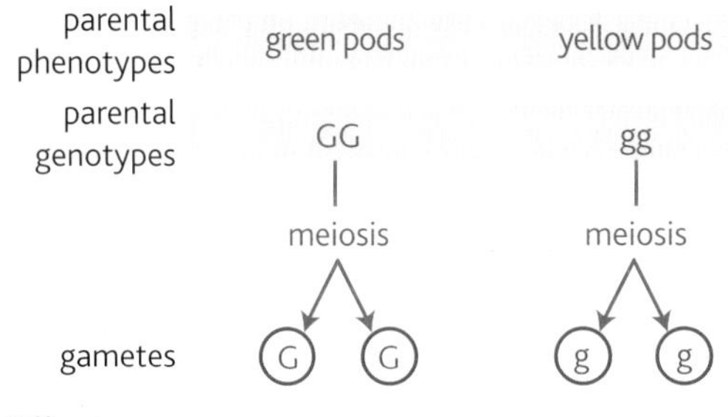

Offspring:

	♂ Gametes	
♀ **Gametes**	(G)	(G)
(g)	Gg	Gg
(g)	Gg	Gg

All (100%) green pods (Gg).

b **Where plant is heterozygous:**

parental phenotypes: green pods, yellow pods

parental genotypes: Gg, gg

meiosis, meiosis

gametes: (G) (g), (g) (g)

Offspring:

	♂ Gametes	
♀ **Gametes**	(G)	(g)
(g)	Gg	gg
(g)	Gg	gg

Half (50%) green pods (Gg).

Half (50%) yellow pods (gg).

2 a homozygous dominant (GG)

b We cannot be absolutely certain because if the unknown genotype were heterozygous (Gg) the gametes produced would contain alleles of two types: either dominant (G) or recessive (g). It is a matter of chance which of these gametes fuses with those from our recessive parent – all these gametes have a recessive allele (g). It is just possible that, in every case, it is the gametes with the dominant allele that fuse and so all the offspring show the dominant character. Provided the sample of offspring is large enough, however, we can be reasonably sure that the unknown genotype is homozygous dominant.

3 a heterozygous (Gg)

b We can be certain because 7 of the offspring display the recessive character (in our case yellow pods). These plants are homozygous recessive and must have obtained one recessive allele from each parent. Our unknown parental genotype must therefore have a recessive allele and be heterozygous (in our case Gg). It is theoretically possible that the plants with yellow pods were due to a mutation but this is most unlikely. The unexpectedly low number of plants with yellow pods is the result of random fusion of the gametes.

8.3

1 E = XX

F = XY

2 A = not colour blind/normal vision

B = not colour blind/normal vision

D = colour blind

3 G = X^RX^r

H = X^RY

I = X^RX^R

J = X^rY

4 0% – because sons inherit their X chromosome from their mother and she has only alleles for normal vision (X^R).

5 By mutation (of the R allele).

A right royal disease

1 Because the ancestors from whom they are descended (Edward VII and Victoria) did not have, or carry, alleles for haemophilia.

2 a The disease of haemophilia only occurs in males and not females.

b Parents without the disease are shown to have children with the disease. Alexandria and Tsar Nicholas II do not have the disease but their son Tsarevitch Alexis does. (Note: There are many other examples.)

3 a X^HX^H

b X^hY

c X^HX^h

4 Anastasia could have either genotype X^HX^h or X^HX^H, depending on whether she inherited an X^H or an X^h from her mother Alexandra. Waldemar's genotype must be X^hY. Therefore:

a Sons would inherit a Y from Waldemar and either an X^H or X^h from Anastasia (mother). Therefore the possible genotypes are X^HY or X^hY.

b Daughters must inherit X^h from Waldemar (father) and either X^H or X^h from Anastasia (mother). Therefore the possible genotypes are X^hX^h or X^HX^h.

8.4

1. C^AC^A C^AC^{Ch} C^AC^H C^AC^a
2. The man is not the father.

   Reasons – child has blood group AB and therefore has alleles I^AI^B. The mother is blood group A and therefore either I^AI^O or I^AI^A. In either case she could have provided the I^A alleles to the child but not the I^B allele. The I^B allele must have come from the real father. The supposed father is blood group O and therefore has alleles I^OI^O. He cannot provide an I^B allele and so cannot be the father.
3.

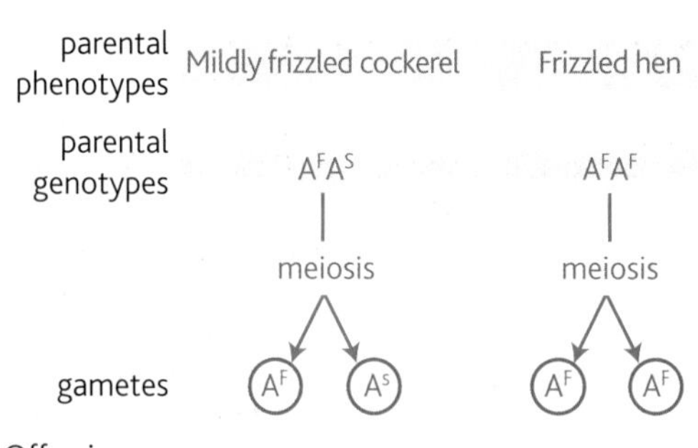

Offspring:

	Cockerel gametes	
Hen gametes	A^F	A^S
A^F	A^FA^F	A^FA^S
A^F	A^FA^F	A^FA^S

Half (50%) frizzled fowl (A^FA^F)
Half (50%) mildly frizzled fowl (A^FA^S)

Coat of many colours

1. The gene for coat colour is on the X chromosome and is therefore sex-linked.

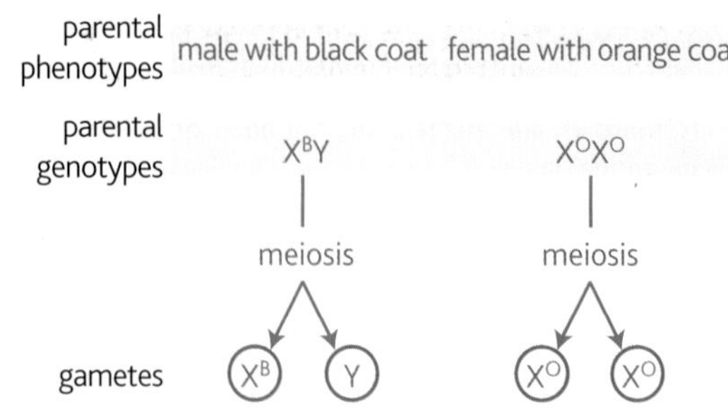

Offspring:

	Male gametes	
Female gametes	X^B	Y
X^O	X^BX^O	X^OY
X^O	X^BX^O	X^OY

Half (50%) tortoiseshell females (X^BX^O)
Half (50%) orange (ginger) males (X^OY)

2. A tortoiseshell coat requires both a B and an O allele. These alleles are on the X chromosome. Males (XY) have only one X chromosome and so cannot have both B and O alleles.
3. Kittens develop inside their mother and so are kept warm / at a uniform temperature.

   As the kitten's coat is light-coloured, tyrosinase must have been denatured / inactivated at this warm temperature.

   After birth, a kitten is exposed to cooler environmental temperatures and its extremities (ears, face, feet and tail) will be the coolest as they are furthest from the main body, where heat is generated and have a large surface area to volume ratio.

   Cooler temperature means tyrosine is activated / not denatured.

   Tyrosinase therefore catalyses the production of dark pigment in these areas.

8.5

1. a Gene pool is all the alleles of all the genes (or one particular gene) of all the individuals in a population at any one time.

   b Allelic frequency is the number of times an allele occurs within the gene pool.
2. The proportion of dominant and recessive alleles of any gene in the population remains the same from one generation to the next.
3. No mutations arise.

   The population is isolated / no flow of alleles into, or out of, the population.

   No natural selection occurs / all alleles are equally advantageous.

   The population is large.

   Mating within the population is random.
4. $p + q = 1.0$ and $p = 0.942$

   Therefore $q = 1.0 - 0.942 = 0.058$

   Frequency of the heterozygous genotype = $2pq$

   $= 2 \times 0.942 \times 0.058$

   $= 0.109$

   As a percentage = $0.109 \times 100 = 10.9\%$.

Not as black and white as it seems

1. It is not sex-linked because the number of males and females of each wing colour are approximately equal – the small difference is due to statistical error.
2. 0.254 (25.4%)

   Number of moths with dark-coloured wings (having two recessive alleles) = 562. Total sample = 1653 + 562 = 2215.

   Proportion with two recessive alleles = 562 ÷ 2215 = 0.254.
3. a 0.254 of the sample population has two recessive alleles

   Therefore $q^2 = 0.254$ and $q = \sqrt{0.254} = 0.504$.

   b $q = 0.504$ and $p + q = 1.0$.

   Therefore $p = 1.0 - 0.504 = 0.496$.

   c Frequency of heterozygotes = $2pq = 2 \times 0.496 \times 0.504 = 0.5$

   Therefore % of heterozygotes = 50%
4. Capture a sample of moths and mark them in some way. Release them back into the population.

Some time later randomly recapture a given number of moths.
Record the number of marked and unmarked moths in this second sample.
Calculate the size of the population as follows:

$$\frac{\text{total number of moths in first sample} \times \text{total number of moths in second sample}}{\text{number of marked moths recaptured}}$$

8.6

1 Selection is the process by which organisms that are better adapted to their environment survive and breed, while those less well adapted fail to do so.

2

Directional selection	Stabilising selection
• Favours/selects phenotypes at one extreme of a population	• Favours/selects phenotypes around the mean of a population
• Changes the characteristics of a population	• Preserves the characteristics of a population
• Distribution curve remains the same shape but the mean shifts to the left or right	• Distribution curve becomes narrower and higher but the mean does not change

3 Directional selection – because birds to one side of the mean (heavier birds) were being selected for, while those to the other side of the mean (lighter birds) were being selected against. The population's characteristics are being changed, not preserved.

Early selection

1 a Few infants are born with a low birth mass (below 2 kg) and few infants are born with a high birth mass (above 4 kg). The majority of infants have a birth mass between 2 kg and 4 kg with the highest percentage having a birth mass of 3.25 kg.

b As infant birth mass increases up to 3.25 kg, the infant mortality rate decreases. At infant birth masses above 3.25 kg, the infant mortality rate increases again.

2 Stabilising selection – because the mortality rate is greater at the two extremes. The infants with the highest and lowest birth masses are more likely to die (are being selected against) while those around the mean are less likely to die (are being selected for / favoured). The population's characteristics are being preserved rather than changed.

They must be cuckoo!

1 Removing cuckoo eggs means there will be more food for the magpie's own chicks. These chicks have a greater probability of being successfully raised to adulthood.

2 Alleles for this type of behaviour are obviously present in the adult birds. There is a high probability that some of the chicks will inherit these alleles. Removing cuckoo eggs increases the probability of more of these chicks surviving to breed and therefore passing on the alleles for this behaviour to subsequent generations.

3 Displaying this behaviour has previously been of no advantage to magpies and so no selection for this behaviour has taken place. Although cuckoos have now arrived, it will take many generations for selection to operate and for allele frequencies to change.

4 Directional selection – because the population's characteristics are being changed, not preserved.

8.7

1 A species is a group of individuals that share similar genes and are capable of breeding with one another to produce fertile offspring. In other words they belong to the same gene pool.

2 Speciation is the evolution of new species from existing species.

3 Geographical isolation occurs when a physical barrier, such as mountains or oceans, prevents two populations from breeding with one another.

4 Geographically isolated populations may experience different environmental conditions. In each population, phenotypes that are best suited to the particular environmental conditions are selected. The composition of the alleles in each gene pool therefore changes as they pass to subsequent generations. The composition of the gene pool of each population becomes increasingly different over time. Being geographically isolated, individuals of each population cannot breed with one another and so the two gene pools remain separate and different.

9.1

1 (Negative chemo-) taxis – wastes are often removed from an organism because they are harmful. Moving away prevents the waste harming the organism and so increases its chance of survival.

2 (Positive chemo-) taxis – increases the chances of sperm cells fertilising the egg cells of other mosses and so helps to produce more moss plants / future generations. Cross-fertilisation increases genetic variability, making species better able to adapt to future environmental changes.

3 (Negative geo-) tropism – takes the seedlings above the ground and into the light, where they can photosynthesise. More photosynthesis means more carbohydrate and so a better chance of survival.

9.2

1 brain / spinal cord
2 brain / spinal cord
3 motor
4 sensory
5 involuntary
6 stimulus
7 (temperature) receptor
8 sensory
9 intermediate
10 motor
11 effectors

9.3

1 Autonomic nervous system – controls the involuntary activities of internal muscles and glands.

2 Sympathetic nervous system stimulates effectors and so speeds up an activity; prepares for stressful situations, e.g. the fight or flight response.

Parasympathetic nervous system inhibits effectors and slows down an activity; controls activities under resting conditions, conserving energy and replenishing the body's reserves.

3 Blood pressure remains high because the parasympathetic system is unable to transmit nerve impulses to the SA node, which decreases heart rate and so lowers blood pressure.

4 a Heart rate remains as it was before taking exercise – after exercise, blood pressure increases and CO_2 concentration of blood rises (causing blood pH to be lowered). The changes are detected by pressure and chemical receptors in the wall of the carotid arteries. As the nerve from here to the medulla oblongata is cut, no nerve impulse can be sent to the centres that control heart rate.

b Blood CO_2 concentration increases as a result of increased respiration during exercise.

9.4

1 Stretch-mediated sodium channel – a special type of sodium channel that changes its permeability to sodium when it changes shape / is stretched.

2 pressure on Pacinian corpuscle → corpuscle changes shape → stretches membrane of neurone → widens stretch-mediated sodium ion channels → allows sodium ions into neurone → changes potential of (depolarises) membrane → produces generator potential

3 Only rod cells are stimulated by low-intensity (dim) light. Rod cells cannot distinguish between different wavelengths / colours of light, therefore the object is perceived only in a mixture of black and white, i.e. grey.

4 Light reaching Earth from a star is of low intensity. Looking directly at a star, light is focused on to the fovea, where there are only cone cells. Cone cells respond only to high light intensity so they are not stimulated by the low light intensity from the star and it cannot be seen. Looking to one side of the star means that light from the star is focused towards the outer regions of the retina, where there are mostly rod cells. These are stimulated by low light intensity and therefore the star is seen.

10.1

1 Hormone response is slow, widespread and long-lasting.

Nervous response is rapid, localised and short-lived.

2 Histamine and prostaglandins – both cause dilation of small arteries and arterioles and increased permeability of capillaries.

3 Response ensures that roots grow downwards into the soil, thus anchoring the plant firmly and bringing them closer to water (needed for photosynthesis).

4 Animal hormones are made in particular organs and affect other organs some distance away.

Plant growth factors are made by cells located throughout the plant and have localised effects.

Discovering the role of IAA in tropisms

1 experiment 1

2 As mica conducts electricity it will not prevent electrical messages passing from the shoot tip but it will prevent chemical messages passing. As there is no response, the message must be chemical and must pass down the shaded side.

3 Displacement of the tip means that the chemical initially only moves down the side of the shoot that is in contact with the tip. This side grows more rapidly, causing bending away from that side.

4 It prevents chemicals / IAA, but not light, passing from one side to the other.

5 Results support the hypothesis that IAA is transported from the lighter side to the darker side of the shoot.

Experiment 8 shows that the total IAA produced and collected is the same whether the shoot is in the light or the dark. This discounts the theory that light destroys IAA or inhibits its production.

Experiment 9 shows that the amount of IAA produced at either side of the tip is the same. The glass plate prevents any sideways transfer.

Experiment 10 shows that the IAA is transferred from the light to the dark side of the shoot soon after it is produced because more than twice as much IAA is found on the dark side of the shoot than on the light side.

10.2

1 (nerve) impulses / action potentials

2 nucleus

3 rough endoplasmic reticulum

4 dendrites

5 Schwann cells

6 insulation

7 myelin

8 motor

9 sensory

10 intermediate

10.3

1 Active transport of sodium ions out of the axon by sodium–potassium pumps is faster than active transport of potassium ions into the axon. Potassium ions diffuse out of the axon but few, if any, sodium ions diffuse into the axon because the sodium 'gates' are closed. Overall, there are more positive ions outside than inside and therefore the outside is positive relative to the inside.

2 A = closed B = open C = closed
D = closed E = closed F = open

Measuring action potentials

1 sodium and potassium ions

2 At resting potential (0.5 ms) there is a positive charge on the outside of the membrane and a negative charge inside, due to the high concentration of sodium ions outside the membrane. The energy of the stimulus causes the sodium voltage-gated channels in the axon membrane to open and therefore sodium ions diffuse in through the channels, along their electrochemical gradient. Being positively charged, they begin a reversal in the potential difference across the membrane. As sodium ions enter, so more sodium ion channels open, causing an even greater influx of sodium ions and an even greater reversal of potential difference: from –70 mv up to +40 mv at 2.0 ms.

3 Two action potentials take place in 10 ms.

Each action potential takes 10 ÷ 2 = 5 ms / action potentials are 5 ms apart.

There are 1000 ms in 1 second.

Therefore there are 1000 ÷ 5 = 200 action potentials in 1 second.

10.4

1 a node of Ranvier

b Because the remainder of the axon is covered by a myelin sheath that prevents ions being exchanged / prevents a potential difference being set up.

c It moves along in a series of jumps from one node of Ranvier to the next.

d saltatory (conduction)

e It is faster than in an unmyelinated axon.

2 It remains the same / does not change.

10.5

1 During the refractory period the sodium voltage-gated channels are closed so no sodium ions can move inwards and no action potential is possible. This means there must be an interval between one impulse and the next.

2 All-or-nothing principle – There is a particular level of stimulus that triggers an action potential. At any level above this threshold, a stimulus will trigger an action potential that is the same regardless of the size of the stimulus (the 'all' part). Below the threshold, no action potential is triggered (the 'nothing' part).

Different axons different speeds

1 The greater the diameter of an axon the faster the speed of conductance. Comparing the data for the two myelinated axons shows that the 20 mm diameter axon conducts at 120 m s^{-1} while the 10 mm diameter axon conducts at only 50 m s^{-1}. Likewise, the data for the two unmyelinated axons show that the 500 mm diameter axon conducts at 25 m s^{-1} while the 1 mm diameter axon conducts at 2 m s^{-1}.

2 In myelinated axons, the myelin acts as an electrical insulator. Action potentials can only form where there is no myelin (at nodes of Ranvier). The action potential therefore jumps from node to node (= saltatory conduction) which makes its conductance faster.

3 Schwann cells

4 The presence of myelin has the greater effect because a myelinated human sensory axon conducts an action potential at twice the speed of the squid giant axon, despite being only 1/50th of its diameter. (Note: Similar comparisons can be made between other types of axon, e.g. squid and human motor axons.)

5 Temperature affects the speed of conductance of action potentials. The higher the temperature, the faster the conductance. The conductance of action potentials in the squid will therefore change as the environmental temperature changes. It will react more slowly at lower temperatures.

10.6

1 It possesses many mitochondria and large amounts of endoplasmic reticulum.

2 It has receptor molecules on its membrane.

3 Neurotransmitter is released from vesicles in the presynaptic neurone into the synaptic cleft when an action potential reaches the synaptic knob. The neurotransmitter diffuses across the synapse to receptor molecules on the postsynaptic neurone to which it binds, thereby setting up a new action potential.

4 Only one end can produce neurotransmitter and so this end alone can create a new action potential in the neurone on the opposite side of the synapse. At the other end there is no neurotransmitter that can be released to pass across the synapse and so no new action potential can be set up.

5 a The relatively quiet background noise of traffic produces a low-level frequency of action potentials in the sensory neurones from the ear. The amount of neurotransmitter released into the synapse is insufficient to exceed the threshold in the postsynaptic neurone and to trigger an action potential and so the noise is 'filtered out' / ignored. Louder noises create a higher frequency and the amount of neurotransmitter released is sufficient to trigger an action potential in the postsynaptic neurone and so there is a response. This is an example of temporal summation. (Note: An explanation in terms of spatial summation is also valid: many sound receptors with a range of thresholds → more receptors respond to the louder noise → more neurotransmitter → response.)

b Reacting to low-level stimuli (background traffic noise) that present little danger can overload the (central) nervous system and so organisms may fail to respond to more important stimuli. High-level stimuli (sound of horn) need a response because they are more likely to represent a danger.

6 As the inside of the membrane is more negative than at resting potential, more sodium ions must enter in order to reach the potential difference of an action potential, i.e. it is more difficult for depolarisation to occur. Stimulation is less likely to reach the threshold level needed for a new action potential.

10.7

1 a sodium ions

b acetylcholine

c ATP

d calcium ions

2 To recycle the choline and ethanoic acid; to prevent acetylcholine from continuously generating a new action potential in the postsynaptic neurone.

Effects of drugs on synapses

1 They will reduce pain.

2 They act like endorphins by binding to the receptors and therefore preventing action potentials being created in the neurones of the pain pathways.

3 Prozac might prevent the elimination of serotonin from the synaptic cleft (Note: Any biologically accurate answer that results in more serotonin in the synaptic cleft is acceptable.)

4 By increasing the concentration of serotonin in the synaptic cleft, its activity is increased, reducing depression, which is caused by reduced serotonin activity.

5 It will reduce muscle contractions (cause muscles to relax).

6 Valium increases the inhibitory effects of GABA so therefore there are fewer action potentials on the nerve pathways that cause muscles to contract.

7 The molecular structure of Vigabatrin is similar to GABA so it may be a competitive inhibitor (compete) for the active site of the enzyme that breaks down GABA. As less GABA is broken down by the enzyme, more of it is available to inhibit neurone activity. Or Vigabatrin might bind to GABA receptors on the neurone membrane and mimic its action, thereby inhibiting neuronal activity.

11.1

1 Muscles require much energy for contraction. Most of this energy is released during the Krebs cycle and electron transport chain in respiration. Both these take place in mitochondria.

2 A = Z-line B = H-zone C = I-band (isotropic band) D = A-band (anisotropic band).

3 The actin and myosin filaments lie side by side in a myofibril and overlap at the edges where they meet. Where they overlap, both filaments can be seen. Where they do not overlap, we see one or other filament only.

4 Slow-twitch fibres contract more slowly and provide less powerful contractions over a longer period.
Fast-twitch fibres contract more rapidly and produce powerful contractions but only for a short duration.

5 Slow-twitch fibres have myoglobin to store oxygen, much glycogen to provide a source of metabolic energy, a rich supply of blood vessels to deliver glucose and oxygen, and numerous mitochondria to produce ATP.

Fast-twitch fibres have thicker and more numerous myosin filaments, a high concentration of enzymes involved in anaerobic respiration and a store of phosphocreatine to rapidly generate ATP from ADP in anaerobic conditions.

11.2

1 Myosin is made of two proteins. The fibrous protein is long and thin in shape, which enables it to combine with others to form a long thick filament along which the actin filament can move. The globular protein forms two bulbous structures (the head) at the end of a filament (the tail). This shape allows it to exactly fit recesses in the actin molecule, to which it can become attached. Its shape also means it can be moved at an angle. This allows it to change its angle when attached to actin and so move it along, causing the muscle to contract.

2 Phosphocreatine stores the phosphate that is used to generate ATP from ADP in anaerobic conditions. A sprinter's muscles often work so strenuously that the oxygen supply cannot meet the demand. The supply of ATP from mitochondria during aerobic respiration therefore ceases. Sprinters with the most phosphocreatine have an advantage because ATP can be supplied to their muscles for longer, and so they perform better.

3 A single ATP molecule is enough to move an actin filament a distance of 40 nm.

Total distance moved by actin filament = 0.8 μm (= 800 nm).

Number of ATP molecules required = 800 ÷ 40 = 20.

4 One role of ATP in muscle contraction is to attach to the myosin heads, thereby causing them to detach from the actin filament and making the muscle relax. As no ATP is produced after death, there is none to attach to the myosin, which therefore remains attached to actin, leaving the muscle in a contracted state, i.e. *rigor mortis*.

12.1

1 Homeostasis is the maintenance of a constant internal environment in organisms.

2 Maintaining a constant temperature is important because enzymes function within a narrow range of temperatures. Fluctuations from the optimum temperature mean enzymes function less efficiently. If the variation is extreme, the enzyme may be denatured and cease to function altogether. A constant temperature means that reactions occur at a predictable and constant rate.

3 Maintaining a constant blood glucose concentration is important in ensuring a constant water potential. Changes to the water potential of the blood and tissue fluids may cause cells to shrink and expand (even to bursting point), due to water leaving or entering by osmosis. In both instances the cells cannot operate normally. A constant blood glucose concentration also ensures a reliable source of glucose for respiration by cells.

12.2

1 conduction, convection, radiation

2 sweat glands, hair (erector) muscle, fat tissue, arterioles (shunt vessels)

3 Food is a source of heat produced by its metabolism in the body. A large, compact mammal will have a smaller surface area to volume ratio than a smaller, less compact one. Because heat is lost from the 'surface' and produced in the 'volume', the larger mammal will both produce more heat and lose less heat per gram of body mass compared to the smaller mammal. It will therefore need less food in order to maintain its body temperature.

4 Explanation – On a cold day, the arterioles in the skin are made smaller in diameter (vasoconstriction) to help conserve heat, thus reducing the flow of blood to the body surface. This lack of surface blood makes the skin appear pale. On a hot day, vasodilation occurs and so more blood flows near to the body surface to help lose heat and the skin appears redder (less pale).

Significance – Vasoconstriction conserves body heat, while vasodilation increases heat loss – they are therefore mechanisms of thermoregulation.

Comparing heat gain and loss in ectotherms and endotherms

1 It allows accurate comparisons to be made even though the animals have different body masses. An increase in body size or body mass means there is increased heat generation.

2 a Both increase proportionally up to 25 °C. Above 25 °C, heat generation increases more rapidly (gradient / slope of line increases), whereas evaporative heat loss increases at the same rate (gradient / slope of line remains the same).

b In a mammal, the relationship is the inverse / opposite, i.e. as evaporative heat loss increases, heat generation decreases.

3 Above 25 °C, the metabolic heat generation in reptiles becomes much more rapid. They therefore generate heat faster than they can lose it. As a result, their body temperature increases and enzymes may be denatured, leading to death. As reptiles have no physiological means of cooling, they must seek shade in order to reduce their body temperature.

4 Sweating or panting increases.

The role of the hypothalamus in temperature regulation

1 The temperature of the hypothalamus is inversely proportional to both the metabolic rate and the body temperature. As the temperature of the hypothalamus increases, the other two factors decrease. As the temperature of the hypothalamus decreases, the other two factors increase.

2 As the hypothalamus is cooled, it causes a rise in metabolic activity in order to increase heat production. As a result, the body temperature rises. As the hypothalamus is warmed, it causes a fall in metabolic activity in order to reduce heat production. As a result, the body temperature falls.

12.3

1 respiration
2 brain
3 osmotic / water potential
4 carbohydrate
5 glycogen
6 muscles
7 amino acids
8 gluconeogenesis
9 glycogen
10 respiration
11 islets of Langerhans
12 insulin
13 glucagon
14 adrenaline

12.4

1 Type I is caused by an inability to produce insulin.
Type II is caused by receptors on body cells losing their responsiveness to insulin.

2 Type I is controlled by the injection of insulin.
Type II is controlled by regulating the intake of carbohydrate in the diet and matching this to the amount of exercise taken.

3 Diabetes is a condition in which insulin is not produced by the pancreas. This leads to fluctuations in the blood glucose level. If the level is below normal, there may be insufficient glucose for the release of energy by cells during respiration. Muscle and brain cells in particular may therefore be less active, leading to tiredness.

4 Match your carbohydrate intake to the amount of exercise that you take. Avoid becoming overweight by not consuming excessive quantities of carbohydrate and by taking regular exercise.

Effects of diabetes on substance levels in the blood

1 adrenaline

2 The rise in insulin level is both greater and more rapid in group Y than in group X.

3 Glucose is removed from blood by cells using it during respiration.

4 Glucose level rises at first because the glucose that is drunk is absorbed into the blood (glucose line on the graph rises). This rise in blood glucose causes insulin to be secreted from cells (β cells) in the pancreas (insulin line rises steeply). Insulin causes increased uptake of glucose into liver and muscle cells, activates enzymes that convert glucose into glycogen and fat, and increases cellular respiration. The effect of all these actions is to reduce glucose levels (glucose line falls from 2.5 hours onwards). As the glucose level rises after 1 hour, so the glucagon level falls. The reduction in glucagon level decreases glucose production from other sources (glycogen, amino acids and glycerol) and so also helps to reduce blood glucose levels. As the blood glucose level falls (after 2.5 hours) so the glucagon level increases to help maintain the blood glucose at its normal level.

5 Group X has diabetes and therefore the glucose intake does not stimulate insulin production (insulin level shown on the graph is low). The glucose level in the blood therefore continues to rise (glucose line rises steeply) as there is no insulin to reduce its level. Blood glucose level remains high, falling only slightly as it is respired by cells.

6 As it is respired by cells, the glucose level will decrease steadily until it falls below the normal level.

13.1

1 Positive feedback occurs when the feedback causes the corrective measures to remain turned on. In doing so, it causes the system to deviate even more from the original (normal) level.
Negative feedback occurs when the feedback causes the corrective measures to be turned off. In doing so, it returns the system to its original (normal) level.

2 If the information is not fed back once an effector has corrected any deviation and returned the system to the set point, the receptor will continue to stimulate the effector and an over-correction will lead to a deviation in the opposite direction from the original one.

3 It gives a greater degree of homeostatic control.

Control of blood water potential

1 As sweating involves a loss of water from the blood, its water potential will decrease (be lower or more negative).

2 a osmotic cells (in the hypothalamus)
b kidney

3 Being a hormone, it is transported in the blood plasma.

4 Absorption (taking in or consumption or drinking) of water because water has been lost during sweating. As the water potential of the blood returns to normal, the lost water must have been replaced. However, the kidney only excretes less water, it does not replace it. Therefore process X must be the way in which water is replaced.

5 negative feedback

13.2

1 a pituitary gland
b corpus luteum in the ovary
c follicle in the ovary

2 There are many examples, the shortest of which include:
more oestrogen → more inhibition of FSH production → less FSH→ less stimulation of oestrogen production → less oestrogen
more progesterone → more inhibition of luteinising hormone (LH) production → less LH → less stimulation of progesterone production → less progesterone

3 Oestrogen is produced in developing follicles. Menopausal women have few follicles left so produce less oestrogen. Oestrogen inhibits the production of FSH so the reduction in oestrogen level means that there is less inhibition of FSH. Therefore more FSH is produced, i.e. FSH levels rise.

4 High progesterone levels inhibit the production of FSH and LH. When progesterone is withdrawn from the diet, its level falls and so the production of FSH and LH resumes. FSH causes follicles to develop in the ovary and LH causes eggs to be released (ovulation).

The oestrous cycle in pigs

1 Day 9 – because ovulation is caused by a high concentration of LH and its level is highest (it peaks on day 9).

2 Days 0–5: oestrogen levels are rising but relatively low, inhibiting the production of FSH (negative feedback) and so FSH levels are low.
Days 5–6: oestrogen levels rise to a critical point, stimulating the production of FSH (positive feedback) and so FSH levels rise.
Days 9–12: oestrogen levels fall below the same critical level, inhibiting FSH production again (negative feedback) and so FSH levels also fall.

3 Progesterone inhibits the production of FSH, and FSH causes follicles in the ovary to develop and mature. If FSH levels are low, follicles will not develop and so no mature eggs will be available for release.

14.1

1 a Universal – because it is the same in all organisms.
b Degenerate – because most amino acids have more than one codon.
c Non-overlapping – because each base in the sequence is read only once.

2 Any 3 from: RNA is smaller than DNA / RNA is usually a single strand and DNA a double helix / the sugar in RNA is ribose while the sugar in DNA is deoxyribose / in RNA the base uracil replaces the base thymine found in DNA

3 A codon is the triplet of bases on messenger RNA that codes for an amino acid.
An anticodon is the triplet of bases on a transfer RNA molecule that is complementary to the codon.

Differences between DNA, mRNA and tRNA

1 a The amount of DNA in a gamete is half that in a body cell.
b It allows gametes to fuse during sexual reproduction without doubling the total amount of DNA at each generation. In so doing it increases genetic variety by allowing the genetic information of two parents to be combined in the offspring.

2 a DNA needs to be stable to enable it to be passed from generation to generation unchanged and thereby allow offspring to be very similar to their parents. Any change to the DNA is a mutation and is normally harmful.
b mRNA is produced to help manufacture a protein, e.g. an enzyme. It would be wasteful to produce the protein continuously when it is only needed periodically. mRNA therefore breaks down once it has been used and is produced again only when next required.

14.2

1 The enzyme RNA polymerase moves along the template DNA strand, causing the bases on this strand to join with the individual complementary nucleotides from the pool that is present in the nucleus. The RNA polymerase adds the nucleotides one at a time, to build a strand of pre-mRNA until it reaches a particular sequence of bases on the DNA that it recognises as a 'stop' code.

2 DNA helicase – This acts on a specific region of the DNA molecule to break the hydrogen bonds between the bases, causing the two strands to separate and expose the nucleotide bases in that region.

3 Splicing is necessary because pre-mRNA has nucleotide sequences derived from introns in DNA. These introns are non-functional and, if left on the mRNA, would lead to the production of non-functional polypeptides or no polypeptides at all. Splicing removes these non-functional introns from pre-mRNA.

4 a UACGUUCAGGUC
b 4 amino acids (1 amino acid is coded for by 3 bases so 12 bases code for 4 amino acids)

5 Some of the base pairs in the genes are introns (non-functional DNA). These introns are spliced from pre-mRNA so the resulting mRNA has fewer nucleotides.

14.3

1 ribosome

2 a UAG on tRNA
b TAG on DNA

3 A tRNA molecule attaches an amino acid at one end and has a sequence of 3 bases, called an anticodon, at the other end. The tRNA molecule is transferred to a ribosome on an mRNA molecule. The anticodon on tRNA pairs with the complementary codon sequence on mRNA. Further tRNA molecules, with amino acids attached, line up along the mRNA in the sequence determined by the mRNA bases. The amino acids are joined by peptide bonds. Therefore the tRNA helps to ensure the correct sequence of amino acids in the polypeptide.

4 One of the codons is a stop codon that indicates the end of polypeptide synthesis. Stop codons do not code for any amino acid so there is one less amino acid than there are codons.

Interpreting the genetic code

1 a GUA
b TCA

2 leucine

3 glutamic acid-alanine-proline-tyrosine-alanine

4 GTAGGACTGGAT

Cracking the code

1 The DNA of the cell would produce its own mRNA so it would be impossible to determine which of the many polypeptides produced was due to the synthetic DNA.

2 The codon UUU – because the very radioactive polypeptide (39 800 counts min^{-1}) was only produced from the mixture containing poly U. This polypeptide must be made up of phenylalanine because this is the only radioactive amino acid present. As the synthetic mRNA contains only the base sequence UUUUUUU, etc., one codon for phenylalanine must be UUU.

3 As a control experiment to show that the radioactivity was due to the labelled phenylalanine rather than some other factor, e.g. background radiation.

4 It may not be possible to say whether the mRNA sequence starts with U and therefore reads UGU GUG UGU, etc. or starts with G and reads GUG UGU GUG. Equally it may not be possible to say whether the polypeptide sequence begins with cysteine or valine. It may therefore be impossible to relate a particular codon to a particular amino acid.

5 It means that most amino acids have more than one codon.

6 Because they do not code for any amino acid – they are 'stop' codons that mark the end of a polypeptide.

14.4

1. A deletion because the fifth nucleotide (A) has been lost. The sequence prior to and after this is the same. (Note: The last base in the mutant version was previously the 13th in the sequence and therefore not shown in the normal version.)
2. In a deletion, all codons after the deletion are affected (frame-shift). Therefore most amino acids coded for by these codons will be different and the polypeptide will be significantly affected. In a substitution, only a single codon, and therefore a single amino acid, will be affected. The effect on the polypeptide is likely to be less severe.
3. The mutation may result from the substitution of one base in the mRNA with another. Although the codon affected will be different, as the genetic code is degenerate, the changed codon may still code for the same amino acid. The polypeptide will be unchanged and there will be no effect.
4. These errors may be inherited and may therefore have a permanent affect on the whole organism. Errors in transcription usually affect only specific cells, are temporary and are not inherited . They are therefore less damaging.
5. Proto-oncogenes, which stimulate cell division, and tumour suppressor genes, which inhibit cell division.

Mutagenic agents

1. The codons in mRNA will be CAU AAA UAA (Note: In mRNA, guanine is coded for by cytosine in DNA, adenine by thymine and uracil by adenine as usual, but after the change, cytosine becomes uracil in DNA and this codes for adenine in mRNA.)
2. substitution gene mutation
3. The active site of DNA polymerase can no longer fit the DNA molecule because the shapes of some DNA bases have been altered by X-rays.
4. The replication of DNA requires DNA polymerase and so the process cannot continue.
5. Public opinion, special interest groups such as the owners of shops selling or using sunbeds, manufacturers, consumers, professional bodies (e.g. members of the medical profession), the media and other scientists.

15.1

1. Totipotent cells are cells with the ability to develop into any other cell of the organism.
2. In animals, only a few cells are totipotent. In humans these are known as stem cells and are found in the embryo, the inner lining of the intestine, skin and bone marrow. In plants, many of the cells throughout a plant are totipotent.
3. In skin cells, the gene that codes for keratin is expressed, but not the gene for myosin. The genetic code for keratin is translated into the protein keratin, which the cell therefore produces, but the genetic code for myosin is not translated. In muscle cells, the gene for myosin is expressed but not the gene for keratin. In the same way, the genetic code for myosin rather than keratin is translated and so only myosin is produced.

Growth of plant tissue cultures

1. differentiation
2. IAA and 2,4-D
3. In test tube 1 the low concentration of IAA produces moderate shoot development but when a high concentration of cytokinin is added (test tube 3) the presence of cytokinin influences the effects of the IAA by reducing shoot development to a 'little'.

Human embryonic stem cells and the treatment of disease

1. Any properly structured and evaluated accounts that make scientifically accurate points in a reasoned fashion are acceptable, e.g.:

For	Against
• Huge potential to cure many debilitating diseases	• It is wrong to use humans, including potential humans, as a means to an end
• Wrong to allow suffering when it can be relieved	• Embryos are human, they have human genes, and deserve the same respect and treatment as adult humans
• Embryos are created for other purposes (IVF) so why not stem cells	• It is the 'slippery slope' to the use of older embryos and fetuses for research
• Embryos of less than 14 days are not recognisably human and so do not command the same respect as adults or fetuses	• It could lead to research and development of human cloning and, although banned in the UK, the information gained could be used elsewhere
• There is no risk of research escalating or including fetuses because current legislation prevents this	• It undermines respect for life
• Adult stem cells are not as suitable as embryonic stem cells and it may be many years before they are, in the meantime many people suffer unnecessarily	• Adult stem cells are an available alternative and energies should be directed towards developing these

15.2

1. Transcriptional factors stimulate transcription of a gene.
2. Oestrogen diffuses through the phospholipid portion of a cell-surface membrane into the cytoplasm of a cell, where it combines with a site on a receptor portion of the transcriptional factor. Oestrogen changes the shape of the receptor molecule, releasing an inhibitor molecule from the DNA binding site on the transcription factor.

The transcriptional factor now enters the nucleus through a nuclear pore and combines with DNA, stimulating transcription of the gene that makes up that portion of DNA, i.e. it stimulates gene expression.

3 The other strand would have complementary bases (i.e. GCUA instead of CGAU respectively). It is unlikely that these opposite base pairings would complement a sequence on the mRNA. The siRNA, with enzyme attached, would therefore not bind to the mRNA and so would be unaffected.

Cancer – the 'two hit' hypothesis

1 A person with a family history of cancer may already have one mutated allele for the inactivation of the tumour suppressor gene. As X-rays increase mutation rates they might advance the likelihood of cancer in these patients. Patients with no family history of cancer are less at risk because they are less likely to have inherited a mutant allele.

2 The proto-oncogene mutant allele might be dominant whereas the tumour suppressor mutant allele might be recessive. If so, it requires just one dominant proto-oncogene allele to cause cancer where as it will take two recessive tumour suppressor alleles (homozygous state) to cause cancer.

3 Tumour repressor genes inhibit cell division. Mutated forms of these genes are inactive and so cell division increases and a tumour forms. The introduction of normal tumour repressor genes means that the inhibition of cell division will be resumed and the tumour growth will stop.

4 Oncogenes cause cancer by permanently activating protein receptors on cells and so they stimulate cell division. By destroying these receptors on cancer cells, division will be halted and tumour growth will stop.

Gene expression in haemoglobin

1 It allows the fetus to load its haemoglobin with oxygen from the mother's haemoglobin where the two blood supplies come close to each other (at the placenta).

2 alpha = 50% beta = 20% gamma = 30%

3 The gene for gamma-globulin is expressed less while the gene for beta-globulin is expressed more.

4 Expression of the gene for gamma-globulin is progressively reduced as a result of either preventing transcription, and hence preventing the production of mRNA, or by the breakdown of mRNA before its genetic code can be translated.

5 A possible therapy would be to express (switch on) the gene for gamma-globulin and prevent the expression of (switch off) the gene for beta-globulin. This would result in haemoglobin being of the fetal rather than the adult type.

16.1

1 recombinant
2 reverse transcriptase
3 complementary (cDNA)
4 DNA polymerase
5 restriction endonucleases
6 blunt
7 sticky
8 CTTAAG

16.2

1 A vector transfers genes (DNA) from one organism into another.

2 To show which cells (bacteria) have taken up the plasmid (gene).

3 Results can be obtained more easily and more quickly – because, with antibiotic-resistance markers, the bacterial cells with the required gene are killed, so replica plating is necessary to obtain the cells with the gene. With fluorescent gene markers, the bacterial cells are not killed and so there is no need to carry out replica plating.

4 a B, C, D, J, K and L – because those that did not take up the plasmid will not have taken up the gene for ampicillin resistance and so will be the ones that are killed on the ampicillin plate, i.e. the colonies that have disappeared.

b E, F and I – because those with the plasmid containing gene X will have lost the gene for tetracycline resistance and therefore the colonies will have been killed on the tetracycline plate, i.e. the colonies will have disappeared.

16.3

1 Primers are short pieces of DNA that have a set of bases complementary to those at the end of the DNA fragment to be copied.

2 Primers attach to the end of a DNA strand that is to be copied and provide the starting sequences for DNA polymerase to begin DNA cloning. DNA polymerase can only attach nucleotides to the end of an existing chain. They also prevent the two separate strands from rejoining.

3 Because the sequences at the opposite ends of the two strands of DNA are different.

4 hydrogen bonds

5 Biological contaminants may contain DNA and this DNA would also be copied.

16.4

1 The tomatoes do not soften when they ripen and so they can be harvested, transported and stored more easily and without damage and yet the flavour is unimpaired.

2 Advantage – The crop yield is greater and so food prices are not as high. This is because the herbicide kills only the weeds that are competing for light, water and minerals.

Disadvantage – The herbicide (or its breakdown products) might accumulate further up the food chain and might be toxic to other organisms / The herbicide-resistant gene might pass to other plants which will then be unaffected by the herbicide, rendering it useless.

3 Insulin produced by recombinant DNA technology is identical to human insulin and so has no side-effects / it does not induce an immune response / there is no need to slaughter animals to obtain it / there is less risk of transferring infection or disease because there is no donor animal.

Evaluation of DNA technology

Arguments should be reasoned, logical and based on sound science, and should use specific examples rather than vague references.

1 Whichever aspects are chosen the beneficial aspects to humans should be clear, e.g.:

Genetically modified crops that can be grown in extreme conditions – greater productivity; more food; less poverty and hunger in some human populations.

2 Arguments must relate directly to the aspects chosen in question 1 and should oppose the use of the technology, e.g.:

Genetically modified crops that can be grown in extreme conditions – risk of damaging the ecological balance; risk of the gene passing to other organisms; dangers from unforeseen by-products of the plants' metabolism; dangers from possible mutations of the genes; the economic consequences for developing countries.

16.5

1 If both parents are heterozygous for the CFTR gene, then each would carry one dominant and one recessive allele for the condition. They would not suffer from CF as they have the dominant allele. If an offspring inherits one recessive allele from each parent, he/she would suffer from CF.

2 Somatic-cell gene therapy targets just the affected tissues, e.g. lung tissue, and the additional gene is not present in sperm or eggs and is therefore not passed on to future generations. As the cells of the lung tissues are continually dying and being replaced, the treatment needs to be repeated periodically – as often as every few days.

3 Using a harmless virus (called an adenovirus) as a vector; wrapping the gene in lipid molecules to enable it to pass through the cell-surface membranes of lung epithelial cells. In both cases the gene preparation is sprayed into the nostrils and drawn into the lungs during breathing.

Effectiveness of gene therapy

1 Because somatic-cell gene therapy requires regular repeat treatments, the secondary response means that the immune response is enhanced on the second and subsequent occasions.

2 Germ-line gene therapy means that all body cells have the additional gene permanently. Therefore the effect lasts for an individual's lifetime and no repeat treatments are required, there is no immune response and no need for a means of delivery.

16.6

1 A DNA probe is a short, single-stranded section of DNA that has some label attached that makes it easily identifiable.

2 It starts the process of DNA synthesis by making the DNA double stranded (DNA polymerase only works on double-stranded DNA). It carries the radioactive label for later identification of the DNA fragment produced.

3 a two

b cytosine

c CACTGTTCAT

Compiling a restriction map

1

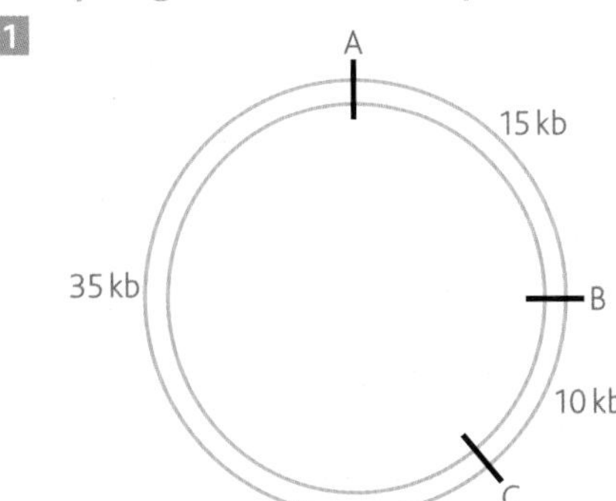

16.7

1 Determine the order of nucleotides on the mutated gene by DNA sequencing – produce a fragment of DNA that has complementary bases to the mutated portion of the gene – label the fragments to form a DNA probe – make multiple copies of the DNA probe using PCR techniques – add the probe to DNA fragments from the individual being tested.

If the donor has a mutant allele, the probe will bind to the complementary bases on the donor DNA. These fragments will now be labelled and can be distinguished from the rest of the DNA.

2 a Tumour suppressor genes inhibit cell division.

b He/she might change their lifestyle to reduce the risk of cancer, e.g. by giving up smoking, losing weight, eating more healthily and avoiding mutagens as far as possible; checking more regularly for early symptoms of cancer; choosing to undergo gene therapy.

Sickle-cell anaemia

1 mRNA has changed from GAG to GUG.

2 The change in an amino acid may affect how it bonds with other amino acids in the same, or other, polypeptide chains. This can alter the tertiary structure of the protein, changing it from a compact globular shape to a more fibrous structure. Some amino acids may bind to amino acids on different haemoglobin molecules rather than those within their own molecule – hence the haemoglobin molecules stick together.

3 Sufferers become tired because the haemoglobin S does not carry oxygen and capillaries become blocked. Cells, especially those making up the muscle and the brain, are unable to produce sufficient energy from respiration due to low oxygen availability.

4 Co-dominance is the expression of both alleles in the phenotype.

5 1 in 4 (25%)

6 a They will be the same.

b The Hardy-Weinberg equation is $p^2 + 2pq + q^2 = 1.0$

H^A frequency = 0.6 = p

H^S frequency = 0.4 = q

$0.6^2 + (2 \times 0.6 \times 0.4) + 0.4^2 = 1.0$

$0.36 + 0.48 + 0.16 = 1.0$

Individuals with $H^AH^A = p^2 = (0.36 \div 1.0 \times 175) = 63$

Individuals with $H^AH^S = 2pq = (0.48 \div 1.0 \times 175) = 84$

Individuals with $H^SH^S = q^2 = (0.16 \div 1.0 \times 175) = 28$.

7 Stabilising selection – because it favours average individuals rather than those at the extremes (i.e. those that are disadvantaged by being either susceptible to malaria or having sickle-cell anaemia).

8 The sickle-cell trait confers some resistance to malaria. Therefore, if individuals who are heterozygous for this condition live in a region where malaria is prevalent, the advantage of not developing malaria may more than offset the disadvantage of having sickle-cell trait. Individuals who live in non-malarial regions remain at a disadvantage because they have the sickle-cell trait.

Implications of genetic screening

1 Any 2 from each list:

For	Against
• Allows parents the choice of whether to have children who may have a genetic disorder	• Cures may be found for genetic diseases so that those who suffer from them can lead normal lives
• Allows patients with oncogenes to change their lifestyle or choose a particular treatment in order to increase their chances of living longer	• It might lead to the elimination of some genes and so reduce the genetic diversity that is essential to evolution
• Can prevent the birth of children with sometimes painful and debilitating diseases	• Difficult to define what counts as a disease – could lead to 'designer babies'
	• All life is sacrosanct and we should not decide which type of baby is acceptable and which is not – could lead to eugenics

16.8

1 PCR is used to increase the quantity of DNA because the quantity available, e.g. at a crime scene, is often very small.

2 a Suspect B – because the bands on this suspect's genetic fingerprint match those of the genetic fingerprint of blood found at the crime scene.

b To eliminate the victim as the source of the blood sample found at the scene.

Other uses of genetic fingerprinting

1 The chemicals may inhibit some of the restriction endonucleases, which would then fail to cut some sections of DNA. There would therefore be a greater number of longer DNA fragments than normal and the fingerprint would be different.

2 In a person with the allele for Huntington's disease, some of the DNA fragments will be larger than those in a person without the allele because of the extra repeating units on the gene. These will travel a shorter distance in the electrophoresis gel and so there will be more thicker bands nearest the start of the fingerprint (where the initial sample was located).

3 Genetic fingerprints can determine how closely any two individuals are related. The closer the match between their fingerprints, the closer they are related. Therefore, to avoid the problems caused by inbreeding, it is advisable to mate animals whose fingerprints differ the most.

Locating DNA fragments

1 restriction endonuclease

2 A = 2 B = 4 C = 5 D = 1 E = 6 F = 3

Explanation – the shorter the fragments (those with fewer base pairs) the further they travel and the longer the fragments the less distance they travel.

3 5 times – because 5 cuts produce 6 fragments.

Index

Acknowledgements

Photograph Acknowledgements

The authors and publisher are grateful to the following for permission to reproduce photographs and other copyright material in this book.

Alamy/22DigiTal: 75; Alamy/Acro Images GmbH: 110 (left); Alamy/AGStockUSA, Inc.: 73 (bottom); Alamy/Bruce Coleman Inc.: 231; Alamy/Chris Gomersall: 11 (bottom), 97; Alamy/Chris Johnson: 8; Alamy/Chris Knapton: 257; Alamy/Christopher Leggett: 213; Alamy/David Hosking: 131; Alamy/David Levenson: 59; Alamy/David Noton Photography: 92, 93; Alamy/David R. Frazier Photolibrary, Inc.: 82 (bottom); Alamy/F1online digitale Bildagentur GmbH: 253; Alamy/geogphotos: 101; Alamy/guatebrian: 89; Alamy/Hawkeye: 81 (top); Alamy/Janine Wiedel Photolibrary: 273 (bottom); Alamy/Juniors Bildharchiv: 11 (top); Alamy/Leslie Garland Picture Library: 7; Alamy/Martin Shields: 157; Alamy/Mike Hill: 73 (top); Alamy/Nigel Cattlin: 98; Alamy/Paul Glendell: 77 (bottom); Alamy/Phototake Inc.: 273 (top); Alamy/Stefan Sollfors: 143 (top); Alamy/STOCKFOLIO: 268; Alamy/Victorio Castellani: 254; Alamy/WoodyStock: 233; Digital Vision JA (NT): 195 (top); Getty/LWA/Dann Tardiff: viii; Glenn and Susan Toole: 5, 90, 100 (top), 100 (bottom), 104, 106; iStockphoto.com: 3, 4, 10, 12, 15 (top), 15 (bottom), 21, 29 (top), 29 (bottom), 121, 191, 197 (top), 197 (bottom), 201; Oxford Scientific/David Fox: 127; Oxford Scientific/Eyal Bartov: 198 (right); Oxford Scientific/Herve Conge Phototake Inc.: 34; Oxford Scientific/Kathie Atkinson: 77 (top); Oxford Scientific/Norbert Rosing: 198 (left); Oxford Scientific/Phototake Inc.: 72; Oxford Scientific/Raymond Blythe: 68; Photodisc 6 (NT): 195 (bottom), 197 (middle); Photolibrary: 18; Science Photo Library/Alfred Pasieka: 141, 222; Science Photo Library/Anatomical Travelogue: 150; Science Photo Library/Astrid & Hanns-Frieder Michler: 186 (top), 203; Science Photo Library/Biology Media: 186 (bottom); Science Photo Library/Biophoto Associates: 118; Science Photo Library/Bob Gibbons: 108; Science Photo Library/BSIP, ERMAKOFF: 270; Science Photo Library/Chris Martin Bahr: 143 (middle); Science Photo Library/Cordelia Molloy: 124; Science Photo Library/CRNI: 168, 175; Science Photo Library/David Aubrey: 82 (top); Science Photo Library/Dr Gopal Murti: 246, 251; Science Photo Library/Dr Jeremy Burgess: 33; Science Photo Library/Dr Kenneth R. Miller: 36; Science Photo Library/Dr Linda Stannard, UCT: 265; Science Photo Library/Dr P. Marazzi: 157; Science Photo Library/Dr Yorgos Nikas: 239; Science Photo Library/Duncan Shaw: 110 (right); Science Photo Library/Eye of Science: 58, 274; Science Photo Library/Hugh Spencer: 96; Science Photo Library/ISM: 56; Science Photo Library/J.C. Revy: 204; Science Photo Library/J.C. Revy & A, Goujeon, ISM: 216; Science Photo Library/Jean Clauderevy, ISM: 161; Science Photo Library/Jerome Wexler: 143 (bottom); Science Photo Library/Mark Burnett: 113; Science Photo Library/Martin Dohrn: 79; Science Photo Library/Martyn Chillmaid: 44; Science Photo Library/Mauro Fermariello: 263; Science Photo Library/Medi-Mation: 148; Science Photo Library/Omikron: 152; Science Photo Library/Rosenfeld Images Ltd: 237; Science Photo Library/Saturn Stills: 206; Science Photo Library/Scott Sinklier/AGStockUSA: 81 (bottom); Science Photo Library/Simon Fraser: 105, 107 (top), 107 (middle), 107 (bottom); Science Photo Library/Steve Gschmeissner: 163, 171; Science Photo Library/Steve Percival: 178; Science Photo Library/Suzanne L. & Joseph T. Collins: 65.

Every effort has been made to trace and contact all copyright holders and we apologise if any have been overlooked. The publisher will be pleased to make the necessary arrangements at the first opportunity.